Protein Degradation

Edited by
R. John Mayer,
Aaron J. Ciechanover, and
Martin Rechsteiner

Further Titles of Interest

Mayer, R. J., Ciechanover, A. J., Rechsteiner, M. (Eds.)

Protein Degradation, Vol. 3

ISBN: 978-3-527-31435-5

Sipe, J. D. (Ed.)

Amyloid Proteins

The Beta Sheet Conformation and Disease

ISBN: 978-3-527-31072-2

Mayer, R. J., Ciechanover, A. J., Rechsteiner, M. (Eds.)

Protein Degradation, Vol. 2

ISBN: 978-3-527-31130-9

Buchner, J., Kiefhaber, T. (Eds.)

Protein Folding Handbook

ISBN: 978-3-527-30784-5

Mayer, R. J., Ciechanover, A. J., Rechsteiner, M. (Eds.)

Protein Degradation, Vol. 1

ISBN: 978-3-527-30837-8

Protein Degradation

Vol. 4: The Ubiquitin-Proteasome System and Disease

Edited by
R. John Mayer, Aaron J. Ciechanover, and Martin Rechsteiner

WILEY-VCH Verlag GmbH & Co. KGaA

The Editors

R. John Mayer
University of Nottingham
Queens Medical Center
Nottingham NG7 2UH
United Kingdom

Aaron Ciechanover
Technion-Israel Institute
Dept. of Biochemistry
Afron Street, Bat Galim
31096 Haifa 31096
Israel

Dr. Martin Rechsteiner
University of Utah
Dept. of Biochemistry
50 N. Medical Drive
Salt Lake City, UT 84132
USA

Library of Congress Card No.:
applied for

British Library Cataloguing-in-Publication Data
A catalogue record for this book is available from the British Library.

Bibliographic information published by the Deutsche Nationalbibliothek
Die Deutsche Nationalbibliothek lists this publication in the Deutsche Nationalbibliografie; detailed bibliographic data are available in the Internet at <http://dnb.d-nb.de>.

Typesetting SNP Best-set Typesetter Ltd., Hong Kong
Printing Betz-Druck GmbH, Darmstadt
Binding Litges & Dopf GmbH, Heppenheim
Cover Design Grafik-Design Schulz, Fußgönheim

Printed in the Federal Republic of Germany
Printed on acid-free paper

ISBN: 978-3-527-31436-2

Contents

Protein Degradation, Vol. 4: The Ubiquitin-Proteasome System and Disease.
Edited by R. J. Mayer, A. Ciechanover, M. Rechsteiner

ISBN: 978-3-527-31436-2

Preface

There is an incredible amount of current global research activity devoted to understanding the chemistry of life. The genomic revolution means that we now have the basic genetic information in order to understand in full the molecular basis of the life process. However, we are still in the early stages of trying to understand the specific mechanisms and pathways that regulate cellular activities. Occasionally discoveries are made that radically change the way in which we view cellular activities. One of the best examples would be the finding that reversible phosphorylation of proteins is a key regulatory mechanism with a plethora of downstream consequences. Now the seminal discovery of another post-translational modification, protein ubiquitylation, is leading to a radical revision of our understanding of cell physiology. It is becoming ever more clear that protein ubiquitylation is as important as protein phosphorylation in regulating cellular activities. One consequence of protein ubiquitylation is protein degradation by the 26S proteasome. However, we are just beginning to understand the full physiological consequences of covalent modification of proteins, not only by ubiquitin, but also by ubiquitin-related proteins.

Because the Ubiquitin Proteasome System (UPS) is a relatively young field of study, there is ample room to speculate on possible future developments. Today a handful of diseases, particularly neurodegenerative ones, are known to be caused by malfunction of the UPS. With perhaps as many as 1000 human genes encoding components of ubiquitin and ubiquitin-related modification pathways, it is almost certain that many more diseases will be found to arise from genetic errors in the UPS or by pathogen subversion of the system. This opens several avenues for the development of new therapies. Already the proteasome inhibitor Velcade is producing clinical success in the fight against multiple myeloma. Other therapies based on the inhibition or activation of specific ubiquitin ligases, the substrate recognition components of the UPS, are likely to be forthcoming. At the fundamental research level there are a number of possible discoveries especially given the surprising range of biochemical reactions involving ubiquitin and its cousins. Who would have guessed that the small highly conserved protein would be involved in endocytosis or that its relative Atg8 would form covalent bonds to a phospholipid during autophagy? We suspect that few students of ubiquitin will be surprised if it or a ubiquitin-like protein is one day found to be covalently attached to a nucleic acid for some biological purpose.

Protein Degradation, Vol. 4: The Ubiquitin-Proteasome System and Disease.
Edited by R. J. Mayer, A. Ciechanover, M. Rechsteiner

ISBN: 978-3-527-31436-2

We are regularly informed by the ubiquitin community that the initiation of this series of books on the UPS is extremely timely. Even though the field is young, it has now reached the point at which the biomedical scientific community at large needs reference works in which contributing authors indicate the fundamental roles of the ubiquitin proteasome system in all cellular processes. We have attempted to draw together contributions from experts in the field to illustrate the comprehensive manner in which the ubiquitin proteasome system regulates cell physiology. There is no doubt then when the full implications of protein modification by ubiquitin and ubiquitin-like molecules are fully understood we will have gained fundamental new insights into the life process. We will also have come to understand those pathological processes resulting from UPS malfunction. The medical implications should have considerable impact on the pharmaceutical industry and should open new avenues for therapeutic intervention in human and animal diseases. The extensive physiological ramifications of the ubiquitin proteasome system warrant a series of books of which this is the forth one. The focus of this book is on the role of the UPS in disease.

Aaron Ciechanover
Marty Rechsteiner
John Mayer

List of Contributors

Jessica M. Boname
University of Cambridge
Department Medicine
Cambridge Institute for Medical Research
Addenbrooke's Hospital
Hills Road
Cambridge, CB2 2XY
UK

Philip Coffino
University of California
Department of Microbiology and Immunology
San Francisco, CA 94143-0414
USA

Ted M. Dawson
Johns Hopkins University School of Medicine
Institute for Cell Engineering
Departments of Neurology
733 North Broadway, Suite 719
Baltimore, MD 21205
USA

Valina L. Dawson
Johns Hopkins University School of Medicine
Institute for Cell Engineering
Departments of Neurology
733 North Broadway, Suite 711
Baltimore, MD 21205
USA

Ivan Dikic
Goethe University School of Medicine
Institute of Biochemistry II
University Hospital, Bldg 75
Theodor-Stern-Kai 7
60590 Frankfurt (Main)
Germany

Kaisa Haglund
Goethe University Hospital
The Norwegian Radium Hospital
Department of Biochemistry
Institute for Cancer Research
Montebello
0310 Oslo
Norway

Martin A. Hoyt
University of California
Department of Microbiology and Immunology
San Francisco, CA 94143-0414
USA

Protein Degradation, Vol. 4: The Ubiquitin-Proteasome System and Disease.
Edited by R. J. Mayer, A. Ciechanover, M. Rechsteiner

ISBN: 978-3-527-31436-2

Zlatka Kostova
National Cancer Institute at Frederick
Laboratory of Protein Dynamics and Signaling
Center for Cancer Research
1050 Boyles Str., Building 560, Room 12–39
Frederick, MD 21702
USA

Paul J. Lehner
University of Cambridge
Cambridge Institute for Medical Research
Department of Medicine
Addenbrooke's Hospital
Hills Road
Cambridge, CB2 2XY
UK

Maria G. Masucci
Karolinska Institutet
Department of Cell and Molecular Biology
Berzelius väg, 35
S-17177 Stockholm
Sweden

Kevin St. P. McNaught
Mount Sinai School of Medicine
Department of Neurology
Annenberg 14-73
One Gustave L. Levy Place
New York , NY 10029
USA

Antje Schäfer
Universität Stuttgart
Institut für Biochemie
Pfaffenwaldring 55
70569 Stuttgart
Germany

Martin Scheffner
University of Konstanz
Department of Biology
Laboratory of Cellular Biochemistry
78457 Konstanz
Germany

Sathya R. Sriram
Johns Hopkins University School of Medicine
Institute for Cell Engineering
Department of Neurology
733 North Broadway, Suite 711
Baltimore, MD 21205
USA

Olivier Staub
University of Lausanne
Department of Pharmacology and Toxicology (Cellular)
Rue du Bugnon 27
1005 Lausanne
Switzerland

Ryosuke Takahashi
Kyoto University Graduate School of Medicine
Department of Neurology
54 Shogoin Kawahara-cho
Sakyo-ku
Kyoto, 606-8507
Japan

Dieter H. Wolf
Universität Stuttgart
Institut für Biochemie
Pfaffenwaldring 55
70569 Stuttgart
Germany

Dimitris P. Xirodimas
University of Dundee
Sir James Black Centre
Division of Gene Regulation and
Expression
Dow Street
Dundee, DD1 5EH
Scotland
UK

1
Ubiquitin Signaling and Cancer Pathogenesis

Kaisa Haglund and Ivan Dikic

1.1
Introduction

Post-translational modifications of proteins allow cells to respond dynamically to intra- and extracellular stimuli to control cellular processes [1]. A modification that has been given special attention among all possible modifications is protein ubiquitination, due to the frequency of its occurrence and the key role it plays in the inducible and reversible control of signaling pathways which regulate cellular homeostasis [2–4]. Tagging of proteins with ubiquitin occurs in a three-step process through the sequential action of the ubiquitin activating (E1), conjugating (E2) and ligase (E3) enzymes [5, 6]. Ubiquitination is a dynamic and reversible modification, and the rapid removal of ubiquitin from substrates and the processing of ubiquitin chains is catalyzed by de-ubiquitinating enzymes (DUBs) [7]. The regulation of DUBs is attracting increasing interest, since they serve to switch off the ubiquitin signal or to initiate a shift between different modifications of the same lysine residue. Moreover, there seems to be an interesting interplay between E3 ubiquitin ligases and DUBs. Interactions between E3s and DUBs have been shown to regulate the stability of E3s which undergo autoubiquitination. This type of interaction also leads the DUB to its substrate and regulates the target stability [7].

Ubiquitin modification can occur in multiple ways, making it a very diverse modification with distinct cellular functions (Figure 1.1). In its simplest form, a single ubiquitin molecule is attached to a single lysine residue in a substrate, which is defined as monoubiquitination [8, 9]. Alternatively, several single ubiquitin molecules can be attached to several different lysines, which is referred to as multiple monoubiquitination or multiubiquitination [10, 11]. Moreover, ubiquitin contains seven lysines itself that can be used to form various types of ubiquitin chain in an iterative process known as polyubiquitination [5, 12]. Interestingly, all seven lysines (Lys6, Lys11, Lys27, Lys29, Lys33, Lys48 and Lys63) have the potential to be used in chain formation, giving rise to chains with different linkages or branches [13].

Monoubiquitination is involved in endocytosis of plasma membrane proteins, the sorting of proteins into the multivesicular body (MVB), budding of

Protein Degradation, Vol. 4: The Ubiquitin-Proteasome System and Disease.
Edited by R. J. Mayer, A. Ciechanover, M. Rechsteiner

ISBN: 978-3-527-31436-2

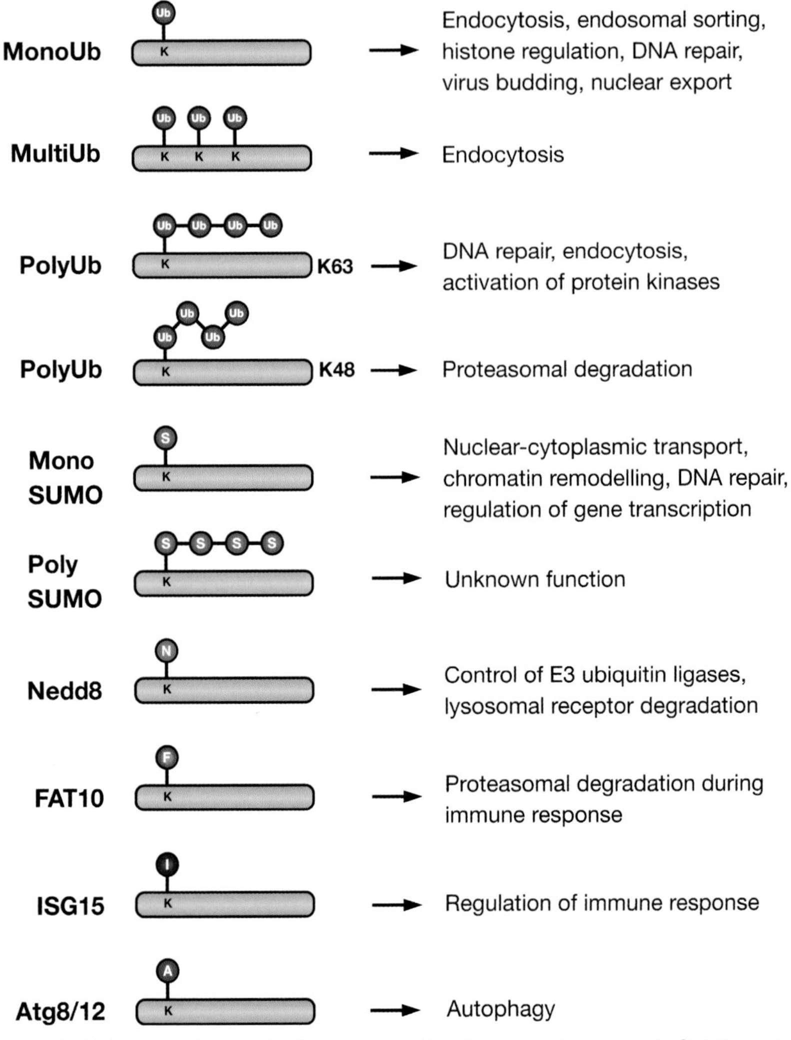

Fig. 1.1. Ubiquitin and ubiquitin-like protein (Ubl) modifications regulate a wide variety of cellular processes. Ubiquitin and Ubls share the same ubiquitin superfold and are collectively referred to as ubiquitons. All ubiquitons are attached via their C-terminal glycine residue to lysine residues in target proteins via a covalent isopeptide bond. Monoubiquitination (MonoUb) is essential for endocytosis and/or endosomal sorting of a variety of receptors, regulation of histones, DNA repair, virus budding and nuclear export. Tagging of several lysines with single ubiquitin molecules (MultiUb), is involved in endocytosis of certain RTKs and regulation of p53 localization. Polyubiquitination (PolyUb), the formation of ubiquitin chains via different lysines of ubiquitin, targets proteins for degradation in the 26S proteasome when linked via lysine 48, and has non-proteolytic functions, including control of DNA repair, endocytosis and activation of protein kinases when linked via lysine 63. Sumolyation controls several processes in the cell nucleus, including DNA repair, protein localization, chromatin remodeling and gene transcription. Neddylation regulates the activity of several E3 ubiquitin ligases, including Cbl, Mdm2 and cullins, and cooperates with ubiquitin to target EGFRs for lysosomal degradation [33, 34]. ISG15 and FAT10 are dimeric ubiquitons implicated in immune response [33, 34]. Atg8 and Atg12 play important roles in autophagy, the degradation of bulk cytoplasmic components, by contributing to the formation of autophagosomes during nutrient starvation of cells [33, 34]. Ub, ubiquitin; K, lysine; S, SUMO; N, Nedd8; F, FAT10; I, ISG15; A, Atg8/12.

retroviruses, DNA repair, histone activity and transcriptional regulation [8, 9, 14–16]. Multiple monoubiquitination is also involved in endocytosis of receptor tyrosine kinases (RTKs) and in nuclear export of p53 [10, 11]. In the case of polyubiquitination, the functions of polyubiquitin chains linked via lysines 48 and 63 have been best characterized. Proteins that are polyubiquitinated with Lys48-linked chains are recognized by ubiquitin-binding subunits of the 26S proteasome and are targeted for proteasomal degradation [5, 17]. Chains linked via Lys63, on the other hand, are involved in regulating endocytosis, DNA repair and activation of NF-κB [2, 14, 18–20]. Thus, whereas Lys48-linked polybiquitination was the first proteolytic signal described, it is becoming clear that monoubiquitination and Lys63-linked polyubiquitination function in several non-proteolytic cellular processes to regulate signaling networks.

1.1.1 Ubiquitin Signaling Networks

Ubiquitination is similar to phosphorylation and functions as a signaling device in cellular signaling networks. First, ubiquitination is an inducible event, which can be triggered by signals such as extracellular stimuli, phosphorylation and DNA damage [2]. This is associated with the fact that E3 ubiquitin ligases are tightly regulated by signal-induced mechanisms, such as post-translational modifications, compartmentalization, degradation and oligomerization [21, 22]. A prominent example is the ubiquitin ligase Cbl, which is recruited to a particular phosphotyrosine residue in the epidermal growth factor receptor (EGFR) following its ligand-induced activation, and subsequently tyrosine phosphorylation of Cbl itself promotes its ubiquitin ligase activity and consequently ubiquitination of the EGFR [23–25].

Second, ubiquitination is a reversible signal that is modulated by the action of DUBs, which is critical for the dynamic regulation of ubiquitin networks in the cell. The regulation of DUB activity is only beginning to be understood, and structural data indicate that these enzymes are in an active conformation only when bound to ubiquitin. Some DUBs require formation of complexes with other proteins in order to become active, and it has been reported that some are inhibited by phosphorylation or degradation [7]. For example, CYLD, an important DUB in the NF-κB pathway, undergoes inhibitory phosphorylation after TNF-α stimulation, leading to the accumulation of one of its substrates, Lys63-ubiquitinated TRAF2 [26].

Ubiquitin mediates many of its functions by interacting with highly specialized ubiquitin-binding domains (UBDs) in downstream effector proteins. More than 15 UBDs (UBA, UIM, IUIM, UEV, GAT, CUE, PAZ, NZF, GLUE, UBM, UBZ, VHS etc.) have been discovered so far [13, 27–31]. The structures of most of these domains have been elucidated when they are complexed with ubiquitin and it appears that they have many different tertiary structures and bind ubiquitin with relatively low affinity (50–100 μM) [13, 30]. The low affinity of UBD–ubiquitin interactions allows rapid assembly and disassembly of interaction networks, which facilitates dynamic biochemical processes [9, 13]. Moreover, it is thought that a local increase in the concentration of UBD-containing proteins and UBDs, for example by the formation of multimeric complexes or the presence of several UBDs within the same protein, might increase the rate at which UBD–ubiquitin interactions occur [9, 13, 30]. Furthermore, some UBDs can bind several ubiquitin molecules simultaneously, as has been reported for the UIM of the endocytic sorting protein Hrs (hepatocyte growth factor-regulated tyrosine kinase substrate) [32]. Due to its versatility, the numerous substrates that can be tagged with ubiquitin and the various proteins containing UBDs, ubiquitination is thus involved in complex networks of interactions in time and space that regulate key cellular functions, such as signaling, endocytosis, cell cycle and DNA repair.

1.1.2
Ubiquitin-like Proteins

The complexity of cellular signaling networks is further increased by modifications with ubiquitin-like (Ubl) proteins, including the small ubiquitin-related modifier (SUMO), Neural precursor cell-expressed developmentally downregulated 8 (Nedd8), interferon-stimulated gene 15 (ISG15), FAT10, Atg8 and Atg12 [33, 34], all of which regulate a variety of physiological processes (Figure 1.1). All Ubls share a similar three-dimensional structure, the ubiquitin superfold which is a β-grasp fold. Despite the varying degrees of sequence similarity, all proteins containing this fold are collectively known as ubiquitons [34].

In a manner similar to that involved in the tagging of proteins with ubiquitin, Ubls are covalently attached to their target proteins via a cascade of three enzymes (E1, E2, E3) which are partially specific for each of the Ubls [33]. As with ubiquitin, Ubls most frequently attach to lysines, although the free N-terminus can be an attachment site for both for ubiquitin and Ubls. In contrast to the ubiquitin system, Ubls generally form mono-conjugates with the substrates and not polymeric chains (Figure 1.1). SUMO conjugates have been observed, however, but their function is not yet known [35]. It is very likely that there are specialized interaction domains for all the Ubls, although they have only been described for a subset. SUMO-interacting motifs (SIMs) have been assigned [36–39], and some known UBDs interact not only with ubiquitin, but also with Nedd8 [40]. Moreover, it is interesting to note that UBDs and SIMs bind at distinct surface locations on ubiquitin and SUMO, respectively, resulting in highly specific interactions which provide some insights into the different cellular functions of these two proteins [1].

In many cases, there is an active interplay between ubiquitin and Ubls in the regulation of individual proteins and/or cellular pathways. For example, the same lysine residue can be modified with either ubiquitin or SUMO, leading to the activation of completely different downstream pathways. The modification of PCNA (proliferating cell nuclear antigen), that forms a clamp that recruits DNA polymerases to the replication fork, with either ubiquitin or SUMO induces error-prone DNA repair or DNA synthesis, respectively [14]. Moreover, there is apparent cooperation between ubiquitin and Nedd8 during downregulation of the epidermal growth factor receptor (EGFR). EGF stimulation triggers Cbl-mediated neddylation of the EGFR, which in turn promotes the subsequent Cbl-mediated ubiquitination of the receptor and its degradation [40].

Further complexity in Ubl signaling networks results from the fact that Ubl domains can be found within the genetically-encoded sequence of proteins. Many proteins containing Ubl domains interact with the proteasome, but there are also several examples in which the ubiquitin fold is involved in mediating protein–protein interactions in signal transduction cascades, consistent with the important role of ubiquitin and Ubls in both degradation and signaling pathways [34].

1.2 Ubiquitin in Cancer Pathogenesis

The development of cancer is a multi-step process which results from mutations in the cellular pathways that control signaling, endocytosis, cell-cycle and cell-death and interactions between the tumor and its surrounding tissue [41]. Deregulation of components of the ubiquitination machinery appears to be a common theme in the development of cancers [4, 42–44]. Mutations or overexpression of numerous E3 ubiquitin ligases can convert them to potent oncogenes and some E3s and DUBs act as tumor suppressors (Table 1.1). Several substrates that are affected by alterations in E3 and DUB activity play key roles in the cell cycle, DNA repair, NF-κB signaling, RTK signaling and angiogenesis and their levels or activity are precisely regulated by ubiquitination (Table 1.1; Figure 1.2). In the following sections we will highlight the nature of role that the ubiquitin system plays in maintaining the homeostatic balance of these processes and why its deregulation promotes the development of different types of tumors.

1.2.1 Ubiquitin in Cell Cycle Control

Deregulation of cell-cycle control is a fundamental characteristic of cancer. Uncontrolled proliferation of cancer cells occurs because the precise regulation of the cell cycle has been disrupted [41]. Progression through the cell cycle is mediated by cyclin-dependent kinases (CDKs) whose activity is regulated by cyclins and CDK inhibitors (CDKIs) [43]. These undergo ubiquitin-mediated proteolysis which results in their periodic expression, ensuring that the cell cycle proceeds at normal speed. Cyclins act as accelerators of the cell cycle, whereas CDKIs function as brakes. Therefore, cyclins (D1 and E) are frequently overexpressed in human cancers and the CDKI p27 is a prominent tumor suppressor [43, 45, 46].

Three structurally-related cullin-dependent E3 ubiquitin ligases, SKP1-CUL1-F-box-protein (SCF)/Skp2, SCF/Fbw1 and anaphase-promoting complex/cyclosome (APC/C), are involved in regulating the levels of cyclins and CDK inhibitors by promoting their polyubiquitination and degradation in the proteasome [43].

Table 1.1. Summary of pathways and proteins regulated by ubiquitination and whose deregulation leads to the development of cancer.

Pathway	Deregulated protein	Function	Type of deregulation	Substrate	Ubiquitin modification	Cancer type	References
Cell cycle	SCF/Skp2	E3 ubiquitin ligase subunit, oncogene	Overexpression	p27, cyclin E	Lys48-linked polyubiquitination	Lung cancer, malignant melanoma, lymphoma	43
	SCF/Fbw7	E3 ubiquitin ligase subunit, tumor suppressor	Mutation	Cyclin E	Lys48-linked polyubiquitination	Ovarian cancer, breast cancer, endometrial cancer	43
	APC/C	E3 ubiquitin ligase, tumor suppressor	Mutation	Cyclin B, securin	Lys48-linked polyubiquitination	Colorectal cancer	46
DNA repair	Mdm2	E3 ubiquitin ligase, oncogene	Overexpression	p53	Lys48-linked polyubiquitination	Non-small cell lung cancer, soft-tissue carcinoma, colorectal cancer	59
	HAUSP	DUB, tumor suppressor	Mutation	p53, Mdm2	Deubiquitination	Non-small-cell lung cancer	67
	BRCA1	E3 ubiquitin ligase, tumor suppressor	Germline mutation	γ-tubulin	Polyubiquitination	Breast cancer, ovarian cancer	68, 69
	FANCL	E3 ubiquitin ligase, tumor suppressor	Mutation	FANCD2	Monoubiquitination	Fanconi anemia-related cancers	68, 74
NF-κB signaling	CYLD	DUB, tumor suppressor	Mutation	NEMO, TRAF2, TRAF6, Bcl-3	Deubiquitination of Lys63-linked ubiquitin chains	Cylindromatosis	53
RTK signaling	Cbl	E3 ubiquitin ligase, proto-oncogene	Mutation	RTKs	Multiple monoubiquitination	Breast cancer, glioblastoma, head and neck cancer, lymphoma	81–84, 86, 87
Angiogenesis	SCF/VHL	E3 ubiquitin ligase subunit, tumor suppressor	Mutation	HIF1α	Lys48-linked polyubiquitination	Kidney cancer, blood vessel tumors in the CNS	78

A. Cell cycle

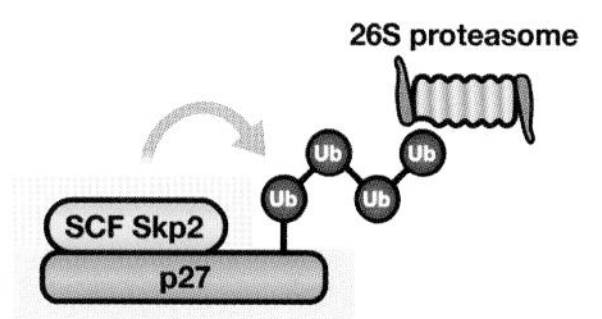

B. DNA repair/cell cycle

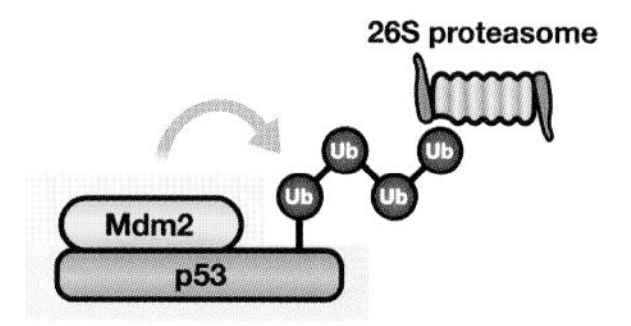

C. DNA repair

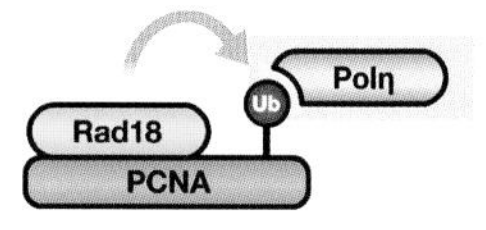

D. RTK degradation

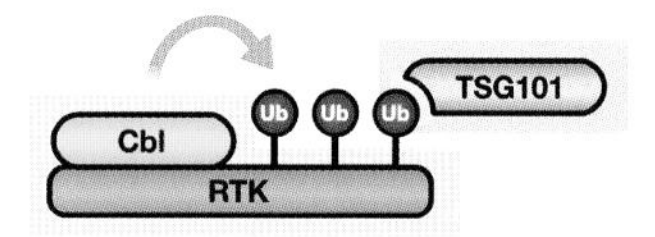

E. Angiogenesis

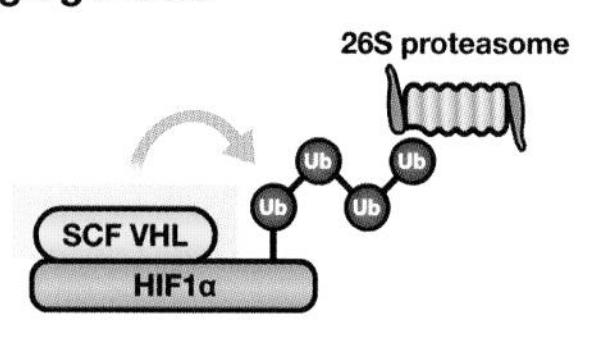

F. NF-κB signaling

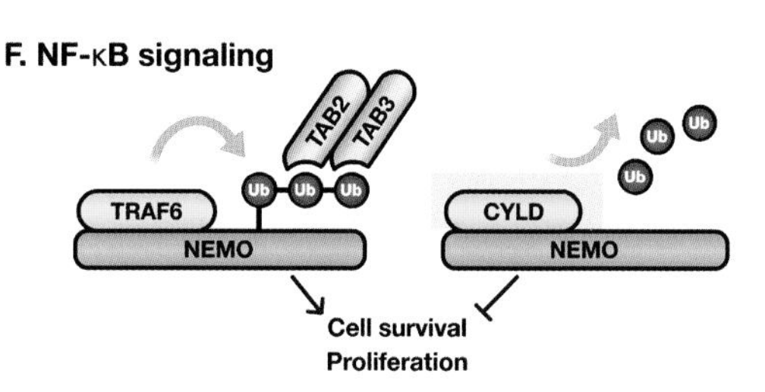

Fig. 1.2. Overview of cancer-relevant ubiquitin-dependent pathways. SCF/Skp2, Mdm2, Rad18, Cbl, SCF/VHL and TRAF6 all are E3 ubiquitin ligases (yellow) that mediate specific types of ubiquitination of their respective substrates which are indicated in the figure (p27, p53, PCNA, RTKs, HIF1α and NEMO) (green). The proteasome, which has UBD-containing subunits, and UBD-containing proteins (Polη, TSG101, TAB2/3) are shown in blue. (A, B, E) Lys48-linked polyubiquitination of p27, p53 and HIF1α leads to their proteasomal degradation, promoting cell cycle progression (p27, p53) or block of production of pro-angiogenic factors (HIF1α). SCF/Skp2 and Mdm2 act as oncogenes, because their overexpression leads to increased proliferation and the development of cancer. SCF/VHL, on the other hand, acts as a tumor suppressor, since its mutation leads to the accumulation of HIF1α, aberrant angiogenesis and tumorigenesis. (C) Rad18 mediates monoubiquitination of PCNA, a modification responsible for recruiting ubiquitin binding domain (UBD)-containing TLS polymerases to the site of DNA damage. Mutation of TLS polymerase Polη leads to a variant of a skin tumor syndrome called Xeroderma pigmentosum. (D) Cbl mediates multiple monoubiquitination of RTKs, which is recognized by ubiquitin-binding domains in proteins of the endocytic sorting machinery, including TSG101. Mutation of the Cbl binding site in RTKs, mutations of Cbl that abolish its ubiquitin ligase activity, or mutation in TSG101 all lead to defective receptor sorting and degradation, causing constitutive signaling and tumorigenesis. (F) TRAF6 mediates Lys63-linked polyubiquitination of NEMO, which recruits the UBD-containing proteins TAB2/3, leading to activation of the protein kinase TAK1 that is required for NF-κB activation. CYLD, the DUB that removes Lys63-linked chains from NEMO, is mutated in a cancer syndrome called cylindromatosis. Tumor suppressors are indicated in turquoise and oncogenes or proto-oncogenes in pink.

SCF/Skp2 targets among others p27 and cyclin E, and SCF/Fbw1 targets cyclin E for polyubiquitination and proteasomal degradation, events that regulate the G1–S transition (Figure 1.2) [47]. APC/C, on the other hand, promotes polyubiquitination and degradation of mitotic cyclins and securin, which are required for termination of the mitotic cycle and separation of the sister chromatids, respectively [46]. In this way APC/C maintains the normal chromosome number, alterations of which are a prevalent form of genetic instability in human cancers. These E3 ubiquitin ligases thus act at different time points during the cell cycle and importantly they appear to interplay in a regulatory loop [43].

Due to their central function in cell cycle progression, aberrant expression or mutations of SCF/Skp2, Fbw1 or APC/C have been found in several human cancers (Table 1) [43, 45, 46]. Skp2 has oncogenic properties in transgenic mouse models, is frequently overexpressed in lung cancers and its overexpression is correlated with poor prognosis in a wide range of cancer types [43]. Fbw1, on the other hand, acts as a tumor suppressor. Mutations in the *FBW1* gene have been reported in ovarian, breast and endometrial cancer, often correlated with increased cyclin E levels [43]. APC/C also functions as a tumor suppressor and is mutated in more than 70% of colorectal carcinomas [46]. Thus, cumulative evidence indicates that deregulation of the ubiquitin system in cell-cycle control is closely linked to the development of cancer.

1.2.2 **Ubiquitin in the NF-κB Pathway**

The NF-κB family of transcription factors triggers the expression of genes that are central mediators of cell survival, proliferation, and innate and adaptive immune responses. The role of NF-κB in cancer is connected to its constitutive activation of anti-apoptotic signals in both pre-neoplastic and malignant cells, and its emerging role in regulating tumor angiogenesis and invasion [48]. NF-κB activation is controlled by ubiquitination of several of the components of the NF-κB pathway [2, 18, 49]. A key step in the activation of NF-κB is its release from the inhibitor IκB and its subsequent translocation from the cytoplasm to the nucleus where it triggers the expression of its target genes. A central regulator of this process is the IκB kinase (IKK) complex, which consists of two catalytic subunits (IKKα and IKKβ) and a regulatory subunit (IKKγ/NEMO). IKK promotes IκB phosphorylation which recruits the E3 ubiquitin ligase SCF-βTRCP to IκB which in turn promotes Lys48-linked ubiquitination and proteasomal degradation, thereby releasing NF-κB [18, 49].

Another type of ubiquitin modification is exemplified by Lys63-linked polyubiquitination which also plays a central role in NF-κB activation by activating protein kinases. Both IKK and the kinase that activates IKK, TGFβ-activated kinase (TAK1), require Lys63-linked chains synthesized by the E3 ubiquitin ligase TNF receptor associated factor 6 (TRAF6) for their activation [18]. IKK activation requires the modification of the regulatory subunit NEMO with Lys63-linked chains [50]. TAK1 activation depends on the interaction between the UBDs of the TAK1-binding proteins TAB1, 2 and 3 with substrates modified with Lys63-linked polyubiquitin chains, and ubiquitinated NEMO is a likely interaction partner (Figure 1.2) [51].

Since ubiquitination plays a central role in NF-κB activation, its removal by DUBs is critical to the downregulation of the NF-κB signal. To date, two DUBs have been identified to have important roles in regulating the NF-κB pathway, A20 and cylindromatosis (CYLD). A20 has a dual role in downregulating NF-κB signaling. First, A20 specifically removes Lys63-linked ubiquitin chains from the receptor-interacting protein (RIP), an essential mediator of TNF receptor 1 (TNFR1) signaling, and subsequently it attaches Lys48-linked ubiquitin chains to promote its proteasomal degradation [52]. Whether there is a genetic link between A20 and the risk of cancer still needs to be established [4].

CYLD was originally identified as a tumor suppressor gene that is mutated in familial cylindromatosis, an autosomal dominant disease characterized by multiple tumors of the skin appendages [53]. CYLD contains a ubiquitin C-terminal hydrolase (UCH) domain and acts as a DUB that removes Lys63-linked chains from several NF-κB pathway members, including the ubiquitin ligases TRAF2 and TRAF6, the IKK subunit NEMO and the transcriptional co-activator Bcl-3 (Figure 1.2) [54–58]. In this way CYLD regulates the duration of NF-κB activation and its loss thus correlates with tumorigenesis.

These examples illustrate that modification of pathway components containing Lys63-linked ubiquitin chains (NEMO, TRAFs, RIP, Bcl-3) triggers the activation of NF-κB, whereas ubiquitin removal is a common theme in its inactivation, thereby preventing excessive cell proliferation and tumor development.

1.2.3
Ubiquitin as a Signal in DNA Repair

The maintenance of DNA integrity is pivotal to the prevention of cancer-promoting mutations in the genome. Cells have therefore developed elaborate DNA repair systems to respond to DNA damage. Emerging data show that ubiquitin modification plays a major role in DNA repair response both by regulating cell cycle arrest (p53, Mdm2, HAUSP, BRCA1 and FANCD2) and by controlling trans-lesion DNA synthesis (PCNA and TLS polymerases).

1.2.3.1 p53 Pathway

The *p53* gene is mutated in more than 50% of human cancers. p53 is a transcription factor with an essential role in promoting cell-cycle arrest, apoptosis and DNA repair when cells encounter DNA damage. In this way, p53 hinders proliferation of damaged cells and acts as a tumor suppressor [59, 60]. In order to maintain cellular homeostasis, the levels of p53 are highly regulated in cells. In unstressed cells, the levels of p53 are kept low and this is mediated by ubiquitin-dependent proteasomal degradation. Mdm2 is a RING-type E3 ubiquitin ligase responsible for promoting both monoubiquitination and Lys48-linked polyubiquitination of p53 in a dose-dependent manner (Figure 1.2) [11]. Monoubiquitination of p53 promotes its nuclear export and polyubiquitination, its degradation by nuclear proteasomes. Normally, the interaction between Mdm2 and p53 is disrupted when cells encounter DNA damage or other stresses, promoting an accumulation of p53 in the nucleus, cell-cycle arrest and DNA repair [61]. Overexpression of Mdm2, on the other hand, leads to aberrant deactivation of p53, which is observed in many types of tumors (Table 1.1) [59, 62, 63].

Herpes simplex-associated ubiquitin-specific protease (HAUSP) is involved in p53 deubiquitination and stabilization [64, 65]. Importantly, its overexpression is sufficient to promote cell-cycle arrest and apoptosis, suggesting that it could act as a tumor suppressor [65]. On the other hand, disruption of the *HAUSP* gene in human cancer cell lines by targeted homologous recombination, also leads to p53 stabilization and activation [66]. These contradictory results could be explained by the presence of other targets of HAUSP, such as Mdm2, which determine p53 levels [64]. Nevertheless, mutations of the *HAUSP* gene are associated with an increased risk for non-small-cell lung cancer [67].

1.2.3.2 **BRCA1 and FANCD2**

The breast cancer susceptibility genes *BRCA1* and *BRCA2* and products of the Fanconi anemia (FA) gene act as tumor suppressors. They function in a network of interconnected biological processes and have important roles in cell-cycle checkpoint control and DNA repair of double strand breaks by mediating homologous recombination [60]. Germline mutations in one allele of either *BRCA1* or *BRCA2* cause hereditary breast and ovarian cancer syndrome and mutations in FA genes (*FANCA, FANCB, FANCC, FANCD1, FANCD2* etc.) can cause FA, a genetic disorder associated with increased susceptibility to cancer [68].

BRCA1 acts as a RING-type E3 ubiquitin ligase and its activity is increased when it is complexed with the structurally and functionally related BRCA1-associated RING domain 1 (BARD1) ubiquitin ligase [69]. Specific mutations in the RING domain of BRCA1 abolish its ubiquitin ligase activity and tumor suppression capabilities. Interestingly, BRCA1 and BARD1 preferentially promote formation of Lys6-linked chains, a chain type that seems to be primarily involved in substrate stabilization [69].

BRCA1- and *BARD1*-deficient mice show centrosome amplification, defective G2–M checkpoint control and genetic instability [69]. Among the ubiquitinated targets of BRCA1/BARD1 is the centrosome component γ-tubulin [70]. Following their duplication during cell division, centrosomes help to form the spindle apparatus that segregates the duplicated chromosomes into daughter cells. Mutation of the ubiquitination site in γ-tubulin leads to amplification of centrosome numbers, a defect associated with chromosome missegregation and the development of cancer [70].

When DNA is damaged, BRCA1 binds to FANCD2 in nuclear foci that are required for cell-cycle checkpoint control and DNA repair [71]. The localization of FANCD2 to these foci is promoted by its monoubiquitination, suggesting that monoubiquitin-binding proteins might be involved in its recruitment [72]. FANCD2 undergoes monoubiquitination in BRCA1–/– cells, indicating that another E3 ubiquitin ligase promotes this modification [73]. Indeed, a component of the nuclear FA–protein complex, FANCL, possesses E3 ubiquitin ligase activity against FANCD2 via its RING-finger-like plant domain (PHD) [74]. Deubiquitination of FANCD2 by ubiquitin-specific protease 1 (USP1), on the other hand, may play an important role when cells restart the cell cycle after DNA damage [75].

1.2.3.3 PCNA and TLS Polymerases

DNA damage blocks the progression of the replication fork and in order to avoid stalling the replication process and circumventing the damaged sites, cells replace the high-fidelity replicative polymerase Polδ with one of the five specialized low stringency DNA polymerases which are able to perform trans-lesion DNA synthesis (TLS) across different types of damage [76]. That TLS is crucial for cells is emphasized by the fact that defects in TLS polymerases can cause disease. Mutations in TLS polymerase Polη are found in patients suffering from a variant of Xeroderma pigmentosum, a UV-induced skin tumor syndrome [77].

Proliferating cell nuclear antigen (PCNA) functions to recruit different polymerases to the site of DNA replication or repair, and its ubiquitination and deubiquitination plays a major role in the polymerase switch. Non-ubiquitinated PCNA recruits the replicative polymerase Polδ during DNA replication. Certain types of DNA damage, on the other hand, induce Rad18-mediated monoubiquitination of PCNA. This modification triggers the recruitment of TLS polymerases, all of which contain UBDs, the so-called ubiquitin-binding motif (UBM) or ubiquitin-binding zinc finger (UBZ) (Figure 1.2) [14, 27]. After trans-lesion synthesis has taken place, the low fidelity TLS polymerases are exchanged for Polδ to ensure accurate continued replication. Therefore, the DUB USP1 removes ubiquitin from PCNA during normal replication to allow recruitment of Polδ and is degraded once the DNA becomes damaged, again allowing monoubiquitination of PCNA and recruitment of the TLS polymerases [20].

1.2.4 Ubiquitin Networks in Angiogenesis

Rapidly growing tumors require efficient blood and nutrient supply and therefore secrete growth factors, such as vascular endothelial growth factor (VEGF) and platelet derived growth factor (PDGF), to promote angiogenesis, the formation of new capillaries. Therefore, it is not surprising that an anti-angiogenic protein, such as VHL (von Hippel-Lindau), would be a tumor suppressor [78]. The *VHL* gene encodes a component of an SCF-like ubiquitin ligase and is mutated in patients suffering from the familial cancer susceptibility, von Hippel-Lindau syndrome, that is associated with cancer of the kidney and tumors in the blood vessels of the central nervous system [78, 79]. Under normoxic conditions, VHL binds to the hydroxylated α-subunits of the hypoxia-inducible factor (HIF) heterodimeric transcription factors and targets them for polyubiquitination and proteasomal degradation (Figure 1.2) [80]. During hypoxic conditions, HIF1α is not hydroxylated and can thus not be bound by VHL, leading to its stabilization. HIF1α then triggers the transcription of several genes encoding pro-angiogenic growth factors, including VEGF, PDGFβ and transforming growth factor α (TGFα) [78]. Mutation of VHL is thought to lead to constantly increased levels of HIF1α and its target growth factors even under normoxic conditions [78], thus stimulating the formation of new blood vessels and tumors.

1.2.5
Ubiquitin Networks in Receptor Endocytosis

Constitutive receptor tyrosine kinase (RTK) signaling, resulting from receptor overexpression, autocrine growth factor loops and activating mutations, can cause cell transformation and cancer [81]. Moreover, loss of negative regulation of RTKs is an important factor contributing to enhanced receptor signaling [82–84]. RTKs are downregulated by endocytosis and lysosomal degradation, which requires ligand-induced Cbl-mediated receptor multiple monoubiquitination, Lys63-linked polyubiquitination and neddylation (Figure 1.2) [9, 10, 40, 85]. Ubiquitin attached to RTKs serves as a sorting tag that is recognized by UBD-containing endocytic proteins along the endocytic pathway, ensuring that they targeted into the inner vesicles of the multivesicular body (MVB), which destines them for lysosomal degradation [9, 15, 16]. Therefore, RTK mutations that lead to the loss of the binding site for the ubiquitin ligase Cbl in addition to Cbl mutants lacking ubiquitin ligase activity, cause defective downregulation of the receptor [83, 84]. Prominent examples of RTKs that have been found mutated in tumors and have escaped Cbl-mediated ubiquitination and degradation include EGFR (EGFRvV, v-erbB and EGFRvIII), MET (TRP-MET) and c-Kit (v-Kit) [83, 86, 87]. Oncogenic forms of Cbl (v-Cbl, Cbl-70Z, ΔY368-Cbl, ΔY371-Cbl) all lack ubiquitin ligase activity and are thought to act as dominant negative proteins and to compete with endogenous Cbl for binding to activated RTKs [24, 25]. Deletions of the extracellular area of the EGFR (EGFRvIII) are found in approximately 40% of glioblastomas and the EGFR family member ErbB2 is frequently overexpressed in breast cancer [87]. Overexpression of ErbB2 favors the formation of EGFR/ErbB2 heterodimers which recruit Cbl less efficiently, and are thus not degraded, but rather recycled back to the cell surface [88–90].

Interestingly, components of the endosomal sorting complex required for transport (ESCRT) machinery that sorts ubiquitinated cargo into the MVB [15, 91], are also linked to the development of tumors. Mutations in the components of the ESCRT-I tumor susceptibility gene 101 (TSG101) and hepatocellular carcinoma-related protein 1 (HCRP1) have been implicated in tumor development [84, 92, 93]. TSG101 contains a ubiquitin-binding UEV domain that binds to ubiquitinated cargo and is required for effective receptor sorting into the MVB (Figure 1.2) [15]. Moreover, mutations of *erupted* (TSG101) and *Vps25* (an ESCRT-II component) have been shown to cause neoplastic tumor growth in the fruit fly [94–98]. Thus, proper ubiquitin-dependent lysosomal degradation of activated RTKs prevents constitutive receptor signaling and carcinogenesis.

1.3 Targeting Ubiquitin Networks in Cancers

Due to its common deregulation in the development of cancers, targeting the ubiquitin system in cancer therapeutics emerges as a promising approach. The major challenge is to develop drugs that specifically act on the desired ubiquitin system component or substrate without affecting other pathways. Possible strategies involve inhibiting ubiquitin activation or conjugation, ubiquitin ligase activity of oncogenic E3s, by blocking either E2 or substrate binding, or inhibiting the degradation of cancer-preventing tumor suppressors [99]. Since the ubiquitin activation and proteasomal degradation steps involve ATP-dependent and proteolytic enzymes, respectively, which are classical drug targets, they represent therapeutically attractive points of intervention [99]. The major concern with these strategies, however, is their wide action on numerous substrates and pathways within the cell which may produce severe side effects. Intervening in the E3–substrate interaction therefore represents a more selective approach which could lead to more effective treatment and fewer nonspecific effects (Figure 1.3).

1.3.1 Targeting Interactions between E3s and their Substrates

This strategy has been successfully applied when targeting the interaction between the oncogenic E3 ubiquitin ligase Mdm2 and the tumor suppressor p53 with two

A. E3-substrate interaction

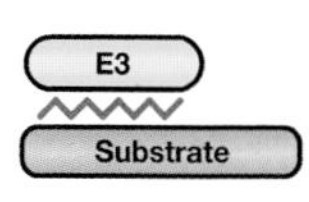

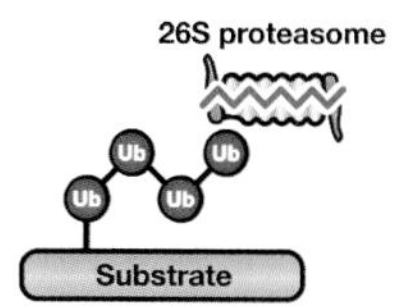

C. Proteasome recognition

D. Ubiquitin-UBD interaction

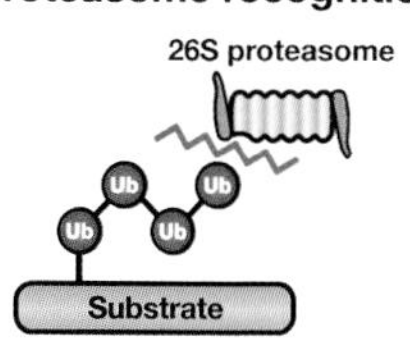

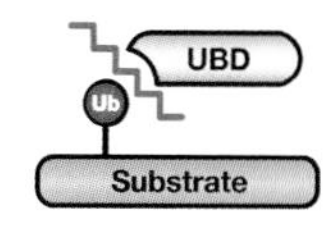

Fig. 1.3. Different approaches for targeting the ubiquitin system in cancer therapy. (A) Interference with the interaction between the E3 ubiquitin ligase and the substrate. Examples: Nutlin and RITA. (B) Inhibition of the catalytic activity of the proteasome. Example: Bortezomib. (C) Interference with the interaction between Lys48-polyubiquitinated substrates and the proteasome. Example: Ubistatins. (D) Interference with the interaction between ubiquitinated substrates and the corresponding ubiquitin-binding domain (UBD)-containing proteins. The main difficulty with all these strategies is in achieving optimum specificity and selectivity. See the main text for more details.

types of small molecule inhibitors that were identified in anti-cancer drug screens, Nutlins (cis-imidazole derivatives) and RITA (2,5-bis(5-hydroxymethyl-2-thienyl)furan). Nutlins occupy the p53 binding pocket of Mdm2 and RITA binds p53 and in this way they both prevent the p53–Mdm2 interaction [99]. Consequently, both compounds stabilize p53, leading to p53-dependent cell-cycle arrest in cancer cell lines and the inhibition of tumor growth in nude mice [99]. Although the initial studies on these inhibitors seem promising, their bioavailability and usefulness in the treatment of human cancer need to be thoroughly studied. Despite higher specificity, major concerns still remain: do these interventions yield unwanted effects such as affecting other substrates of Mdm2, other interaction partners of p53 or p53-related proteins such as p63 and p73? These basic concerns apply in each of the cases where the interaction surface between the E3 ubiquitin ligase and the substrate is targeted.

1.3.2 Targeting the Proteasome

Surprisingly, the biggest success so far in targeting the ubiquitin system in cancer therapy has been the development of Bortezomib, a small molecule proteasome inhibitor that binds reversibly to the active site of the 20S proteasome subunit [100]. Despite concerns regarding the lack of specificity due to the inhibition of the entire proteasomal protein degradation system, this inhibitor is being successfully used clinically in the treatment of relapsed, refractory multiple myeloma, and is being studied in a variety of hematological cancers and solid tumors, including non-Hodgkin's lymphoma, prostate, breast and non-small-cell lung cancers. Bortezomib is thought to inhibit cell proliferation by blocking the degradation of proteins involved in cell-cycle control and apoptosis (including p53, cyclins and IκB) [101]. Interestingly, Bortezomib shows selective cytotoxicity against cancer cells compared to normal cells both *in vitro* and *in vivo* [99]. Therefore, development of strategies involving proteasome inhibitors may be useful in the therapy of certain types of tumors (Figure 1.3).

1.3.3 Other Approaches

Apart from targeting the ubiquitin system itself, some clinically-effective monoclonal antibodies and small molecule inhibitors have been shown to promote ubiquitin-dependent degradation of oncogenic proteins. Such an example is Herceptin, a monoclonal antibody used for the treatment of breast cancer tumors overexpressing ErbB2 which increases Cbl-mediated ErbB2 ubiquitination and degradation [102].

Another promising and challenging approach to targeting the ubiquitin system in cancer therapy is to alter the ubiquitin-induced protein–protein interactions in cells [4]. Although this approach offers more specificity than any of the above-mentioned strategies due to the fact that there are more ubiquitin-induced interactions than existing enzymes, there are still many difficulties related to this strategy. The main obstacles include targeting the flat and hydrophobic interaction surface between ubiquitin and UBDs and dealing with the low affinities of such interactions. Modulators of polyubiquitin chain recognition, the ubistatins, have been shown to bind specifically to the interfaces between Lys48-linked ubiquitin molecules, and to inhibit ubiquitin-dependent proteasomal degradation of certain substrates (Figure 1.3) [103]. The ubistatins are however not cell-permeable, but it may be possible to use them to target interaction surfaces between ubiquitin and UBDs after further developing them and increasing their bioavailability. Other types of inhibitor of ubiquitin–UBD interactions could target either the hydrophobic surface of ubiquitin containing Ile44, with which most UBDs interact, or specific UBDs (Figure 1.3). However, these approaches are also associated with issues of specificity, since ubiquitin and ubiquitin chains are attached to numerous proteins and UBDs are found in a vast variety of proteins. Despite these drawbacks, the increasing interest and knowledge gained in this field will ensure that several novel strategies for targeting the ubiquitin system with higher specificity will be developed in the near future.

1.4 Conclusions and Future Perspectives

In conclusion, we note that there is a strong link between alterations in ubiquitin signaling networks and the hallmarks of cancer, including uncontrolled proliferation and cell cycle divisions (SCF ligases, NF-κB), increased cellular signaling (RTKs, NF-κB), defective endocytosis (Cbl, RTKs, TSG101), increased cell survival (NF-κB), defective DNA repair (p53, BRCA1, TLS polymerases) and increased angiogenesis (VHL). A complete understanding of the interplay between ubiquitination and other post-translational modifications such as phosphorylation and acetylation, between ubiquitin ligases and DUBs, as well as ubiquitin and Ubls in cellular networks will have a great impact on our insight into cancer-promoting mechanisms and our ability to design smart drugs for the treatment of cancer.

References

1 Seet, B.T., Dikic, I., Zhou, M.M. and Pawson, T. (2006) Reading protein modifications with interaction domains. *Nat Rev Mol Cell Biol* **7**, 473–483.

2 Haglund, K. and Dikic, I. (2005) Ubiquitylation and cell signalling. *EMBO J* **24**, 3353–3359.

3 Sigismund, S., Polo, S. and Di Fiore, P.P. (2004) Signaling through monoubiquitination. *Curr Top Microbiol Immunol* **286**, 149–185.

4 Hoeller, D., Hecker, C.M. and Dikic, I. (2006) Ubiquitin and ubiquitin-like proteins in cancer pathogenesis. *Nat Rev Cancer* **6**, 776–788.

5 Hershko, A. and Ciechanover, A. (1998) The ubiquitin system. *Annu Rev Biochem* **67**, 425–479.

6 Pickart, C.M. and Eddins, M.J. (2004) Ubiquitin: structures, functions, mechanisms. *Biochim Biophys Acta* **1695**, 55–72.

7 Nijman, S.M., Luna-Vargas, M.P., Velds, A., Brummelkamp, T.R., Dirac, A.M., Sixma, T.K. and Bernards, R. (2005) A genomic and functional inventory of deubiquitinating enzymes. *Cell* **123**, 773–786.

8 Hicke, L. (2001) Protein regulation by monoubiquitin. *Nat Rev Mol Cell Biol* **2**, 195–201.

9 Haglund, K., Di Fiore, P.P. and Dikic, I. (2003) Distinct monoubiquitin signals in receptor endocytosis. *Trends Biochem Sci* **28**, 598–603.

10 Haglund, K., Sigismund, S., Polo, S., Szymkiewicz, I., Di Fiore, P.P. and Dikic, I. (2003) Multiple monoubiquitination of RTKs is sufficient for their endocytosis and degradation. *Nat Cell Biol* **5**, 461–466.

11 Li, M., Brooks, C.L., Wu-Baer, F., Chen, D., Baer, R. and Gu, W. (2003) Mono- versus polyubiquitination: differential control of p53 fate by Mdm2. *Science* **302**, 1972–1975.

12 Pickart, C.M. and Fushman, D. (2004) Polyubiquitin chains: polymeric protein signals. *Curr Opin Chem Biol* **8**, 610–616.

13 Hicke, L., Schubert, H.L. and Hill, C.P. (2005) Ubiquitin-binding domains. *Nat Rev Mol Cell Biol* **6**, 610–621.

14 Hoege, C., Pfander, B., Moldovan, G.L., Pyrowolakis, G. and Jentsch, S. (2002) RAD6-dependent DNA repair is linked to modification of PCNA by ubiquitin and SUMO. *Nature* **419**, 135–141.

15 Katzmann, D.J., Odorizzi, G. and Emr, S.D. (2002) Receptor downregulation and multivesicular-body sorting. *Nat Rev Mol Cell Biol* **3**, 893–905.

16 Hicke, L. and Dunn, R. (2003) Regulation of membrane protein transport by ubiquitin and ubiquitin-binding proteins. *Annu Rev Cell Dev Biol* **19**, 141–172.

17 Pickart, C. and Cohen, R.E. (2004) Proteasomes and their kin: proteases in the machine age. *Nat Rev Mol Cell Biol* **5**, 177–187.

18 Krappmann, D. and Scheidereit, C. (2005) A pervasive role of ubiquitin conjugation in activation and termination of IkappaB kinase pathways. *EMBO Rep* **6**, 321–326.

19 Galan, J.M. and Haguenauer-Tsapis, R. (1997) Ubiquitin lys63 is involved in ubiquitination of a yeast plasma membrane protein. *EMBO J* **16**, 5847–5854.

20 Huang, T.T., Nijman, S.M., Mirchandani, K.D., Galardy, P.J., Cohn, M.A., Haas, W., Gygi, S.P., Ploegh, H.L., Bernards, R. and D'Andrea, A.D. (2006) Regulation of monoubiquitinated PCNA by DUB autocleavage. *Nat Cell Biol* **8**, 339–347.

21 D'Azzo, A., Bongiovanni, A. and Nastasi, T. (2005) E3 ubiquitin ligases as regulators of membrane protein trafficking and degradation. *Traffic* **6**, 429–441.

22 Gao, M. and Karin, M. (2005) Regulating the regulators: control of protein ubiquitination and ubiquitin-like modifications by extracellular stimuli. *Mol Cell* **19**, 581–593.

23 Levkowitz, G., Waterman, H., Ettenberg, S.A., Katz, M., Tsygankov, A.Y., Alroy, I., Lavi, S., Iwai, K., Reiss, Y., Ciechanover, A., et al. (1999) Ubiquitin ligase activity and tyrosine phosphorylation underlie suppression of growth factor signaling by c-Cbl/Sli-1. *Mol Cell* **4**, 1029–1040.

24 Thien, C.B. and Langdon, W.Y. (2001) Cbl: many adaptations to regulate protein tyrosine kinases. *Nat Rev Mol Cell Biol* **2**, 294–307.

25 Schmidt, M.H. and Dikic, I. (2005) The Cbl interactome and its functions. *Nat Rev Mol Cell Biol* **6**, 907–919.

26 Reiley, W., Zhang, M., Wu, X., Granger, E. and Sun, S.C. (2005) Regulation of the deubiquitinating enzyme CYLD by IkappaB kinase gamma-dependent phosphorylation. *Mol Cell Biol* **25**, 3886–3895.

27 Bienko, M., Green, C., Crosetto, N., Rudolf, F., Zapart, G., Coull, B., Kannouche, P., Wider, G., Peter, M., Lehmann, A.R., et al. (2005) Ubiquitin-binding domains in Y-family polymerases regulate tranlesion synthesis. *Science* **310**, 1821–1824.

28 Lee, S., Tsai, Y.C., Mattera, R., Smith, W.J., Kostelansky, M.S., Weissman, A.M., Bonifacino, J.S. and Hurley, J.H. (2006) Structural basis for ubiquitin recognition and autoubiquitination by Rabex-5. *Nat Struct Mol Biol* **13**, 264–271.

29 Penengo, L., Mapelli, M., Murachelli, A.G., Confalonieri, S., Magri, L., Musacchio, A., Di Fiore, P.P., Polo, S. and Schneider, T.R. (2006) Crystal structure of the ubiquitin binding domains of rabex-5 reveals two modes of interaction with ubiquitin. *Cell* **124**, 1183–1195.

30 Harper, J.W. and Schulman, B.A. (2006) Structural complexity in ubiquitin recognition. *Cell* **124**, 1133–1136.

31 Slagsvold, T., Aasland, R., Hirano, S., Bache, K.G., Raiborg, C., Trambaiolo, D., Wakatsuki, S. and Stenmark, H. (2005) Eap45 in mammalian ESCRT-II binds ubiquitin via a phosphoinositide-interacting GLUE domain. *J Biol Chem* **280**, 19600–19606.

32 Hirano, S., Kawasaki, M., Ura, H., Kato, R., Raiborg, C., Stenmark, H. and Wakatsuki, S. (2006) Double-sided ubiquitin binding of Hrs-UIM in endosomal protein sorting. *Nat Struct Mol Biol* **13**, 272–277.

33 Kerscher, O., Felberbaum, R. and Hochstrasser, M. (2006) Modification of proteins by ubiquitin and ubiquitin-like proteins. *Annu Rev Cell Dev Biol* **22**, 159–180.

34 Welchman, R.L., Gordon, C. and Mayer, R.J. (2005) Ubiquitin and ubiquitin-like proteins as multifunctional signals. *Nat Rev Mol Cell Biol* **6**, 599–609.

35 Bylebyl, G., Belichenko, I. and Johnson, E. (2003) The SUMO isopeptidase Ulp2 prevents accumulation of SUMO chains in yeast. *J Biol Chem* **278**, 44113–44120.

36 Hecker, C.M., Rabiller, M., Haglund, K., Bayer, P. and Dikic, I. (2006) Specification of SUMO1- and SUMO2-interacting motifs. *J Biol Chem* **281**, 16117–16127.

37 Hannich, J.T., Lewis, A., Kroetz, M.B., Li, S.J., Heide, H., Emili, A. and Hochstrasser, M. (2005) Defining the SUMO-modified proteome by multiple approaches in *Saccharomyces cerevisiae*. *J Biol Chem* **280**, 4102–4110.

38 Song, J., Durrin, L.K., Wilkinson, T.A., Krontiris, T.G. and Chen, Y. (2004) Identification of a SUMO-binding motif that recognizes SUMO-modified proteins. *Proc Natl Acad Sci U S A* **101**, 14373–14378.

39 Minty, A., Dumont, X., Kaghad, M. and Caput, D. (2000) Covalent modification of p73alpha by SUMO-1. Two-hybrid screening with p73 identifies novel SUMO-1-interacting proteins and a SUMO-1 interaction motif. *J Biol Chem* **275**, 36316–36323.

40 Oved, S., Mosesson, Y., Zwang, Y., Santonico, E., Shtiegman, K., Marmor, M.D., Kochupurakkal, B.S., Katz, M., Lavi, S., Cesareni, G. and Yarden, Y. (2006) Conjugation to Nedd8 instigates ubiquitylation and down-regulation of activated receptor tyrosine kinases. *J Biol Chem* **281**, 21640–21651.

41 Hanahan, D. and Weinberg, R.A. (2000) The hallmarks of cancer. *Cell* **100**, 57–70.

42 Fang, S., Lorick, K.L., Jensen, J.P. and Weissman, A.M. (2003) RING finger ubiquitin protein ligases: implications for tumorigenesis, metastasis and for molecular targets in cancer. *Semin Cancer Biol* **13**, 5–14.

43 Nakayama, K.I. and Nakayama, K. (2006) Ubiquitin ligases: cell-cycle control and cancer. *Nat Rev Cancer* **6**, 369–381.

44 Jiang, Y.H. and Beaudet, A.L. (2004) Human disorders of ubiquitination and proteasomal degradation. *Curr Opin Pediatr* **16**, 419–426.

45 Mani, A. and Gelmann, E.P. (2005) The ubiquitin-proteasome pathway and its role in cancer. *J Clin Oncol* **23**, 4776–4789.

46 Wasch, R. and Engelbert, D. (2005) Anaphase-promoting complex-dependent proteolysis of cell cycle regulators and genomic instability of cancer cells. *Oncogene* **24**, 1–10.

47 Guardavaccaro, D. and Pagano, M. (2004) Oncogenic aberrations of cullin-dependent ubiquitin ligases. *Oncogene* **23**, 2037–2049.

48 Karin, M. (2006) Nuclear factor-kappaB in cancer development and progression. *Nature* **441**, 431–46.

49 Karin, M. and Ben-Neriah, Y. (2000) Phosphorylation meets ubiquitination: the control of NF-[kappa]B activity. *Annu Rev Immunol* **18**, 621–663.

50 Zhou, H., Wertz, I., O'Rourke, K., Ultsch, M., Seshagiri, S., Eby, M., Xiao, W. and Dixit, V.M. (2004) Bcl10 activates the NF-kappaB pathway through ubiquitination of NEMO. *Nature* **427**, 167–171.

51 Kanayama, A., Seth, R.B., Sun, L., Ea, C.K., Hong, M., Shaito, A., Chiu, Y.H., Deng, L. and Chen, Z.J. (2004) TAB2 and TAB3 activate the NF-kappaB pathway through binding to polyubiquitin chains. *Mol Cell* **15**, 535–548.

52 Wertz, I.E., O'Rourke, K.M., Zhou, H., Eby, M., Aravind, L., Seshagiri, S., Wu, P., Wiesmann, C., Baker, R., Boone, D.L. et al. (2004) De-ubiquitination and ubiquitin ligase domains of A20 downregulate NF-kappaB signalling. *Nature* **430**, 694–699.

53 Bignell, G.R., Warren, W., Seal, S., Takahashi, M., Rapley, E., Barfoot, R., Green, H., Brown, C., Biggs, P.J., Lakhani, S.R., et al. (2000) Identification of the familial cylindromatosis tumour-suppressor gene. *Nat Genet* **25**, 160–165.

54 Trompouki, E., Hatzivassiliou, E., Tsichritzis, T., Farmer, H., Ashworth, A. and Mosialos, G. (2003) CYLD is a deubiquitinating enzyme that negatively regulates NF-kappaB activation by TNFR family members. *Nature* **424**, 793–796.

55 Brummelkamp, T.R., Nijman, S.M., Dirac, A.M. and Bernards, R. (2003) Loss of the cylindromatosis tumour suppressor inhibits apoptosis by activating NF-kappaB. *Nature* **424**, 797–801.

56 Kovalenko, A., Chable-Bessia, C., Cantarella, G., Israel, A., Wallach, D. and Courtois, G. (2003) The tumour suppressor CYLD negatively regulates NF-kappaB signalling by deubiquitination. *Nature* **424**, 801–805.

57 Ikeda, F. and Dikic, I. (2006) CYLD in ubiquitin signaling and tumor pathogenesis. *Cell* **125**, 643–645.

58 Massoumi, R., Chmielarska, K., Hennecke, K., Pfeifer, A. and Fassler, R. (2006) Cyld inhibits tumor cell proliferation by blocking Bcl-3-dependent NF-kappaB signaling. *Cell* **125**, 665–677.

59 Bode, A.M. and Dong, Z. (2004) Post-translational modification of p53 in tumorigenesis. *Nat Rev Cancer* **4**, 793–805.

60 Sherr, C.J. (2004) Principles of tumor suppression. *Cell* **116**, 235–246.

61 Brooks, C.L. and Gu, W. (2006) p53 ubiquitination: Mdm2 and beyond. *Mol Cell* **21**, 307–315.

62 Oliner, J.D., Kinzler, K.W., Meltzer, P.S., George, D.L. and Vogelstein, B. (1992) Amplification of a gene encoding a p53-associated protein in human sarcomas. *Nature* **358**, 80–83.

63 Leach, F.S., Tokino, T., Meltzer, P.S., Burrel, M., Oliner, J.D., Smith, S., Hill, D.E., Sidransky, D., Kinzler, K.W. and Vogelstein, B. (1993) p53 mutation and Mdm2 amplification in human soft tissue sarcomas. *Cancer Res* **53**, 2231–2234.

64 Li, M., Brooks, C.L., Kon, N. and Gu, W. (2004) A dynamic role of HAUSP in the p53-Mdm2 pathway. *Mol Cell* **13**, 879–886.

65 Li, M., Chen, D., Shiloh, A., Luo, J., Nikolaev, A.Y., Qin, J. and Gu, W. (2002) Deubiquitination of p53 by HAUSP is an important pathway for p53 stabilization. *Nature* **416**, 648–653.

66 Cummins, J.M., Rago, C., Kohli, M., Kinzler, K.W., Lengauer, C. and Vogelstein, B. (2004) Tumour suppression: disruption of HAUSP gene stabilizes p53. *Nature* **428**, 1 p following 486.

67 Masuya, D., Huang, C., Liu, D., Nakashima, T., Yokomise, H., Ueno, M., Nakashima, N. and Sumitomo, S. (2006) The HAUSP gene plays an important role in non-small cell lung carcinogenesis through p53-dependent pathways. *J Pathol* **208**, 724–732.

68 Venkitaraman, A.R. (2004) Tracing the network connecting BRCA and Fanconi

anaemia proteins. *Nat Rev Cancer* **4**, 266–276.

69 Irminger-Finger, I. and Jefford, C.E. (2006) Is there more to BARD1 than BRCA1? *Nat Rev Cancer* **6**, 382–391.

70 Starita, L.M., Machida, Y., Sankaran, S., Elias, J.E., Griffin, K., Schlegel, B.P., Gygi, S.P. and Parvin, J.D. (2004) BRCA1-dependent ubiquitination of gamma-tubulin regulates centrosome number. *Mol Cell Biol* **24**, 8457–8466.

71 D'andrea, A.D. and Grompe, M. (2003) The Fanconi anaemia/BRCA pathway. *Nat Rev Cancer* **3**, 23–34.

72 Gregory, R.C., Taniguchi, T. and D'andrea, A.D. (2003) Regulation of the Fanconi anemia pathway by monoubiquitination. *Semin Cancer Biol* **13**, 77–82.

73 Vandenberg, C.J., Gergely, F., Ong, C.Y., Pace, P., Mallery, D.L., Hiom, K. and Patel, K.J. (2003) BRCA1-independent ubiquitination of FANCD2. *Mol Cell* **12**, 247–254.

74 Meetei, A.R., de Winter, J.P., Medhurst, A.L., Wallisch, M., Waisfisz, Q., Van De Vrugt, H.J., Oostra, A.B., Yan, Z., Ling, C., Bishop, C.E., et al. (2003) A novel ubiquitin ligase is deficient in Fanconi anemia. *Nat Genet* **35**, 165–170.

75 Nijman, S.M., Huang, T.T., Dirac, A.M., Brummelkamp, T.R., Kerkhoven, R.M., D'andrea, A.D. and Bernards, R. (2005) The deubiquitinating enzyme USP1 regulates the Fanconi anemia pathway. *Mol Cell* **17**, 331–339.

76 Lehmann, A.R. (2006) Translesion synthesis in mammalian cells. *Exp Cell Res* **312**, 2673–2676.

77 Lehmann, A.R. (2005) Replication of damaged DNA by translesion synthesis in human cells. *FEBS Lett* **579**, 873–876.

78 Kaelin, W.G., Jr. (2002) Molecular basis of the VHL hereditary cancer syndrome. *Nat Rev Cancer* **2**, 673–682.

79 Pause, A., Lee, S., Worrell, R.A., Chen, D.Y., Burgess, W.H., Linehan, W.M. and Klausner, R.D. (1997) The von Hippel-Lindau tumor-suppressor gene product forms a stable complex with human CUL-2, a member of the Cdc53 family of proteins. *Proc Natl Acad Sci USA* **94**, 2156–2161.

80 Cockman, M.E., Masson, N., Mole, D.R., Jaakkola, P., Chang, G.W., Clifford, S.C., Maher, E.R., Pugh, C.W., Ratcliffe, P.J. and Maxwell, P.H. (2000) Hypoxia inducible factor-alpha binding and ubiquitylation by the von Hippel-Lindau tumor suppressor protein. *J Biol Chem* **275**, 25733–25741.

81 Blume-jensen, P. and Hunter, T. (2001) Oncogenic kinase signalling. *Nature* **411**, 355–365.

82 Dikic, I. and Giordano, S. (2003) Negative receptor signalling. *Curr Opin Cell Biol* **15**, 128–135.

83 Peschard, P. and Park, M. (2003) Escape from Cbl-mediated downregulation: a recurrent theme for oncogenic deregulation of receptor tyrosine kinases. *Cancer Cell* **3**, 519–523.

84 Bache, K.G., Slagsvold, T. and Stenmark, H. (2004) Defective downregulation of receptor tyrosine kinases in cancer. *EMBO J* **23**, 2707–2712.

85 Mosesson, Y., Shtiegman, K., Katz, M., Zwang, Y., Vereb, G., Szollosi, J. and Yarden, Y. (2003) Endocytosis of receptor tyrosine kinases is driven by monoubiquitylation, not polyubiquitylation. *J Biol Chem* **278**, 21323–21326.

86 Citri, A. and Yarden, Y. (2006) EGF-ERBB signalling: towards the systems level. *Nat Rev Mol Cell Biol* **7**, 505–516.

87 Sebastian, S., Settleman, J., Reshkin, S.J., Azzariti, A., Bellizzi, A. and Paradiso, A. (2006) The complexity of targeting EGFR signalling in cancer: from expression to turnover. *Biochim Biophys Acta* **1766**, 120–139.

88 Yarden, Y. and Sliwkowski, M.X. (2001) Untangling the ErbB signalling network. *Nat Rev Mol Cell Biol* **2**, 127–137.

89 Lenferink, A.E., Pinkas-Kramarski, R., Van de Poll, M.L., Van Vugt, M.J., Klapper, L.N., Tzahar, E., Waterman, H., Sela, M., Van Zoelen, E.J. and Yarden, Y. (1998) Differential endocytic routing of homo- and hetero-dimeric ErbB tyrosine kinases confers signaling superiority to receptor heterodimers. *EMBO J* **17**, 3385–3397.

90 Muthuswamy, S.K., Gilman, M. and Brugge, J.S. (1999) Controlled dimerization of ErbB receptors provides evidence for differential signaling by homo- and heterodimers. *Mol Cell Biol* **19**, 6845–6857.

91 Slagsvold, T., Pattni, K., Malerod, L. and Stenmark, H. (2006) Endosomal and non-endosomal functions of ESCRT proteins. *Trends Cell Biol* **16**, 317–326.

92 Xu, Z., Liang, L., Wang, H., Li, T. and Zhao, M. (2003) HCRP1, a novel gene that is downregulated in hepatocellular carcinoma, encodes a growth-inhibitory protein. *Biochem Biophys Res Commun* **311**, 1057–1066.

93 Lee, M.P. and Feinberg, A.P. (1997) Aberrant splicing but not mutations of TSG101 in human breast cancer. *Cancer Res* **57**, 3131–3134.

94 Hariharan, I.K. and Bilder, D. (2006) Regulation of imaginal disc growth by tumor-suppressor genes in Drosophila. *Annu Rev Genet* **40**, 385–361.

95 Moberg, K.H., Schelble, S., Burdick, S.K. and Hariharan, I.K. (2005) Mutations in erupted, the Drosophila ortholog of mammalian tumor susceptibility gene 101, elicit non-cell-autonomous overgrowth. *Dev Cell* **9**, 699–710.

96 Thompson, B.J., Mathieu, J., Sung, H.H., Loeser, E., Rorth, P. and Cohen, S.M. (2005) Tumor suppressor properties of the ESCRT-II complex component Vps25 in Drosophila. *Dev Cell* **9**, 711–720.

97 Herz, H.M., Chen, Z., Scherr, H., Lackey, M., Bolduc, C. and Bergmann, A. (2006) vps25 mosaics display non-autonomous cell survival and overgrowth, and autonomous apoptosis. *Development* **133**, 1871–1880.

98 Vaccari, T. and Bilder, D. (2005) The Drosophila tumor suppressor vps25 prevents nonautonomous overproliferation by regulating notch trafficking. *Dev Cell* **9**, 687–698.

99 Nalepa, G., Rolfe, M. and Harper, J.W. (2006) Drug discovery in the ubiquitin–proteasome system. *Nat Rev Drug Discov* **5**, 596–613.

100 Adams, J. (2004) The development of proteasome inhibitors as anticancer drugs. *Cancer Cell* **5**, 417–421.

101 Richardson, P.G., Mitsiades, C., Hideshima, T. and anderson, K.C. (2005) Proteasome inhibition in the treatment of cancer. *Cell Cycle* **4**, 290–296.

102 Klapper, L.N., Waterman, H., Sela, M. and Yarden, Y. (2000) Tumor-inhibitory antibodies to HER-2/ErbB-2 may act by recruiting c-Cbl and enhancing ubiquitination of HER-2. *Cancer Res* **60**, 3384–3388.

103 Verma, R., Peters, N.R., D'Onofrio, M., Tochtrop, G.P., Sakamoto, K.M., Varadan, R., Zhang, M., Coffino, P., Fushman, D., Deshaies, R.J. and King, R.W. (2004) Ubistatins inhibit proteasome-dependent degradation by binding the ubiquitin chain. *Science* **306**, 117–120.

2
Regulation of the p53 Tumor-suppressor Protein by Ubiquitin and Ubiquitin-like Molecules

Dimitris P. Xirodimas

2.1
Functional Domains of p53

The p53 protein is a sequence-specific transcription factor, which either induces or represses expression of a variety of genes. This change in gene expression leads to either cell cycle arrest or apoptosis depending on cellular conditions. The p53 pathway is activated by a variety of genotoxic agents such as ultraviolet (UV) light [1], ionizing radiation [2], chemotherapeutic drugs [3] as well as by non-genotoxic treatments such as withdrawal of growth factors, hypoxia [4], heat shock, depletion of ribonucleoside triphosphates [5]. All these stimuli induce a nuclear accumulation of p53, whereas in normal unstressed cells the protein is present in immunologically undetectable levels. The kinetics of this response may vary depending on the stimulus applied, for example ionizing radiation results in a fast and transient p53 accumulation, while UV radiation induces more prolonged protein stabilization.

The p53 protein can be divided into three independent functional domains which coordinate and regulate the activity of each other in the complete protein. The N-terminus, which includes the first 100 amino acid residues, has been shown to mediate the transcriptional transactivation function of p53 and to be crucial for p53-mediated apoptosis. Components of the transcriptional machinery such as the TATA-associated factors TAFII70 and TAFII31 (subunits of TFIID) [6, 7], the p62 subunit of the transcription/repair factor TFIIH [8] or the co-activators CBP/p300 [9, 10] have been shown to interact with this region of p53. The N-terminus of p53 also contains a proline-rich domain (amino acids 62–91), which is important for the induction of p53-mediated apoptosis [11–13]. Amino acid residues 100–290 of human p53 form an independently folded protease-resistant domain, which binds to DNA in a sequence-specific manner [14, 15]. The DNA binding domain has been selected as a target in the process of tumor progression, as 90% of the missense point mutations in p53 identified in tumors are located in this domain and are responsible for the loss of the biological activity of wild type p53 [16, 17]. The C-terminus of human p53 contains the nuclear localization signal (amino acids 315–320) and the oligomerization domain (amino acids 324–355), which allows

Protein Degradation, Vol. 4: The Ubiquitin-Proteasome System and Disease.
Edited by R. J. Mayer, A. Ciechanover, M. Rechsteiner

ISBN: 978-3-527-31436-2

Mdm2 binding

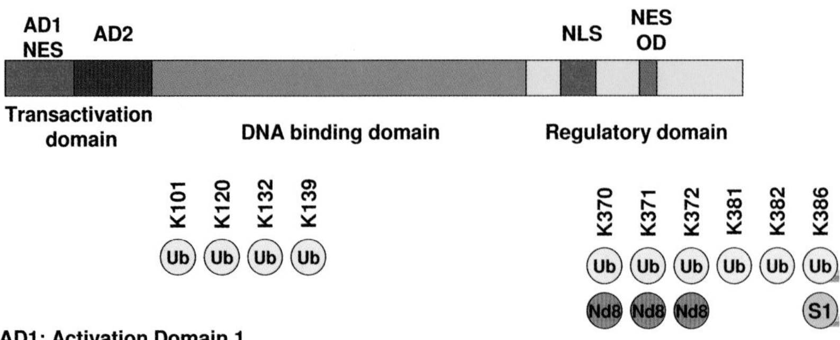

AD1: Activation Domain 1

AD2: Activation Domain 2 / Proline rich domain

NLS: Nuclear localisation signal

NES: Nuclear export signal

OD: Oligomerisation Domain

Fig. 2.1. Functional domains of p53. [6, 7, 11–13, 14–15, 18, 19, 20, 41–44, 72–74]

the formation of tetrameric p53 structures [18, 19]. The last 30 C-terminal p53 amino acid residues (363–393) are involved in the non-specific DNA and RNA binding activity of p53 as well as negatively regulating the sequence-specific DNA binding activity of the core of p53 [20]. Up until recently, the structure of the p53 gene was considered to be simple with the use of a single promoter and the production of two in mouse and three in human splice variants. It appears that the structure of the p53 gene is more complex than at first thought and through the use of an internal promoter in intron 4 and alternative splicing, six new p53 isoforms can be synthesized. Biologically, some of the p53 isoforms can differentially bind to p53 promoters, modulating the expression of p53 target genes [21].

2.2
The Family of Ubiquitin-like Molecules

Ubiquitin is the senior member of a growing family of small polypeptides, the ubiquitin-like family. These molecules are related to ubiquitin but are standing as unique pathways in controlling diverse biological processes. This family includes SUMO/sentrin/Smpt3 (SUMO-1, -2 and -3), NEDD8/Rub1, ISG15/UCRP, Fub, Fat 10 and Apg12 [22, 23]. Mechanistically, their conjugation is very similar to that of ubiquitin. The immature molecule is initially processed by a specific protease to expose at the C-terminus a di-glycine motif. Conjugation of the mature molecule on the ε amino group of a substrate lysine residue involves three main enzymatic activities. An E1 activating enzyme through an ATP-dependent step forms a high energy thiolester bond between an internal cysteine residue and the C-terminal glycine of the ubiquitin-like modifier. In the second step through transesterification, a thiolester bond is formed between the ubiquitin-like molecule and the E2 conjugating enzyme. Finally, through the action of an E3-ligase the lysine of the substrate is covalently modified [24–26]. However, at least for ubiquitin there is evidence that a fourth activity may also be involved (E4-ligase) which stimulates the formation of poly-ubiquitin chains [27]. Biochemical studies have demonstrated that at least for SUMO-1, -2 and -3 covalent modification of the substrate requires the activity of only the first two enzymes mentioned above. It is believed that an E3 ligase increases the efficiency of the transfer of the modifier from the E2 enzyme onto the substrate. For SUMO-1, -2, -3 the E1 enzyme is a heterodimer of SAE1/AOS1 and SAE2/Uba2, whereas the E2 conjugating enzyme is Ubc9 [28–31]. Modification of substrates with SUMO typically occurs on a lysine within a consensus sequence ψKXD/E (where ψ is a hydrophobic residue) [32]. The conjugating enzyme Ubc9 interacts with this motif, possibly explaining the conjugation of substrates in the absence of an E3-ligase activity [33–35]. SUMO-2 and -3 but not -1 have internal consensus sequences and can therefore form poly-SUMO chains [36]. For NEDD8 the E1 is also a heterodimer of APPBP1 (or Ula1 in yeast) and Uba3 proteins and the E2 conjugating enzyme is Ubc12 [37, 38]. These set of enzymes are unique for each pathway and they cannot use ubiquitin or other members of the family.

2.3
E3 Ligases for p53

There are two distinct families of E3-ligases. The HECT (Homologous to E6-AP Carboxyl Terminus, see below) family of E3s which make a thiolester bond with ubiquitin and the RING Finger E3s which instead facilitate the transfer of ubiquitin from the E2 to the substrate. A common phenotype for the RING ligases is their autocatalytic activities which enables them to control their own modification and stability [26]. A number of E3 ligases have been identified as regulators of p53 modification with ubiquitin/SUMO/NEDD8 (see Table 2.1). It was during the early 1990s that the first ligase was identified as a p53 regulator by the fact that the human papillomavirus protein E6 directly interacts with p53 and recruits the E6AP (E6 associated protein) E3-ligase. This results in p53 ubiquitination and proteasomal degradation [39]. Adenoviral proteins E1B55K and E4orf6, which also interact with p53, cooperate in a cullin-based E3-ligase complex to stimulate the degradation of p53 [40]. This is one of the mechanisms by which viral infection neutralizes p53 function and promotes cell proliferation. It was not until 1997 that the first cellular E3-ligase was identified which directly interacts with p53 and controls p53 levels through the ubiquitin/proteasome pathway. The Mdm2 oncogene product was known to interact with the N-terminus of p53 inhibiting p53 transcriptional activity by competing for p53 binding with factors of the basal

Table 2.1. E3-ligases which control the function of p53. [39, 40, 50–52, 53, 54, 59, 60–62, 63–67]

Ligase	Type
E6AP	HECT
E1B55k/E4orf6	Cullin based ligase
Mdm2	RING
COP1	RING
Pihr2	RING
CHIP	U-Box
ARFBP1	HECT
Topors	RING
WWR1	HECT
CARPs	RING
PIAS family	RING

transcriptional machinery [7, 41–44]. However, the significance of the p53–Mdm2 interaction was clearly demonstrated from gene "knock out" experiments in mice [45, 46]. Deletion of Mdm2 appeared to be embryonic lethal, but in contrast mice deficient for both Mdm2 and p53 were viable and developed normally, suggesting that a key activity of Mdm2 is to downregulate the growth-suppressing effects of p53. This motivated different groups to look more carefully at the effects of Mdm2 binding on p53. One approach was to create p53 mutants that were unable to interact with Mdm2 or Mdm2 mutants that were deficient for p53 binding and to use them in co-transfection experiments in cells. Interaction of p53 with Mdm2 dramatically reduced the p53 steady state levels, demonstrating that p53 is degraded through its interaction with Mdm2 [47, 48]. In another approach, peptides which were selected using phage-displayed peptide libraries, and which could disrupt the p53–Mdm2 interaction, were shown to stabilize p53 and activate p53-dependent transcription without the administration of any genotoxic stimuli [49]. Further biochemical studies showed that Mdm2 plays a direct role in the process by acting as an RING finger E3-ligase promoting the modification of p53 with ubiquitin [50–52]. As mentioned above p53 as a transcription factor induces the expression of different genes, one of them being the mdm2 gene itself, thus creating a negative feedback loop. Since then, two other RING finger ligases have been identified which also participate in a negative feedback loop with p53. Pihr2 and Cop1 E3-ligases were shown to stimulate p53 ubiquitination and proteasomal degradation and p53 induces transcription of the pihr2 and cop1 genes [53, 54]. Therefore, during the p53 response to stress stimuli, a variety of ubiquitin ligases can be induced which can then suppress the function of p53. Mdm2 is overexpressed in sarcomas, and increased protein levels of Pirh2 and Cop1 were recently detected in lung and breast tumors respectively [55–58]. These tumors contain wild-type p53, suggesting that the overexpression of these ligases could account for an over-silenced p53 function. ARF-BP1/Mule/HectH9 was recently identified as a HECT-type E3 ligase, which also negatively controls p53 levels. It was found to interact with the p14ARF tumor suppressor (see below) and to control p53 independently of Mdm2. However, ARF-BP1 is not a p53-induced gene and it was also shown to have p53-independent functions [59]. The chaperone-associated ubiquitin ligase CHIP can also suppress the levels of p53. CHIP through its interaction with Hsc70 and Hsp90 facilitates ubiquitination and degradation of chaperone-associated proteins [60]. Given the conformational flexibility of p53, its stability could be regulated through a transient association with molecular chaperones. TOPORS ligases were shown to stimulate modification of p53 both with SUMO and ubiquitin but the physiological implications are not known [61, 62]. A more specific role for the PIAS (Protein Inhibitor of Activated STAT) family of RING ligases in the SUMO conjugation of p53 has been demonstrated [63–65]. Recently, the WWP1 (WW domain-containing protein 1) ligase was shown to ubiquitinate p53 but this interaction seems to increase the levels of p53 [66]. In this case the increase was associated with a cytoplasmic accumulation of p53 and reduction in its transcriptional activity. Also, the CARP (caspase-8 and -10 associated RING proteins) family of apoptotic inhibitors were demonstrated to specifically suppress the levels of p53 phosphorylated at Ser15/20 [67].

2.4
Modification of p53 with Ubiquitin

Covalent modification of substrates with ubiquitin can occur in different formats. Ubiquitin can be conjugated as a single moiety or can form ubiquitin chains through modification of a pre-existing ubiquitin via an internal lysine residue. There are seven lysine residues in ubiquitin which can be used as acceptor sites (K6, K11, K27, K29, K33, K48 and K63). Formation of chains through different lysines creates a unique functional signal. For example chain formation through K48 is regarded as a signal for targeting the substrate for proteasomal degradation, whereas K63 linkage is involved in post replicative DNA repair, translation and endocytosis. One of the most extensively studied roles of p53 modification with ubiquitin is targeting p53 for 26S proteasomal degradation [68–71]. Initial studies identified the regulatory C-terminus of p53 as the domain required for p53 modification with ubiquitin. Specifically, mutation of six lysine residues (K370, K372, K373, K381, K382, K386) located in the last 30 amino acids of p53 dramatically reduced the levels of Mdm2-mediated ubiquitinated p53 [72, 73]. This p53 mutant is expressed at higher protein levels compared to wild-type p53 and has increased transcriptional activity. Additional lysines in the DNA binding domain of p53 (K101, K120, K132, K139) were recently identified [74]. The role of the C-terminal lysines in regulating p53 function was further addressed in a mouse model system where the six or seven C-terminal lysines in p53 where mutated to arginine. Interestingly, the mice were viable, developed normally and the stability of the p53 mutants was very similar to that of the wild-type p53. However, the p537KR mutant showed more rapid DNA damage in the thymus compared to the wild type [75] whereas deficiencies in the activation of p53 target genes in the p536KR mutant mice were observed [76]. It has to be noted however, that lysines are target residues for multiple post-translational modifications and therefore, phenotypes observed in *in-vivo* or *in-vitro* model systems are difficult to attribute to a particular modification. Furthermore, since different modifications may have different biological outcomes the overall functional effect of lysine mutants may represent an average of the individual responses. Initial biological studies suggested that the Mdm2–p53 interaction leads to nuclear export and degradation of p53 in the cytoplasm. This model was based on the identification of nuclear export sequences in Mdm2 and p53 and on the observation that inhibitors of the Crm1-dependent nuclear export such as leptomycin B, resulted in nuclear accumulation of p53 and activation of the response. By using heterokaryon assays, where two different types of cells are fused to create a cell with two discrete nuclei and a common cytoplasm, it was shown that p53 and Mdm2 can shuttle from one nuclei to the other through the common cytoplasm [77–79]. However, the interaction of p53 with Mdm2 is not sufficient to mediate export to the cytoplasm. Mutation of a critical cysteine residue in the RING finger of Mdm2 (C464A), which renders Mdm2 inactive as an E3-ligase but still able to interact with p53, severely impaired the ability of Mdm2 to facilitate nuclear export of p53 [80]. These data suggested that ubiquitin may act as the signal for this translocation event, leading to degradation of p53 in the cytoplasm. This step however, does not appear to be the only mechanism by which p53 is targeted to the proteasome as subsequent studies have shown that Mdm2-mediated ubiquitination and degradation of p53 may occur in the nucleus in the absence of any nuclear to cytoplasmic transport [81–83]. Further analysis suggested that the extent of p53 ubiquitination may control this process. Low levels of Mdm2 promote mono-ubiquitination of p53 facilitating p53 nuclear export, whereas high Mdm2 levels were able to promote multi-ubiquitination of p53 and proteasomal degradation in the nucleus. This model was supported by the observation that a p53 protein fused at its C-terminus with ubiquitin (mimicking mono-ubiquitination) was localized in the cytoplasm [84]. This appears to be a specific signal for ubiquitin as fusion of NEDD8 and SUMO-1 failed to change the localization of p53 [85]. This means of regulation could represent physiological conditions where in unstressed cells with low levels of Mdm2, p53 is exported to the cytoplasm for degradation or for transcription-independent p53 functions, such as the induction of apoptosis through interaction with mitochondria. However, at later stages of the p53 response or in malignancies where the levels of Mdm2 are high, p53 is poly-ubiquitinated and degraded in the nucleus.

2.5
Requirements for Mdm2-mediated Ubiquitination of p53

Ubiquitin modification of a substrate requires the direct interaction of the E3-ligase. Clearly, Mdm2 is the most intensively studied p53 E3-ligase and much information has now been amassed with regard to its requirements for the promotion of p53 ubiquitination and proteasomal degradation. Mdm2 is a 90-kDa protein divided into four major conserved domains: an N-terminal domain (amino acids 23–108), a central domain which contains a highly acidic region (amino acids 209–275), a zinc finger domain (amino acids 289–333) and a RING finger domain (amino acids 460–490) [86, 87]. The studies in which the role of Mdm2 as a regulator of p53 stability was discovered, also showed that direct interaction with p53 is necessary for this process [47, 48]. Studies using peptide mimetics, peptide phage display libraries approaches, crystallographic or mutational analysis showed that amino acids 14–27 in p53 interact with a hydrophobic pocket in the N-terminus of Mdm2. More specifically, Phe 19, Try 23 and Leu 26 are crucial for this interaction as their side chains are buried within the Mdm2 hydrophobic pocket and are the main binding contacts [43, 49, 88–90]. The RING finger domain is required for the E3-ligase activity of Mdm2 and for its suppressive role towards the transcriptional activity of p53. Mutations of potential zinc-coordinating residues in this domain showed that an intact RING finger is required for Mdm2 to promote p53 and its own ubiquitination [50, 51, 91]. Furthermore, this domain of Mdm2 is important for its interaction with its homolog, Mdmx and it is believed that heterodimer formation is mediated via interaction of their RING finger domains [92]. Much interest has been focussed on the central domain of Mdm2 and its role in the regulation of p53 and Mdm2 stability. This domain is responsible for many of the Mdm2 interactions with regulatory proteins such as p14ARF, p300, YY1, Rb, ribosomal proteins, Kap1 and TAFII250 (see below). In particular, deletion of the acidic domain inhibits the ability of Mdm2 to promote p53 degradation. Experiments with hybrid mutants containing Mdm2 and Mdmx domains also showed the importance of the acidic domain in targeting p53 for degradation [93, 94]. Further mutational analysis of potential phosphorylation of serine residues in mouse Mdm2 (Ser238, 240, 244, 251, 254, 258, 260) showed that these mutants were deficient in promoting p53 degradation. However, their ability to ubiquitinate p53 was not affected [95]. Lack of p53 degradation despite efficient ubiquitination was previously shown with an in-frame Mdm2 deletion mutant (amino acids 217–246) deficient for p300 binding and with p53 mutants [82, 96]. This highlights that proteasomal degradation of p53 involves a post-ubiqutination step, which could be regulated through phosphorylation of Mdm2 and/or interaction with additional proteins. However, apart from the main binding area for p53 and Mdm2, additional areas of the proteins interact as secondary binding motifs. From NMR studies on Mdm2 fragments bound to N-terminal p53 peptides, it was suggested that Mdm2 is conformationally flexible, and subject to allosteric regulation upon substrate binding [89]. Consistent with this, is the fact that *in-vitro* interaction of Mdm2 with RNA renders Mdm2 capable of binding to p53 lacking the N-terminus [97]. Further biochemical and biological studies have shown that the acidic domain of Mdm2 can interact with a flexible region in p53 within the DNA binding domain, which is frequently found unfolded in human tumors [98, 99]. While interaction of p53 with Mdm2 through their N-termini is important for p53 degradation, the secondary binding interface may control efficient p53 ubiquitination and proteasomal degradation [100]. On the other hand, p53 needs to have the correct oligomerization status for proper association and processing by

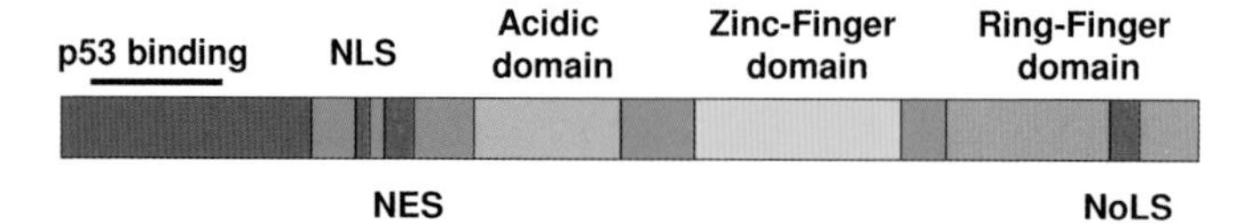

NLS: Nuclear localisation signal

NES: Nuclear export signal

NoLS: Nucleolar localisation signal

Fig. 2.2. Functional domains of Mdm2. [86, 87, 116]

Mdm2. p53 mutants unable to form tetramers are deficient in Mdm2 binding and degradation [101]. These conformational/structural requirements for Mdm2 and p53 to assemble a complex which can be processed by the proteasome, underlie the complexity and precise regulation of these processes.

2.6
Regulation of p53 Ubiquitination

2.6.1
E2 Conjugating Enzymes

As mentioned above, modification of a substrate with ubiquitin or ubiquitin-like molecules involves three well-characterised enzymatic activities. For ubiquitin, there is one E1-activating enzyme and multiple E2-conjugating enzymes; each of the E2 enzymes is capable of cooperating with different E3-ligases to promote modification of the substrates. Different mechanisms exist to modulate the modification status of p53 with regard to ubiquitin. Characterization of the specificity of Mdm2 in the selection of E2-conjugating enzymes for p53 ubiquitination showed that *in vivo* the UbcH5B/C are physiological E2s for Mdm2. These enzymes are involved in the formation of K48-linked ubiquitin chains. Downregulation of these E2s by siRNA led to the accumulation of p53 and Mdm2 proteins in cells. However, the accumulated p53 was transcriptionally inactive and this was due, at least in part, to the concomitant accumulation of the Mdm2 protein. Interestingly, this study also showed that known activators of the p53 response, such as doxorubicin and actinomycin D reduced the levels of expression of UbcH5B/C. This raises the possibility that the levels of E2-conjugating enzymes could be affected by signaling pathways that control the levels of p53 [102]. Ubc13 an E2-conjugating enzyme, which is involved in the formation of complexes with ubiquitin variant proteins (Uevs) to stimulate the formation of K63-dependent ubiquitin chains, was reported to promote the cytoplasmic localization of p53. As with UbcH5B/C the levels of Ubc13 were modulated by stress stimuli, in this case ionizing radiation caused reduction of Ubc13 levels. Ubc13 was found to directly interact with p53 and it still not known whether an E3-ligase complex is required to produce this effect [103].

2.6.2 Interacting Proteins

The p53 pathway is also controlled through the interaction of p53 and Mdm2 with other cellular proteins. The $p14^{ARF}$ tumor suppressor is now well established as a regulator of the Mdm2 protein. ARF (p14 in human, p19 in mice) was identified as an alternative transcript of the Ink4α/ARF locus, which also expresses the $p16^{Ink4\alpha}$ cyclin dependent kinases inhibitor [104, 105]. ARF has been shown to directly interact with Mdm2 and to protect p53 from Mdm2-mediated proteasomal degradation [106–108]. Mechanistically, initial studies proposed that ARF, which itself is predominantly localized in the nucleolus, sequesters Mdm2 to the same compartment. This disrupts the interaction with p53 allowing it to accumulate in the nucleoplasm [109, 110]. Other studies proposed that ARF was able to inhibit the nuclear export of the p53–Mdm2 complex via a step involving the nucleolus [111]. However, protection of p53 from Mdm2-mediated degradation by ARF through the nucleolus does not seem to be the only mechanism of action. ARF mutants which were localized in the nucleoplasm were reported to stabilize p53 and activate the p53 response [112].

Biochemical studies showed that ARF was able to inhibit the E3-ligase activity of Mdm2 [113, 114]. Further *in vivo* studies showed that ARF has a differential role in controlling the ligase activity of Mdm2. Blockade of the proteasomal degradation of p53 and Mdm2 by ARF was accompanied by the inhibition of p53 ubiqutination but not Mdm2 auto-ubiquitination [115]. Mdm2 contains a cryptic nucleolar sequence at the C-terminus (amino acids 466–473) which becomes exposed on interaction with ARF. Mutation of this signal makes Mdm2 resistant to nucleolar localization when ARF is bound [116]. This Mdm2 mutant (Mdm2NoLS), despite its inability to degrade p53, is still auto-ubiquitinated. Expression of ARF as with wild-type Mdm2, blocked proteasomal degradation of Mdm2NoLS without affecting its auto-ubiquitination [115]. This suggests that Mdm2 can be ubiquitinated and degraded outside the nucleolus and ARF can block this step in the absence of nucleolar sequestration. This is consistent with the idea that the “trans” and “auto” ubiquitination of Mdm2 are differentially modulated. ARF also has p53- and Mdm2-independent functions and recently the ARF-BP1 ligase, which controls p53 stability was also shown to be inhibited by ARF [59].

The role of the nucleolus in controlling the function of p53 was recently expanded. In an elegant study, DNA damage of the nucleolus was shown to be necessary and sufficient to stabilize p53 and activate its response [117]. Furthermore, ribosomal proteins such as L11, L5, L23 whose nucleolar localization is part of the ribosome assembly process, were also shown to protect p53 from proteasomal degradation [118–122]. These studies identified a link between the p53 pathway and ribosomal biogenesis, with p53 sensing any nucleolar stress possibly through Mdm2. Interestingly, ribosomal protein L26 was shown to interact with the 5′ untranslated region of p53 mRNA and increase p53 translation during DNA damage [123]. Therefore, by increasing protein synthesis and decreasing proteasomal degradation of p53 an optimal response is achieved. This is further supported by an elegant study showing that Mdm2 protein is rapidly degraded as a result of DNA damage, which then leads to the accumulation of p53 [124]. The involvement of the translational process in regulating p53 stability had been previously suggested with the demonstration that Mdm2 was able to increase protein synthesis of the full length p53 and a shorter form through a second initiation site at residue 47. This form of p53 (p53/47) lacks the N-terminal Mdm2 binding site and by oligomerizing with the full length p53 regulates the stability and function of the full-length p53 [125]. The Mdm2 homolog Mdmx has proven to be an important regulator of p53 [87]. Knockout of the Mdmx gene in mice caused embryonic lethality which was rescued by the concomitant deletion of p53 [126–128]. Mdmx can inhibit p53 transcriptional activity through its interaction with p53 at its N-terminus [129, 130]. However, despite the presence of a C-terminal RING domain which is structurally similar to Mdm2, the ability of Mdmx to promote p53 ubiquitination is very low compared to that of Mdm2 [131]. When overexpressed, Mdmx was shown to block proteasomal degradation of p53 and Mdm2. Mdmx was shown to localize predominantly in the cytoplasm but Mdm2 was able to recruit Mdmx into the nucleus, which led to the inhibition of proteasomal degradation of p53 and Mdm2. However, in contrast to ARF expression, Mdmx preferentially inhibited the auto-ubiquitination activity of Mdm2 but not the ubiquitination of p53 [132–134]. On the other hand, Mdm2 as an E3-ligase can control the levels of Mdmx and it appears that these two proteins cooperate to control each other's stability and the function of p53 [135–137].

A similar autoregulatory loop was identified between Mdm2 and the tumor susceptibility gene 101 (tsg101). Deletion of tsg101 in mice caused embryonic lethality and the accumulation of p53 and this lethality was delayed by the concomitant deletion of p53. TSG101 was found to interact with both Mdm2 and p53 to control their stability. Expression of TSG101 stabilized Mdm2 and decreased the levels of p53, presumably through the increase in Mdm2 levels. This effect was dependent on the ubiquitin conjugating-like domain (Ubc) found in TSG101, possibly affecting Mdm2 ubiquitination. On the other hand Mdm2 is capable of destabilizing TSG101, creating an autoregulatory loop which controls the stability of p53 [138, 139]. These data suggest that apart from the regulation of Mdm2 ligase activity either towards itself or the substrate, there is control of a post-ubiquitination step, essential for the proteasomal processing of p53 and Mdm2.

Biochemical studies suggested that Mdm2 as an E3-ligase promotes the initial modification of p53 with ubiquitin as multiple single moieties (multi mono-ubiquitination), suggesting that additional factors may be involved in the formation of poly-ubiquitinated p53 [140]. Indeed, the transcriptional co-activator p300 and the transcription factors YY1 (Yin Yang 1) were shown to directly interact with Mdm2 and cooperate in the generation of polyubiquitin chains. Both factors interact with the central domain of Mdm2 which plays an important role in p53 degradation

[141, 142]. Mdm2 showed similar cooperation in stimulating the ubiquitination of p53 through the central acidic domain, with the transcriptional co-repressor Kap1, the transcription factor TAFII250, the Mdm2 binding protein MTBP and gankyrin a protein which is commonly overexpressed in hepatocellular carcinomas [143–146]. Transfer of ubiquitinated substrates to the proteasome is typically mediated through carrier proteins, which can interact with both the ubiquitin chain (through a UBA domain) and the proteasome (through a UBL domain) [23, 147]. Rad23 (*S. cerevisiae*) or the human homolog, hHR23A and hHR23B, proteins are involved in the global nucleotide excision repair (NER) pathway and they contain both UBA and UBL domains. A post-ubiquitination role for hHR23 was proposed in the regulation of p53 degradation. An Mdm2–hHR23 complex was identified and the presence of hHR23 was found to be required for genotoxic activation of the p53 response [148–150]. In addition, the retinoblastoma susceptibility protein (pRb) protects p53 from proteasomal degradation and inhibits the anti-apoptotic effects of Mdm2 by its interaction with Mdm2 (amino acids 273–321). However, pRb did not inhibit the suppressing effects of Mdm2 on p53 transactivation activity but rather on p53-mediated transrepression [151].

2.6.3 By Other Post-translational Modifications

Additional post-translational modifications can regulate the ability of Mdm2 to promote p53 ubiquitination and proteasomal degradation. In response to DNA damage both p53 and Mdm2 are targets for phosphorylation events which control the suppressive function of Mdm2 on p53. Phosphorylation/dephosphorylation events involving Ser15, Thr18 and Ser20 are thought to regulate the interaction of Mdm2 with p53 and therefore its capacity to modulate p53 stability. The ATM/ATR family of kinases activated after DNA damage are thought to be involved in these events either by directly phosphorylating p53 (Ser15) or indirectly through the activation of other kinases such as Chk1, Chk2 and Plk3 (Ser20). These phosphorylation events can also "prime" p53 for additional modifications, such as the recruitment of histone acetyltransferases, which acetylate p53 at Lys382 [152–155]. Acetylation of p53 is implicated in the control of p53 stability and its transcriptional activity [156]. This interplay between different modification events was highlighted by the finding that Mdm2 associated with HDAC1 histone deacetylase, resulting in the deacetylation of p53 lysines known to be present in their acetylated form [157]. Deacetylation of p53 is further facilitated by the action of the Kap1 co-repressor. Kap1 was shown to directly interact with Mdm2 and recruit HDAC1 to p53, resulting in deacetylation and Mdm2-dependent increase in the ubiquitination of p53 [146]. It is conceivable, that these lysines become accessible to other modification pathways that use the same residues in p53, such as ubiquitin and ubiquitin-like molecules. This highlights that different signaling pathways can alter or finely modulate the function of p53 by synergizing or competing in its modification. Not surprisingly, the role of Mdm2 as a regulator of p53 stability is controlled by its degree of phosphorylation. When the DNA is damaged, Ser395 and Tyr394 in Mdm2 are phosphorylated by ATM and its downstream tyrosine kinase c-Abl respectively, which then inhibits the ability of Mdm2 to degrade p53 [158–160]. In the case of Ser395, phosphorylation blocks Mdm2-mediated nuclear export and degradation of p53 [161]. The subcellular localization of Mdm2 is also controlled by other induced phosphorylation events. Survival signaling pathways such as AKT activation leads to the phosphorylation of Mdm2 at Ser166 and Ser186 [162, 163] which stimulates the nuclear import of Mdm2 and the decrease in p53 levels. Activation of c-Abl also leads to the phosphorylation of Tyr276 in Mdm2, which then blocks the Mdm2-mediated degradation of p53. This phosphorylation event also facilitates the interaction between Mdm2 and ARF, which leads to nucleolar localization of Mdm2 [164].

2.7
De-ubiquitination of p53

Modification of substrates with ubiquitin and ubiquitin-like molecules is a very dynamic process and is brought about by the action of de-conjugating enzymes which specifically remove the modified molecule from the substrate. Various de-ubiquitinating enzymes which control the p53 degradation process have been identified. HAUSP/USP7 was first identified as a p53 interacting protein, which could block/reverse Mdm2-mediated ubiquitination of p53 and activate the p53 response [165]. However, this effect appears to be more complex as it depends on the relative amounts of HAUSP in cells. As mentioned previously, Mdm2 as a RING ligase is able to promote its own ubiquitination and proteasomal degradation. HAUSP was also shown to interact with Mdm2 to prevent its auto-ubiqutination and degradation. Therefore, in the total absence of HAUSP (gene knockout in cells) the levels of Mdm2 dramatically decrease thus causing an increase in p53 levels [166]. However, a partial reduction in HAUSP levels (siRNA knockdown in cells) caused the levels of p53 protein to decline, suggesting that under these conditions the ubiquitination of p53 but not of Mdm2, was affected [167]. The role of HAUSP in the p53–Mdm2 pathway was shown to be regulated by the death-associated protein Daxx. By simultaneously interacting with HAUSP and Mdm2, Daxx facilitates the de-ubiquitination and stabilization of Mdm2. On the other hand, Daxx promotes Mdm2-mediated ubiquitination and degradation of p53. In the presence of DNA damage Daxx dissociates from Mdm2, leading to Mdm2 destabilization and the activation of the p53 response [168]. Furthermore, HAUSP was shown to control the levels of Mdmx. The de-ubiquitinating activity of HAUSP is impaired after DNA damage, providing a possible mechanism for the rapid and transient destabilization of Mdm2/Mdmx which has been observed under conditions of stress [124, 169, 170]. Two more de-ubiquitinating enzymes were recently reported to control the p53–Mdm2 pathway. USP2a was identified in a bacterial two-hybrid screen for novel Mdm2 interactors and was shown to specifically de-ubiquitinate Mdm2 but not p53. Overexpression of USP2a caused an accumulation of Mdm2 and a decrease in p53 levels through the proteasome pathway. This may be relevant to 50% of prostate tumors where the USP2 gene (encodes for USP2a and USP2b) is found to be amplified. As many of these tumors contain wild-type p53, overexpression of USP2a could explain the suppression of the p53 pathway [171]. USP28 a de-ubiquitinating enzyme found in complex with one of the DNA damage mediator proteins, 53BP1, was reported to be a critical regulator of the apoptotic response induced by p53 after exposure to ionizing radiation (IR) [172].

2.8 SUMO-1/sentrin/smpt3

In contrast to ubiquitin, SUMO modification cannot act as a degradation signal for the 26S proteasome. In fact, in a few reported cases SUMO has been found to block proteasomal degradation by competing with ubiquitin. One of the emerging roles for SUMO conjugation is to control transcriptional activity. With regard to p53, initial reports suggested that conjugation with SUMO-1 increases its transcriptional activity. Lysine 386 was shown to be the acceptor residue for SUMO-1 and is one of the lysines required for the conjugation of ubiquitin to p53; it is also present in the identified SUMO consensus sequence ψKXD/E [173, 174]. Two-hybrid screens originally identified members of the PIAS family of ligases (see above) as interactors with p53 and soon after biochemical and biological studies demonstrated the role of PIAS as regulators of p53–SUMO conjugation [64]. Different groups have assessed the role of these ligases in p53 function with contradictory data [175]. Based on the initial studies it appears that expression of ubc9 and SUMO-1 increased p53 transcriptional activity [173, 174, 176]. However, expression of the PIAS ligases (PIAS1 and PIASx) enhanced p53 SUMO conjugation but resulted in suppression of p53 transcriptional activity [65]. This effect was dependent on the intact E3-ligase activity, as RING finger point mutants were unable to stimulate p53–SUMO conjugation and suppress p53 function. In a more recent study, PIASy was also shown to stimulate p53–SUMO conjugation but the effect of this PIAS member was to activate p53 transcriptional activity [63]. The effect of the SUMO pathway on p53 function becomes more complex, as the p53 regulators Mdm2 and Mdmx are also SUMO conjugated. Removal of SUMO-1 from Mdm2 through the action of the SUMO protease, SUSP4, resulted in increased auto-ubiquitination of Mdm2 and an increase in p53 stability. The levels of SUSP4 were increased after UV-induced DNA damage, providing another example of cooperation between different conjugation pathways in the activation of the p53 response [177].

2.9 NEDD8/Rub1

NEDD8 shares the greatest homology with ubiquitin amongst the family of ubiquitin-like molecules. However, a distinct set of enzymatic activities are involved in

NEDD8 modification of substrate proteins. Genetic experiments in yeast, plants, *C. elegans*, Drosophila and mice have shown an important role for NEDD8 in cell cycle and cell viability [178–181]. However, up until recently the only well-described substrate for NEDDylation was the cullin family of proteins. Cullins are scaffold proteins for the SCF–ubiquitin ligase complex (Skip-1, Cullin, F-box). The role of cullin NEDDylation is to increase the ubiquitin ligase activity of the complex towards its substrates [182]. Two of the first identified non-cullin substrates for NEDD8 were the p53 and Mdm2 proteins. Mdm2 as a RING finger ligase promoted *in vivo* modification of itself and p53 with NEDD8. This modification required the C-terminus of p53 but in contrast to ubiquitin a p53 mutant with three lysines in that domain (K370, K372, K373) mutated into arginine, was deficient in Mdm2-mediated NEDDylation. The role of NEDD8 in the control of p53 function was assessed in the TS41 system. These Chinese Hamster Ovary (CHO) cells carry a temperature sensitive mutation in the APP-BP1 gene (one of the components of the NEDD8 E1-activating enzyme; see above) [183]. When these cells were grown at restrictive temperatures (where the pathway is switched off) the transcriptional activity of p53 was increased. Furthermore, in the absence of the NEDDylation pathway Mdm2 was more potent in inhibiting p53 activity. These data suggest a suppressive role for the NEDD8 conjugation pathway in the transcriptional activity of p53 [184]. The FBX011 protein, a member of the F-box protein family and a component of the SCF complex, was reported to promote the modification of p53 with NEDD8 but not with ubiquitin. Expression of FBX011 did not affect the protein levels of p53 but inhibited p53 transcriptional activity, consistent with a suppressive role for NEDD8 in p53 function [185].

2.10 Therapeutic Intervention through the Ubiquitin Pathway

As previously mentioned, in many tumors, despite the presence of a wild-type p53 gene the pathway is not functional and in some cases this may result from the overexpression of E3-ligases which over-suppress the function of p53. There is evidence to suggest that tumor cells may be more sensitive to the activation of p53 compared to healthy cells, making the activation of p53 an attractive target for potential anti-cancer drugs [186]. The studies, which identified small peptides that disrupt the p53–Mdm2 interaction, showed that this was sufficient to increase the levels of p53 in the absence of any genotoxic stimuli [49]. This provides a proof of principle for targeting the interaction between p53 and Mdm2 or the function of the E3-ligase Mdm2. In support of the former strategy, small molecules called nutlins were isolated as potent and specific inhibitors of the p53–Mdm2 interaction. Nutlins bind in the hydrophobic pocket of Mdm2, in which the N-terminus of p53 is buried. These drug-like molecules were shown to stabilize and activate p53 only in tumor cells that contained wild-type p53 [187]. The ligase activity of Mdm2 has also been used as a target for small molecule inhibitors. Using an *in vitro* Mdm2–p53 ubiquitination assay, compounds were isolated that inhibited p53 ubiquitination. Interestingly, these molecules did not affect Mdm2 auto-ubiquitination and had a similar effect to ARF expression in cells [188]. In other studies, compounds which inhibited the auto-ubiquitination activity of Mdm2 were isolated *in vitro*. At first glance, this may seem paradoxical as use of inhibitors of this class will lead to an accumulation of the Mdm2 protein, which in principal could still interact with p53 and inhibit its function. However, these compounds were shown to stabilize p53 and Mdm2 in cells, and to activate p53 transcriptional activity [189]. Inhibitors of proteasome function such as PS-341 are now being used as treatment for multiple myeloma. The exact mechanism of the anti-tumor action of these compounds is not known, but is characterized by an accumulation of p53 and the induction of phosphorylation at Ser15 [190]. As previously mentioned, over 50% of human tumors contain mutant p53 with the majority of mutations located in the core DNA-binding domain of the protein. More specifically, these mutations either involve residues which make direct contact with DNA ("contact mutants"), or residues that provide structural stability, and proper positioning of the DNA contact residues ("structural mutants"). Restoring the wild-type activity of p53 mutants could be a valuable approach to suppress the uncontrolled growth of these tumors. This strategy is being tested with small molecules such as PRIMA-1, MIRA-1 and CP-31398, which can restore wild type conformation to some p53 mutants [191]. The role of the ubiquitin pathway in the transcriptional silencing of p53 mutants has not yet been addressed. There is evidence to suggest that structural mutants of p53 are thermodynamically unstable leading to an increased association with the chaperone machinery. Chaperone-associated ligases, such as CHIP can promote proteasomal degradation of both mutant and wild-type p53, but this process is more efficient in the case of the mutant protein. Since wild-type p53 is conformationally flexible, it is possible that it takes part in a dynamic interaction with the chaperone machinery, but this event is more profound for the more thermodynamically unstable p53 mutants [60, 192]. Furthermore, mutations in the DNA-binding domain caused an increase in p53 ubiquitination [98]. Clearly, investigation of the role of the ubiquitin pathway in the regulation of mutant p53 function could provide a new approach to restoring the wild type function of mutant p53 in tumor cells.

Acknowledgments

DX is a Research Fellow of the Association for International Cancer Research (AICR).

References

1 Maltzman, W. and Czyzyk, L. (1984) UV irradiation stimulates levels of p53 cellular tumor antigen in nontransformed mouse cells. *Mol Cell Biol* **4**, 1689–1694.

2 Kastan, M.B., Onyekwere, O., Sidransky, D., Vogelstein, B. and Craig, R.W. (1991) Participation of p53 protein in the cellular response to DNA damage. *Cancer Res* **51**, 6304–6311.

3 Fritsche, M., Haessler, C. and Brandner, G. (1993) Induction of nuclear accumulation of the tumor-suppressor protein p53 by DNA-damaging agents. *Oncogene* **8**, 307–318.

4 Graeber, T.G., Peterson, J.F., Tsai, M., Monica, K., Fornace, A.J. Jr. and Giaccia, A.J. (1994) Hypoxia induces accumulation of p53 protein, but activation of a G1-phase checkpoint by low-oxygen conditions is independent of p53 status. *Mol Cell Biol* **14**, 6264–6277.

5 Linke, S.P., Clarkin, K.C., Di Leonardo, A., Tsou, A. and Wahl, G.M. (1996) A reversible, p53-dependent G0/G1 cell cycle arrest induced by ribonucleotide depletion in the absence of detectable DNA damage. *Genes Dev* **10**, 934–947.

6 Lu, H. and Levine, A.J. (1995) Human TAFII31 protein is a transcriptional coactivator of the p53 protein. *Proc Natl Acad Sci USA* **92**, 5154–5158.

7 Thut, C.J., Chen, J.L., Klemm, R. and Tjian, R. (1995) p53 transcriptional activation mediated by coactivators TAFII40 and TAFII60. *Science* **267**, 100–104.

8 Xiao, H., Pearson, A., Coulombe, B., Truant, R., Zhang, S., Regier, J.L., Triezenberg, S.J., Reinberg, D., Flores, O., Ingles, C.J. et al. (1994) Binding of basal transcription factor TFIIH to the acidic activation domains of VP16 and p53. *Mol Cell Biol* **14**, 7013–7024.

9 Gu, W. and Roeder, R.G. (1997) Activation of p53 sequence-specific DNA binding by acetylation of the p53 C-terminal domain. *Cell* **90**, 595–606.

10 Lill, N.L., Grossman, S.R., Ginsberg, D., DeCaprio, J. and Livingston, D.M. (1997) Binding and modulation of p53 by p300/CBP coactivators. *Nature* **387**, 823–827.

11 Baptiste, N., Friedlander, P., Chen, X. and Prives, C. (2002) The proline-rich domain of p53 is required for cooperation with anti-neoplastic agents to promote apoptosis of tumor cells. *Oncogene* **21**, 9–21.

12 Walker, K.K. and Levine, A.J. (1996) Identification of a novel p53 functional domain that is necessary for efficient growth suppression. *Proc Natl Acad Sci USA* **93**, 15335–15340.

13 Zhu, J., Zhou, W., Jiang, J. and Chen, X. (1998) Identification of a novel p53 functional domain that is necessary for mediating apoptosis. *J Biol Chem* **273**, 13030–13036.

14 Bargonetti, J., Friedman, P.N., Kern, S.E., Vogelstein, B. and Prives, C. (1991) Wild-type but not mutant p53 immunopurified proteins bind to sequences adjacent to the SV40 origin of replication. *Cell* **65**, 1083–1091.

15 Kern, S.E., Kinzler, K.W., Bruskin, A., Jarosz, D., Friedman, P., Prives, C. and Vogelstein, B. (1991) Identification of p53 as a sequence-specific DNA-binding protein. *Science* **252**, 1708–1711.

16 Cho, Y., Gorina, S., Jeffrey, P.D. and Pavletich, N.P. (1994) Crystal structure of a p53 tumor suppressor-DNA complex: understanding tumorigenic mutations. *Science* **265**, 346–355.

17 Pavletich, N.P., Chambers, K.A. and Pabo, C.O. (1993) The DNA-binding domain of p53 contains the four conserved regions and the major mutation hot spots. *Genes Dev* **7**, 2556–2564.

18 Jeffrey, P.D., Gorina, S. and Pavletich, N.P. (1995) Crystal structure of the tetramerization domain of the p53 tumor suppressor at 1.7 angstroms. *Science* **267**, 1498–1502.

19 Lee, W., Harvey, T.S., Yin, Y., Yau, P., Litchfield, D. and Arrowsmith, C.H. (1994) Solution structure of the tetrameric minimum transforming domain of p53. *Nat Struct Biol* **1**, 877–890.

20 Hupp, T.R., Meek, D.W., Midgley, C.A. and Lane, D.P. (1992) Regulation of the specific DNA binding function of p53. *Cell* **71**, 875–886.

21 Bourdon, J.C., Fernandes, K., Murray-Zmijewski, F., Liu, G., Diot, A., Xirodimas, D.P., Saville, M.K. and Lane, D.P. (2005) p53 isoforms can regulate p53 transcriptional activity. *Genes Dev* **19**, 2122–2137.

22 Kerscher, O., Felberbaum, R. and Hochstrasser, M. (2006) Modification of proteins by ubiquitin and ubiquitin-like

proteins. *Annu Rev Cell Dev Biol* **22**, 159–180.

23 Welchman, R.L., Gordon, C. and Mayer, R.J. (2005) Ubiquitin and ubiquitin-like proteins as multifunctional signals. *Nat Rev Mol Cell Biol* **6**, 599–609.

24 Ciechanover, A. (1998) The ubiquitin-proteasome pathway: on protein death and cell life. *EMBO J*, **17**, 7151–7160.

25 Hershko, A. and Ciechanover, A. (1998) The ubiquitin system. *Annu Rev Biochem* **67**, 425–479.

26 Weissman, A.M. (2001) Themes and variations on ubiquitylation. *Nat Rev Mol Cell Biol* **2**, 169–178.

27 Koegl, M., Hoppe, T., Schlenker, S., Ulrich, H.D., Mayer, T.U. and Jentsch, S. (1999) A novel ubiquitination factor, E4, is involved in multiubiquitin chain assembly. *Cell* **96**, 635–644.

28 Desterro, J.M., Rodriguez, M.S., Kemp, G.D. and Hay, R.T. (1999) Identification of the enzyme required for activation of the small ubiquitin-like protein SUMO-1. *J Biol Chem* **274**, 10618–10624.

29 Gong, L., Li, B., Millas, S. and Yeh, E.T. (1999) Molecular cloning and characterization of human AOS1 and UBA2, components of the sentrin-activating enzyme complex. *FEBS Lett* **448**, 185–189.

30 Johnson, E.S. and Blobel, G. (1997) Ubc9p is the conjugating enzyme for the ubiquitin-like protein Smt3p. *J Biol Chem* **272**, 26799–26802.

31 Okuma, T., Honda, R., Ichikawa, G., Tsumagari, N. and Yasuda, H. (1999) *In vitro* SUMO-1 modification requires two enzymatic steps, E1 and E2. *Biochem Biophys Res Commun* **254**, 693–698.

32 Rodriguez, M.S., Dargemont, C. and Hay, R.T. (2001) SUMO-1 conjugation in vivo requires both a consensus modification motif and nuclear targeting. *J Biol Chem* **276**, 12654–12659.

33 Bernier-Villamor, V., Sampson, D.A., Matunis, M.J. and Lima, C.D. (2002) Structural basis for E2-mediated SUMO conjugation revealed by a complex between ubiquitin-conjugating enzyme Ubc9 and RanGAP1. *Cell* **108**, 345–356.

34 Lin, D., Tatham, M.H., Yu, B., Kim, S., Hay, R.T. and Chen, Y. (2002) Identification of a substrate recognition site on Ubc9. *J Biol Chem* **277**, 21740–21748.

35 Sampson, D.A., Wang, M. and Matunis, M.J. (2001) The small ubiquitin-like modifier-1 (SUMO-1) consensus sequence mediates Ubc9 binding and is essential for SUMO-1 modification. *J Biol Chem* **276**, 21664–21669.

36 Tatham, M.H., Jaffray, E., Vaughan, O.A., Desterro, J.M., Botting, C.H., Naismith, J.H. and Hay, R.T. (2001) Polymeric chains of SUMO-2 and SUMO-3 are conjugated to protein substrates by SAE1/SAE2 and Ubc9. *J Biol Chem* **276**, 35368–35374.

37 Liakopoulos, D., Doenges, G., Matuschewski, K. and Jentsch, S. (1998) A novel protein modification pathway related to the ubiquitin system. *EMBO J* **17**, 2208–2214.

38 Osaka, F., Kawasaki, H., Aida, N., Saeki, M., Chiba, T., Kawashima, S., Tanaka, K. and Kato, S. (1998) A new NEDD8-ligating system for cullin-4A. *Genes Dev* **12**, 2263–2268.

39 Scheffner, M., Werness, B.A., Huibregtse, J.M., Levine, A.J. and Howley, P.M. (1990) The E6 oncoprotein encoded by human papillomavirus types 16 and 18 promotes the degradation of p53. *Cell* **63**, 1129–1136.

40 Querido, E., Blanchette, P., Yan, Q., Kamura, T., Morrison, M., Boivin, D., Kaelin, W.G., Conaway, R.C., Conaway, J.W. and Branton, P.E. (2001) Degradation of p53 by adenovirus E4orf6 and E1B55K proteins occurs via a novel mechanism involving a Cullin-containing complex. *Genes Dev* **15**, 3104–3117.

41 Momand, J., Zambetti, G.P., Olson, D.C., George, D. and Levine, A.J. (1992) The mdm-2 oncogene product forms a complex with the p53 protein and inhibits p53-mediated transactivation. *Cell* **69**, 1237–1245.

42 Oliner, J.D., Pietenpol, J.A., Thiagalingam, S., Gyuris, J., Kinzler, K. W. and Vogelstein, B. (1993) Oncoprotein MDM2 conceals the activation domain of tumour suppressor p53. *Nature* **362**, 857–860.

43 Picksley, S.M., Vojtesek, B., Sparks, A. and Lane, D.P. (1994) Immunochemical analysis of the interaction of p53 with MDM2; – fine mapping of the MDM2 binding site on p53 using synthetic peptides. *Oncogene* **9**, 2523–2529.

44 Stephen, C.W., Helminen, P. and Lane, D.P. (1995) Characterisation of epitopes on human p53 using phage-displayed peptide libraries: insights into antibody-peptide interactions. *J Mol Biol* **248**, 58–78.

45 Jones, S.N., Roe, A.E., Donehower, L.A. and Bradley, A. (1995) Rescue of embryonic lethality in Mdm2-deficient mice by absence of p53. *Nature* **378**, 206–208.

46 Montes de Oca Luna, R., Wagner, D.S. and Lozano, G. (1995) Rescue of early embryonic lethality in mdm2-deficient mice by deletion of p53. *Nature* **378**, 203–206.

47 Haupt, Y., Maya, R., Kazaz, A. and Oren, M. (1997) Mdm2 promotes the rapid degradation of p53. *Nature* **387**, 296–299.

48 Kubbutat, M.H., Jones, S.N. and Vousden, K.H. (1997) Regulation of p53 stability by Mdm2. *Nature* **387**, 299–303.

49 Bottger, A., Bottger, V., Sparks, A., Liu, W.L., Howard, S.F. and Lane, D.P. (1997b) Design of a synthetic Mdm2-binding mini protein that activates the p53 response *in vivo*. *Curr Biol* **7**, 860–869.

50 Fang, S., Jensen, J.P., Ludwig, R.L., Vousden, K.H. and Weissman, A.M. (2000) Mdm2 is a RING finger-dependent ubiquitin protein ligase for itself and p53. *J Biol Chem* **275**, 8945–8951.

51 Honda, R. and Yasuda, H. (2000) Activity of MDM2, a ubiquitin ligase, toward p53 or itself is dependent on the RING finger domain of the ligase. *Oncogene* **19**, 1473–1476.

52 Honda, R., Tanaka, H. and Yasuda, H. (1997) Oncoprotein MDM2 is a ubiquitin ligase E3 for tumor suppressor p53. *FEBS Lett* **420**, 25–27.

53 Dornan, D., Wertz, I., Shimizu, H., Arnott, D., Frantz, G.D., Dowd, P., O'Rourke, K., Koeppen, H. and Dixit, V. M. (2004b) The ubiquitin ligase COP1 is a critical negative regulator of p53. *Nature* **429**, 86–92.

54 Leng, R.P., Lin, Y., Ma, W., Wu, H., Lemmers, B., Chung, S., Parant, J.M., Lozano, G., Hakem, R. and Benchimol, S. (2003) Pirh2, a p53-induced ubiquitin-protein ligase, promotes p53 degradation. *Cell* **112**, 779–791.

55 Dornan, D., Bheddah, S., Newton, K., Ince, W., Frantz, G.D., Dowd, P., Koeppen, H., Dixit, V.M. and French, D.M. (2004a) COP1, the negative regulator of p53, is overexpressed in breast and ovarian adenocarcinomas. *Cancer Res* **64**, 7226–7230.

56 Duan, W., Gao, L., Druhan, L.J., Zhu, W.G., Morrison, C., Otterson, G.A. and Villalona-Calero, M.A. (2004) Expression of Pirh2, a newly identified ubiquitin protein ligase, in lung cancer. *J Natl Cancer Inst* **96**, 1718–1721.

57 Leach, F.S., Tokino, T., Meltzer, P., Burrell, M., Oliner, J.D., Smith, S., Hill, D.E., Sidransky, D., Kinzler, K.W. and Vogelstein, B. (1993) p53 Mutation and MDM2 amplification in human soft tissue sarcomas. *Cancer Res* **53**, 2231–2234.

58 Oliner, J.D., Kinzler, K.W., Meltzer, P.S., George, D.L. and Vogelstein, B. (1992) Amplification of a gene encoding a p53-associated protein in human sarcomas. *Nature* **358**, 80–83.

59 Chen, D., Kon, N., Li, M., Zhang, W., Qin, J. and Gu, W. (2005) ARF-BP1/Mule is a critical mediator of the ARF tumor suppressor. *Cell* **121**, 1071–1083.

60 Esser, C., Scheffner, M. and Hohfeld, J. (2005) The chaperone-associated ubiquitin ligase CHIP is able to target p53 for proteasomal degradation. *J Biol Chem* **280**, 27443–27448.

61 Rajendra, R., Malegaonkar, D., Pungaliya, P., Marshall, H., Rasheed, Z., Brownell, J., Liu, L.F., Lutzker, S., Saleem, A. and Rubin, E.H. (2004) Topors functions as an E3 ubiquitin ligase with specific E2 enzymes and ubiquitinates p53. *J Biol Chem* **279**, 36440–36444.

62 Weger, S., Hammer, E. and Heilbronn, R. (2005) Topors acts as a SUMO-1 E3 ligase for p53 *in vitro* and *in vivo*. *FEBS Lett* **579**, 5007–5012.

63 Bischof, O., Schwamborn, K., Martin, N., Werner, A., Sustmann, C., Grosschedl, R. and Dejean, A. (2006) The E3 SUMO ligase PIASy is a regulator of cellular senescence and apoptosis. *Mol Cell* **22**, 783–794.

64 Kahyo, T., Nishida, T. and Yasuda, H. (2001) Involvement of PIAS1 in the sumoylation of tumor suppressor p53. *Mol Cell* **8**, 713–718.

65 Schmidt, D. and Muller, S. (2002) Members of the PIAS family act as SUMO ligases for c-Jun and p53 and repress p53 activity. *Proc Natl Acad Sci USA* **99**, 2872–2877.

66 Laine, A. and Ronai, Z. (2006) Regulation of p53 localization and transcription by the HECT domain E3 ligase WWP1. *Oncogene* **26**, 1477–1483.

67 Yang, W., Rozan, L.M., McDonald, E.R.3rd, , Navaraj, A., Liu, J.J., Matthew, E.M., Wang, W., Dicker, D.T. and El-Deiry, W.S. (2006) Carps are ubiquitin ligases that promote MDM2-independent P53 and phospho-P53ser20 degradation. *J Biol Chem* **282**, 3273–3281.

68 Hicke, L. (2001) Protein regulation by monoubiquitin. *Nat Rev Mol Cell Biol* **2**, 195–201.

69 Hofmann, R.M. and Pickart, C.M. (1999) Noncanonical MMS2-encoded ubiquitin-conjugating enzyme functions in assembly of novel polyubiquitin chains for DNA repair. *Cell* **96**, 645–653.

70 Spence, J., Sadis, S., Haas, A.L. and Finley, D. (1995) A ubiquitin mutant with specific defects in DNA repair and multiubiquitination. *Mol Cell Biol* **15**, 1265–1273.

71 Spence, J., Gali, R.R., Dittmar, G., Sherman, F., Karin, M. and Finley, D. (2000) Cell cycle-regulated modification of the ribosome by a variant multiubiquitin chain. *Cell* **102**, 67–76.

72 Nakamura, S., Roth, J.A. and Mukhopadhyay, T. (2000) Multiple lysine mutations in the C-terminal domain of p53 interfere with MDM2-dependent protein degradation and ubiquitination. *Mol Cell Biol* **20**, 9391–9398.

73 Rodriguez, M.S., Desterro, J.M., Lain, S., Lane, D.P. and Hay, R.T. (2000) Multiple C-terminal lysine residues target p53 for ubiquitin-proteasome-mediated degradation. *Mol Cell Biol* **20**, 8458–8467.

74 Chan, W.M., Mak, M.C., Fung, T.K., Lau, A., Siu, W.Y. and Poon, R.Y. (2006) Ubiquitination of p53 at multiple sites in the DNA-binding domain. *Mol Cancer Res* **4**, 15–25.

75 Krummel, K.A., Lee, C.J., Toledo, F. and Wahl, G.M. (2005) The C-terminal lysines fine-tune P53 stress responses in a mouse model but are not required for stability control or transactivation. *Proc Natl Acad Sci USA* **102**, 10188–10193.

76 Feng, L., Lin, T., Uranishi, H., Gu, W. and Xu, Y. (2005) Functional analysis of the roles of posttranslational modifications at the p53 C terminus in regulating p53 stability and activity. *Mol Cell Biol* **25**, 5389–5395.

77 Roth, J., Dobbelstein, M., Freedman, D.A., Shenk, T. and Levine, A.J. (1998) Nucleo-cytoplasmic shuttling of the hdm2 oncoprotein regulates the levels of the p53 protein via a pathway used by the human immunodeficiency virus rev protein. *EMBO J* **17**, 554–564.

78 Stommel, J.M., Marchenko, N.D., Jimenez, G.S., Moll, U.M., Hope, T.J. and Wahl, G.M. (1999) A leucine-rich nuclear export signal in the p53 tetramerization domain: regulation of subcellular localization and p53 activity by NES masking. *EMBO J* **18**, 1660–1672.

79 Tao, W. and Levine, A.J. (1999a) Nucleocytoplasmic shuttling of oncoprotein Hdm2 is required for Hdm2-mediated degradation of p53. *Proc Natl Acad Sci USA* **96**, 3077–3080.

80 Boyd, S.D., Tsai, K.Y. and Jacks, T. (2000) An intact HDM2 RING-finger domain is required for nuclear exclusion of p53. *Nat Cell Biol* **2**, 563–568.

81 Lohrum, M.A., Woods, D.B., Ludwig, R.L., Balint, E. and Vousden, K.H. (2001) C-terminal ubiquitination of p53 contributes to nuclear export. *Mol Cell Biol* **21**, 8521–8532.

82 Xirodimas, D.P., Stephen, C.W. and Lane, D.P. (2001b) Cocompartmentalization of p53 and Mdm2 is a major determinant for Mdm2-mediated degradation of p53. *Exp Cell Res* **270**, 66–77.

83 Yu, Z.K., Geyer, R.K. and Maki, C.G. (2000) MDM2-dependent ubiquitination of nuclear and cytoplasmic P53. *Oncogene* **19**, 5892–5897.

84 Li, M., Brooks, C.L., Wu-Baer, F., Chen, D., Baer, R. and Gu, W. (2003) Mono- versus polyubiquitination: differential control of p53 fate by Mdm2. *Science* **302**, 1972–1975.

85 Brooks, C.L. and Gu, W. (2006) p53 ubiquitination: Mdm2 and beyond. *Mol Cell* **21**, 307–315.

86 Juven-Gershon, T. and Oren, M. (1999) Mdm2: the ups and downs. *Mol Med* **5**, 71–83.

87 Marine, J.C. and Jochemsen, A.G. (2004) Mdmx and Mdm2: brothers in arms? *Cell Cycle* **3**, 900–904.

88 Bottger, A., Bottger, V., Garcia-Echeverria, C., Chene, P., Hochkeppel, H.K., Sampson, W., Ang, K., Howard, S.F., Picksley, S.M. and Lane, D.P. (1997a) Molecular characterization of the hdm2-p53 interaction. *J Mol Biol* **269**, 744–756.

89 Chi, S.W., Lee, S.H., Kim, D.H., Ahn, M.J., Kim, J.S., Woo, J.Y., Torizawa, T., Kainosho, M. and Han, K.H. (2005) Structural details on mdm2-p53 interaction. *J Biol Chem* **280**, 38795–38802.

90 Kussie, P.H., Gorina, S., Marechal, V., Elenbaas, B., Moreau, J., Levine, A.J. and Pavletich, N.P. (1996) Structure of the MDM2 oncoprotein bound to the p53 tumor suppressor transactivation domain. *Science* **274**, 948–953.

91 Argentini, M., Barboule, N. and Wasylyk, B. (2000) The contribution of the RING finger domain of MDM2 to cell cycle progression. *Oncogene* **19**, 3849–3857.

92 Tanimura, S., Ohtsuka, S., Mitsui, K., Shirouzu, K., Yoshimura, A. and Ohtsubo, M. (1999) MDM2 interacts with MDMX through their RING finger domains. *FEBS Lett* **447**, 5–9.

93 Kawai, H., Wiederschain, D. and Yuan, Z.M. (2003) Critical contribution of the MDM2 acidic domain to p53 ubiquitination. *Mol Cell Biol* **23**, 4939–4947.

94 Meulmeester, E., Frenk, R., Stad, R., de Graaf, P., Marine, J.C., Vousden, K.H. and Jochemsen, A.G. (2003) Critical role for a central part of Mdm2 in the ubiquitylation of p53. *Mol Cell Biol* **23**, 4929–4938.

95 Blattner, C., Hay, T., Meek, D.W. and Lane, D.P. (2002) Hypophosphorylation of Mdm2 augments p53 stability. *Mol Cell Biol* **22**, 6170–6182.

96 Argentini, M., Barboule, N. and Wasylyk, B. (2001) The contribution of the acidic domain of MDM2 to p53 and MDM2 stability. *Oncogene* **20**, 1267–1275.

97 Burch, L.R., Midgley, C.A., Currie, R.A., Lane, D.P. and Hupp, T.R. (2000) Mdm2 binding to a conformationally sensitive domain on p53 can be modulated by RNA. *FEBS Lett* **472**, 93–98.

98 Shimizu, H., Burch, L.R., Smith, A.J., Dornan, D., Wallace, M., Ball, K.L. and Hupp, T.R. (2002) The conformationally flexible S9-S10 linker region in the core domain of p53 contains a novel MDM2 binding site whose mutation increases ubiquitination of p53 *in vivo*. *J Biol Chem* **277**, 28446–28458.

99 Shimizu, H., Saliba, D., Wallace, M., Finlan, L., Langridge-Smith, P.R. and Hupp, T.R. (2006) Destabilizing missense mutations in the tumour suppressor protein p53 enhance its ubiquitination *in vitro* and *in vivo*. *Biochem J* **397**, 355–367.

100 Wallace, M., Worrall, E., Pettersson, S., Hupp, T.R. and Ball, K.L. (2006) Dual-site regulation of MDM2 E3-ubiquitin ligase activity. *Mol Cell* **23**, 251–263.

101 Marston, N.J., Jenkins, J.R. and Vousden, K.H. (1995) Oligomerisation of full length p53 contributes to the interaction with mdm2 but not HPV E6. *Oncogene* **10**, 1709–1715.

102 Saville, M.K., Sparks, A., Xirodimas, D.P., Wardrop, J., Stevenson, L.F., Bourdon, J.C., Woods, Y.L. and Lane, D.P. (2004) Regulation of p53 by the ubiquitin-conjugating enzymes UbcH5B/C *in vivo*. *J Biol Chem* **279**, 42169–42181.

103 Laine, A., Topisirovic, I., Zhai, D., Reed, J.C., Borden, K.L. and Ronai, Z. (2006) Regulation of p53 Localization and Activity by Ubc13. *Mol Cell Biol* **26**, 8901–8913.

104 Quelle, D.E., Zindy, F., Ashmun, R.A. and Sherr, C.J. (1995) Alternative reading frames of the INK4a tumor suppressor gene encode two unrelated proteins capable of inducing cell cycle arrest. *Cell* **83**, 993–1000.

105 Serrano, M., Hannon, G.J. and Beach, D. (1993) A new regulatory motif in cell-cycle control causing specific inhibition of cyclin D/CDK4. *Nature* **366**, 704–707.

106 Pomerantz, J., Schreiber-Agus, N., Liegeois, N.J., Silverman, A., Alland, L., Chin, L., Potes, J., Chen, K., Orlow, I., Lee, H.W., Cordon-Cardo, C. and DePinho, R.A. (1998) The Ink4a tumor suppressor gene product, p19Arf, interacts with MDM2 and neutralizes MDM2's inhibition of p53. *Cell* **92**, 713–723.

107 Stott, F.J., Bates, S., James, M.C., McConnell, B.B., Starborg, M., Brookes, S., Palmero, I., Ryan, K., Hara, E., Vousden, K.H. and Peters, G. (1998) The alternative product from the human CDKN2A locus, p14(ARF), participates in a regulatory feedback loop with p53 and MDM2. *EMBO J* **17**, 5001–5014.

108 Zhang, Y., Xiong, Y. and Yarbrough, W.G. (1998) ARF promotes MDM2 degradation and stabilizes p53: ARF-INK4a locus deletion impairs both the Rb and p53 tumor suppression pathways. *Cell* **92**, 725–734.

109 Weber, J.D., Taylor, L.J., Roussel, M.F., Sherr, C.J. and Bar-Sagi, D. (1999) Nucleolar Arf sequesters Mdm2 and activates p53. *Nat Cell Biol* **1**, 20–26.

110 Zhang, Y. and Xiong, Y. (1999) Mutations in human ARF exon 2 disrupt its nucleolar localization and impair its ability to block nuclear export of MDM2 and p53. *Mol Cell* **3**, 579–591.

111 Tao, W. and Levine, A.J. (1999b) P19(ARF) stabilizes p53 by blocking nucleo-cytoplasmic shuttling of Mdm2. *Proc Natl Acad Sci USA* **96**, 6937–6941.

112 Llanos, S., Clark, P.A., Rowe, J. and Peters, G. (2001) Stabilization of p53 by p14ARF without relocation of MDM2 to the nucleolus. *Nat Cell Biol* **3**, 445–452.

113 Honda, R. and Yasuda, H. (1999) Association of p19(ARF) with Mdm2 inhibits ubiquitin ligase activity of Mdm2 for tumor suppressor p53. *EMBO J* **18**, 22–27.

114 Midgley, C.A., Desterro, J.M., Saville, M.K., Howard, S., Sparks, A., Hay, R.T. and Lane, D.P. (2000) An N-terminal p14ARF peptide blocks Mdm2-dependent ubiquitination in vitro and can activate p53 in vivo. *Oncogene* **19**, 2312–2323.

115 Xirodimas, D., Saville, M.K., Edling, C., Lane, D.P. and Lain, S. (2001a) Different effects of p14ARF on the levels of ubiquitinated p53 and Mdm2 *in vivo*. *Oncogene* **20**, 4972–4983.

116 Lohrum, M.A., Ashcroft, M., Kubbutat, M.H. and Vousden, K.H. (2000) Identification of a cryptic nucleolar-localization signal in MDM2. *Nat Cell Biol* **2**, 179–181.

117 Rubbi, C.P. and Milner, J. (2003) Disruption of the nucleolus mediates stabilization of p53 in response to DNA damage and other stresses. *EMBO J* **22**, 6068–6077.

118 Dai, M.S. and Lu, H. (2004) Inhibition of MDM2-mediated p53 ubiquitination and degradation by ribosomal protein L5. *J Biol Chem* **279**, 44475–44482.

119 Dai, M.S., Zeng, S.X., Jin, Y., Sun, X.X., David, L. and Lu, H. (2004) Ribosomal protein L23 activates p53 by inhibiting MDM2 function in response to ribosomal perturbation but not to translation inhibition. *Mol Cell Biol* **24**, 7654–7668.

120 Jin, A., Itahana, K., O'Keefe, K. and Zhang, Y. (2004) Inhibition of HDM2 and activation of p53 by ribosomal protein L23. *Mol Cell Biol* **24**, 7669–7680.

121 Lohrum, M.A., Ludwig, R.L., Kubbutat, M.H., Hanlon, M. and Vousden, K.H. (2003) Regulation of HDM2 activity by the ribosomal protein L11. *Cancer Cell* **3**, 577–587.

122 Marechal, V., Elenbaas, B., Piette, J., Nicolas, J.C. and Levine, A.J. (1994) The ribosomal L5 protein is associated with mdm-2 and mdm-2-p53 complexes. *Mol Cell Biol* **14**, 7414–7420.

123 Takagi, M., Absalon, M.J., McLure, K.G. and Kastan, M.B. (2005) Regulation of p53 translation and induction after DNA damage by ribosomal protein L26 and nucleolin. *Cell* **123**, 49–63.

124 Stommel, J.M. and Wahl, G.M. (2004) Accelerated MDM2 auto-degradation induced by DNA-damage kinases is required for p53 activation. *EMBO J* **23**, 1547–1556.

125 Yin, Y., Stephen, C.W., Luciani, M.G. and Fahraeus, R. (2002) p53 Stability and activity is regulated by Mdm2-mediated induction of alternative p53 translation products. *Nat Cell Biol* **4**, 462–467.

126 Finch, R.A., Donoviel, D.B., Potter, D., Shi, M., Fan, A., Freed, D.D., Wang, C.Y., Zambrowicz, B.P., Ramirez-Solis, R., Sands, A.T. and Zhang, N. (2002) Mdmx is a negative regulator of p53 activity in vivo. *Cancer Res* **62**, 3221–3225.

127 Migliorini, D., Lazzerini Denchi, E., Danovi, D., Jochemsen, A., Capillo, M., Gobbi, A., Helin, K., Pelicci, P.G. and Marine, J.C. (2002b) Mdm4 (Mdmx) regulates p53-induced growth arrest and neuronal cell death during early embryonic mouse development. *Mol Cell Biol* **22**, 5527–5538.

128 Parant, J., Chavez-Reyes, A., Little, N.A., Yan, W., Reinke, V., Jochemsen, A.G. and Lozano, G. (2001) Rescue of embryonic lethality in Mdm4-null mice by loss of Trp53 suggests a nonoverlapping pathway with MDM2 to regulate p53. *Nat Genet* **29**, 92–95.

129 Little, N.A. and Jochemsen, A.G. (2001) Hdmx and Mdm2 can repress transcription activation by p53 but not by p63. *Oncogene* **20**, 4576–4580.

130 Stad, R., Ramos, Y.F., Little, N., Grivell, S., Attema, J., van Der Eb, A.J. and Jochemsen, A.G. (2000) Hdmx stabilizes Mdm2 and p53. *J Biol Chem* **275**, 28039–28044.

131 Badciong, J.C. and Haas, A.L. (2002) MdmX is a RING finger ubiquitin ligase capable of synergistically enhancing Mdm2 ubiquitination. *J Biol Chem* **277**, 49668–49675.

132 Jackson, M.W. and Berberich, S.J. (2000) MdmX protects p53 from Mdm2-mediated degradation. *Mol Cell Biol* **20**, 1001–1007.

133 Migliorini, D., Danovi, D., Colombo, E., Carbone, R., Pelicci, P.G. and Marine, J. C. (2002a) Hdmx recruitment into the nucleus by Hdm2 is essential for its ability to regulate p53 stability and transactivation. *J Biol Chem* **277**, 7318–7323.

134 Stad, R., Little, N.A., Xirodimas, D.P., Frenk, R., van der Eb, A.J., Lane, D.P., Saville, M.K. and Jochemsen, A.G. (2001) Mdmx stabilizes p53 and Mdm2 via two distinct mechanisms. *EMBO Rep* **2**, 1029–1034.

135 de Graaf, P., Little, N.A., Ramos, Y.F., Meulmeester, E., Letteboer, S.J. and Jochemsen, A.G. (2003) Hdmx protein stability is regulated by the ubiquitin ligase activity of Mdm2. *J Biol Chem* **278**, 38315–38324.

136 Linares, L.K., Hengstermann, A., Ciechanover, A., Muller, S. and Scheffner, M. (2003) HdmX stimulates Hdm2-mediated ubiquitination and degradation of p53. *Proc Natl Acad Sci USA* **100**, 12009–12014.

137 Pan, Y. and Chen, J. (2003) MDM2 promotes ubiquitination and degradation of MDMX. *Mol Cell Biol* **23**, 5113–5121.

138 Li, L., Liao, J., Ruland, J., Mak, T.W. and Cohen, S.N. (2001) A TSG101/MDM2 regulatory loop modulates MDM2 degradation and MDM2/p53 feedback control. *Proc Natl Acad Sci USA* **98**, 1619–1624.

139 RuLand, J., Sirard, C., Elia, A., MacPherson, D., Wakeham, A., Li, L., de la Pompa, J.L., Cohen, S.N. and Mak, T.W. (2001) p53 accumulation, defective cell proliferation, and early embryonic lethality in mice lacking tsg101. *Proc Natl Acad Sci USA* **98**, 1859–1864.

140 Lai, Z., Ferry, K.V., Diamond, M.A., Wee, K.E., Kim, Y.B., Ma, J., Yang, T., Benfield, P.A., Copeland, R.A. and Auger, K.R. (2001) Human mdm2 mediates multiple mono-ubiquitination of p53 by a mechanism requiring enzyme isomerization. *J Biol Chem* **276**, 31357–31367.

141 Grossman, S.R., Deato, M.E., Brignone, C., Chan, H.M., Kung, A.L., Tagami, H., Nakatani, Y. and Livingston, D.M. (2003) Polyubiquitination of p53 by a ubiquitin ligase activity of p300. *Science* **300**, 342–344.

142 Sui, G., Affar El, B., Shi, Y., Brignone, C., Wall, N.R., Yin, P., Donohoe, M., Luke, M.P., Calvo, D., Grossman, S.R. and Shi, Y. (2004) Yin Yang 1 is a negative regulator of p53. *Cell* **117**, 859–872.

143 Allende-Vega, N., Saville, M.K. and Meek, D.W. (2007) Transcription factor TAFII250 promotes Mdm2-dependent turnover of p53. *Oncogene* **26**, 4234–4242.

144 Brady, M., Vlatkovic, N. and Boyd, M.T. (2005) Regulation of p53 and MDM2 activity by MTBP. *Mol Cell Biol* **25**, 545–553.

145 Higashitsuji, H., Higashitsuji, H., Itoh, K., Sakurai, T., Nagao, T., Sumitomo, Y., Masuda, T., Dawson, S., Shimada, Y., Mayer, R.J. and Fujita, J. (2005) The oncoprotein gankyrin binds to MDM2/HDM2, enhancing ubiquitylation and degradation of p53. *Cancer Cell* **8**, 75–87.

146 Wang, C., Ivanov, A., Chen, L., Fredericks, W.J., Seto, E., Rauscher, F. J.3rd, and Chen, J. (2005) MDM2 interaction with nuclear corepressor KAP1 contributes to p53 inactivation. *EMBO J* **24**, 3279–3290.

147 Raasi, S. and Pickart, C.M. (2003) Rad23 ubiquitin-associated domains (UBA) inhibit 26 S proteasome-catalyzed proteolysis by sequestering lysine 48-linked polyubiquitin chains. *J Biol Chem* **278**, 8951–8959.

148 Brignone, C., Bradley, K.E., Kisselev, A.F. and Grossman, S.R. (2004) A post-ubiquitination role for MDM2 and hHR23A in the p53 degradation pathway. *Oncogene* **23**, 4121–4129.

149 Glockzin, S., Ogi, F.X., Hengstermann, A., Scheffner, M. and Blattner, C. (2003) Involvement of the DNA repair protein hHR23 in p53 degradation. *Mol Cell Biol* **23**, 8960–8969.

150 Kaur, M., Pop, M., Shi, D., Brignone, C. and Grossman, S.R. (2006) hHR23B is required for genotoxic-specific activation of p53 and apoptosis. *Oncogene* **26**, 1231–1237.

151 Hsieh, J.K., Chan, F.S., O'Connor, D.J., Mittnacht, S., Zhong, S. and Lu, X. (1999) RB regulates the stability and the apoptotic function of p53 via MDM2. *Mol Cell* **3**, 181–193.

152 Appella, E. and Anderson, C.W. (2001) Post-translational modifications and activation of p53 by genotoxic stresses. *Eur J Biochem* **268**, 2764–2772.

153 Lambert, P.F., Kashanchi, F., Radonovich, M.F., Shiekhattar, R. and Brady, J.N. (1998) Phosphorylation of p53 serine 15 increases interaction with CBP. *J Biol Chem* **273**, 33048–33053.

154 Meek, D.W. (2004) The p53 response to DNA damage. *DNA Repair (Amst)* **3**, 1049–1056.

155 Shieh, S.Y., Ikeda, M., Taya, Y. and Prives, C. (1997) DNA damage-induced phosphorylation of p53 alleviates inhibition by MDM2. *Cell* **91**, 325–334.

156 Gu, W., Luo, J., Brooks, C.L., Nikolaev, A.Y. and Li, M. (2004) Dynamics of the p53 acetylation pathway. *Novartis Found Symp* **259**, 197–205; discussion 205–197, 223–195.

157 Ito, A., Kawaguchi, Y., Lai, C.H., Kovacs, J.J., Higashimoto, Y., Appella, E. and Yao, T.P. (2002) MDM2-HDAC1-mediated deacetylation of p53 is required for its degradation. *EMBO J* **21**, 6236–6245.

158 Goldberg, Z., Vogt Sionov, R., Berger, M., Zwang, Y., Perets, R., Van Etten, R. A., Oren, M., Taya, Y. and Haupt, Y. (2002) Tyrosine phosphorylation of Mdm2 by c-Abl: implications for p53 regulation. *EMBO J* **21**, 3715–3727.

159 Khosravi, R., Maya, R., Gottlieb, T., Oren, M., Shiloh, Y. and Shkedy, D. (1999) Rapid ATM-dependent phosphorylation of MDM2 precedes p53 accumulation in response to DNA damage. *Proc Natl Acad Sci USA* **96**, 14973–14977.

160 Sionov, R.V., Moallem, E., Berger, M., Kazaz, A., Gerlitz, O., Ben-Neriah, Y., Oren, M. and Haupt, Y. (1999) c-Abl neutralizes the inhibitory effect of Mdm2 on p53. *J Biol Chem* **274**, 8371–8374.

161 Maya, R., Balass, M., Kim, S.T., Shkedy, D., Leal, J.F., Shifman, O., Moas, M., Buschmann, T., Ronai, Z., Shiloh, Y., Kastan, M.B., Katzir, E. and Oren, M. (2001) ATM-dependent phosphorylation of Mdm2 on serine 395: role in p53 activation by DNA damage. *Genes Dev* **15**, 1067–1077.

162 Ashcroft, M., Ludwig, R.L., Woods, D.B., Copeland, T.D., Weber, H.O., MacRae, E.J. and Vousden, K.H. (2002) Phosphorylation of HDM2 by Akt. *Oncogene* **21**, 1955–1962.

163 Zhou, B.P., Liao, Y., Xia, W., Zou, Y., Spohn, B. and Hung, M.C. (2001) HER-2/neu induces p53 ubiquitination via Akt-mediated MDM2 phosphorylation. *Nat Cell Biol* **3**, 973–982.

164 Dias, S.S., Milne, D.M. and Meek, D.W. (2006) c-Abl phosphorylates Hdm2 at tyrosine 276 in response to DNA damage and regulates interaction with ARF. *Oncogene* **25**, 6666–6671.

165 Li, M., Chen, D., Shiloh, A., Luo, J., Nikolaev, A.Y., Qin, J. and Gu, W. (2002) Deubiquitination of p53 by HAUSP is an important pathway for p53 stabilization. *Nature* **416**, 648–653.

166 Cummins, J.M., Rago, C., Kohli, M., Kinzler, K.W., Lengauer, C. and Vogelstein, B. (2004) Tumour suppression: disruption of HAUSP gene stabilizes p53. *Nature* **428**, 1 p following 486.

167 Li, M., Brooks, C.L., Kon, N. and Gu, W. (2004) A dynamic role of HAUSP in the p53–Mdm2 pathway. *Mol Cell* **13**, 879–886.

168 Tang, J., Qu, L.K., Zhang, J., Wang, W., Michaelson, J.S., Degenhardt, Y.Y., El-Deiry, W.S., Yang, X. (2006) Critical role for Daxx in regulating Mdm2. *Nat Cell Biol* **8**, 855–862.

169 Meulmeester, E., Maurice, M.M., Boutell, C., Teunisse, A.F., Ovaa, H., Abraham, T.E., Dirks, R.W. and Jochemsen, A.G. (2005a) Loss of HAUSP-mediated deubiquitination contributes to DNA damage-induced destabilization of Hdmx and Hdm2. *Mol Cell* **18**, 565–576.

170 Meulmeester, E., Pereg, Y., Shiloh, Y. and Jochemsen, A.G. (2005b) ATM-mediated phosphorylations inhibit Mdmx/Mdm2 stabilization by HAUSP in favor of p53 activation. *Cell Cycle* **4**, 1166–1170.

171 Stevenson, L.F., Sparks, A., Allende-Vega, N., Xirodimas, D.P., Lane, D.P. and Saville, M.K. (2007) The deubiquitinating enzyme USP2a regulates the p53 pathway by targeting Mdm2. *EMBO J* **26**, 976–986.

172 Zhang, D., Zaugg, K., Mak, T.W. and Elledge, S.J. (2006) A role for the deubiquitinating enzyme USP28 in control of the DNA-damage response. *Cell* **126**, 529–542.

173 Gostissa, M., Hengstermann, A., Fogal, V., Sandy, P., Schwarz, S.E., Scheffner, M. and Del Sal, G. (1999) Activation of p53 by conjugation to the ubiquitin-like protein SUMO-1. *EMBO J* **18**, 6462–6471.

174 Rodriguez, M.S., Desterro, J.M., Lain, S., Midgley, C.A., Lane, D.P. and Hay, R.T. (1999) SUMO-1 modification activates the transcriptional response of p53. *EMBO J* **18**, 6455–6461.

175 Kwek, S.S., Derry, J., Tyner, A.L., Shen, Z. and Gudkov, A.V. (2001) Functional analysis and intracellular localization of p53 modified by SUMO-1. *Oncogene* **20**, 2587–2599.

176 Muller, S., Berger, M., Lehembre, F., Seeler, J.S., Haupt, Y. and Dejean, A. (2000) c-Jun and p53 activity is modulated by SUMO-1 modification. *J Biol Chem* **275**, 13321–13329.

177 Lee, M.H., Lee, S.W., Lee, E.J., Choi, S.J., Chung, S.S., Lee, J.I., Cho, J.M., Seol, J. H., Baek, S.H., Kim, K.I., Chiba, T., Tanaka, K., Bang, O.S. and Chung, C.H. (2006) SUMO-specific protease SUSP4 positively regulates p53 by promoting Mdm2 self-ubiquitination. *Nat Cell Biol* **8**, 1424–1431.

178 Jones, D. and Candido, E.P. (2000) The NED-8 conjugating system in *Caenorhabditis elegans* is required for embryogenesis and terminal differentiation of the hypodermis. *Dev Biol* **226**, 152–165.

179 Osaka, F., Saeki, M., Katayama, S., Aida, N., Toh, E.A., Kominami, K., Toda, T., Suzuki, T., Chiba, T., Tanaka, K. and Kato, S. (2000) Covalent modifier NEDD8 is essential for SCF ubiquitin-ligase in fission yeast. *EMBO J* **19**, 3475–3484.

180 Ou, C.Y., Lin, Y.F., Chen, Y.J. and Chien, C.T. (2002) Distinct protein degradation mechanisms mediated by Cul1 and Cul3 controlling Ci stability in Drosophila eye development. *Genes Dev* **16**, 2403–2414.

181 Tateishi, K., Omata, M., Tanaka, K. and Chiba, T. (2001) The NEDD8 system is essential for cell cycle progression and morphogenetic pathway in mice. *J Cell Biol* **155**, 571–579.

182 Pan, Z.Q., Kentsis, A., Dias, D.C., Yamoah, K. and Wu, K. (2004) Nedd8 on cullin: building an expressway to protein destruction. *Oncogene* **23**, 1985–1997.

183 Handeli, S. and Weintraub, H. (1992) The ts41 mutation in Chinese hamster cells leads to successive S phases in the absence of intervening G2, M, and G1. *Cell* **71**, 599–611.

184 Xirodimas, D.P., Saville, M.K., Bourdon, J.C., Hay, R.T. and Lane, D.P. (2004) Mdm2-mediated NEDD8 conjugation of p53 inhibits its transcriptional activity. *Cell* **118**, 83–97.

185 Abida, W.M., Nikolaev, A., Zhao, W., Zhang, W. and Gu, W. (2006) FBXO11 promotes the neddylation of p53 and inhibits its transcriptional activity. *J Biol Chem* **282**, 1797–1804.

186 Lain, S. and Lane, D. (2003) Improving cancer therapy by non-genotoxic activation of p53. *Eur J Cancer* **39**, 1053–1060.

187 Vassilev, L.T., Vu, B.T., Graves, B., Carvajal, D., Podlaski, F., Filipovic, Z., Kong, N., Kammlott, U., Lukacs, C., Klein, C., Fotouhi, N. and Liu, E.A. (2004) *In vivo* activation of the p53 pathway by small-molecule antagonists of MDM2. *Science* **303**, 844–848.

188 Lai, Z., Yang, T., Kim, Y.B., Sielecki, T.M., Diamond, M.A., Strack, P., Rolfe, M., Caligiuri, M., Benfield, P.A., Auger, K.R. and Copeland, R.A. (2002) Differentiation of Hdm2-mediated p53 ubiquitination and Hdm2 autoubiquitination activity by small molecular weight inhibitors. *Proc Natl Acad Sci USA* **99**, 14734–14739.

189 Yang, Y., Ludwig, R.L., Jensen, J.P., Pierre, S.A., Medaglia, M.V., Davydov, I. V., Safiran, Y.J., Oberoi, P., Kenten, J.H., Phillips, A.C., Weissman, A.M. and Vousden, K.H. (2005) Small molecule inhibitors of HDM2 ubiquitin ligase activity stabilize and activate p53 in cells. *Cancer Cell* **7**, 547–559.

190 Hideshima, T., Mitsiades, C., Akiyama, M., Hayashi, T., Chauhan, D., Richardson, P., Schlossman, R., Podar, K., Munshi, N.C., Mitsiades, N. and Anderson, K.C. (2003) Molecular mechanisms mediating antimyeloma activity of proteasome inhibitor PS-341. *Blood* **101**, 1530–1534.

191 Selivanova, G. and Wiman, K.G. (2007) Reactivation of mutant p53: molecular mechanisms and therapeutic potential. *Oncogene* **26**, 2243–2254.

192 Lane, D.P. and Hupp, T.R. (2003) Drug discovery and p53. *Drug Discov Today* **8**, 347–355.

3
The Ubiquitin–Proteasome System in Epstein-Barr Virus Infection and Oncogenesis

Maria G. Masucci

3.1
Introduction

The modification of intracellular proteins by covalent attachment of ubiquitin or ubiquitin-like polypeptides and the degradation of some of these conjugates by the proteasomes are critical events in the regulation of cellular metabolism, proliferation, differentiation and death (reviewed in [1]). These processes are therefore major targets for pathogens that often devote a significant part of their genomes to genes whose products modify the cellular environment or protect the infected cells from the host immune attack. The involvement of the ubiquitin–proteasome system in the life cycle of viruses includes a role in virus entry, virus exit and maturation, regulation of viral and cellular gene expression and modulation of cellular functions including cell cycle, apoptosis and antiviral responses such as interferon production and antigen processing. This chapter will focus on the first identified human tumor virus Epstein-Barr Virus (EBV), a gamma-herpesvirus that is involved in the pathogenesis of a broad spectrum of malignancies of lymphoid and epithelial cell origin (reviewed in [2]). As a result of its relatively large coding capacity, EBV has evolved unique strategies for persistence in the infected host by parasitizing the complex life cycle of its primary target, the B-lymphocyte. Many of these strategies involve modulation of the ubiquitin–proteasome system.

3.2
Viral Interference with the Ubiquitin–Proteasome System

The ubiquitin–proteasome system plays a pivotal role in viral infection and pathogenesis. As extensively discussed elsewhere in this book and in a number of recent reviews, ubiquitin-dependent proteolysis is achieved through two successive steps: the covalent attachment of ubiquitin to the target protein, and the degradation of the ubiquitinated protein by the 26S proteasome with the release of peptide fragments and reusable ubiquitin [1]. The ubiquitination step involves

Protein Degradation, Vol. 4: The Ubiquitin-Proteasome System and Disease.
Edited by R. J. Mayer, A. Ciechanover, M. Rechsteiner

ISBN: 978-3-527-31436-2

three sequential enzymatic reactions that entail the ATP-dependent activation of the C-terminal glycine of ubiquitin by a ubiquitin-activating enzyme (E1), the transfer of the activated ubiquitin to a ubiquitin-conjugating enzyme (E2) and the subsequent formation of a covalent isopeptide bond between the activate C-terminus of ubiquitin and ε-amino group of a lysine residue in the substrate catalyzed by a ubiquitin–protein ligase (E3). After at least four rounds of ubiquitination involving the Lys48 or, less frequently, Lys29 residue of the previously conjugated ubiquitin, the substrate is recognized and subsequently degraded by the 26S proteasome (reviewed in [3]). The specificity of proteolysis appears to be achieved primarily at the step of ubiquitination, mainly due to the capacity of the E3 enzymes to recognize only one or a few specific substrates. The E3s can be divided into three groups: the homologous to E6-associated protein carboxyl terminus (HECT)-domain subfamily E3s that are themselves ubiquitinated before transferring ubiquitin to the substrate; the single-subunit really interesting new gene (RING)-finger subfamily, where the substrate recognition site and the RING domain involved in E2 binding reside in the same protein; the multi-subunit RING-finger subfamily where the substrate recognition and RING domains are found in separate subunits of the ligase, which allows the construction of a huge variety of enzymes with different substrate specificity and finely tuned activity (reviewed in [4]). The efficiency and stability of ubiquitination is also regulated by the activity of deubiquitinating enzymes (DUBs) that hydrolyze the isopeptide bonds between two adjacent ubiquitins, or between ubiquitin and the substrate. More than 90 DUBs have been identified in the human genome (reviewed in [5]). Based on their structure and function, these can be classified into at least five distinct families: ubiquitin-specific proteases (USP), ubiquitin carboxyl-terminal hydrolases (UCH), ovarian tumor (OTU) domain-containing proteases, Josefines, and the Jab1/MPN domain-associated metalloisopeptidase (JAMM) group of hydrolases. In addition to the recycling of damaged or misfolded proteins, the ubiquitin–proteasome system is responsible for the constitutive and induced turnover of regulatory proteins that control a wide variety of cellular functions, including the cell cycle, transcription, translation, signal transduction, antigen processing and apoptosis [1]. Furthermore, modification of proteins by polyubiquitin chains linked through Lys63, by single ubiquitin, or by a growing family of ubiquitin-like molecules including SUMO, Nedd8, ISG15, FAT10, LC3 and several more, each requiring a dedicated set of specific E1, E2 and E3 enzymes, does not result in proteasomal degradation but regulates essential functions such as DNA repair, endo- and exocytosis, protein trafficking between different cellular compartments, and autophagy [6].

Viruses exploit and manipulate this complex system of protein modification and degradation in many different ways. Viral entry and exit from the infected cell follow physiological routes for uptake and export of macromolecules that are controlled by mono or polyubiquitination of receptors and transport proteins. Likewise, the trafficking of viral regulatory and structural proteins in and out of the nucleus and other cellular compartments requires the same type of ubiquitin and ubiquitin-like modification that guide these processes in uninfected cells. In addition, viruses are obligate intracellular parasites that must exploit cellular metabolic processes for energy production and synthesis of proteins and nucleic acids whose regulators are controlled by the ubiquitin–proteasome system. Interference with these regulatory pathways is achieved through the production of multifunctional viral proteins that either mimic the activity of the cellular enzymes, often E3 ligases, or act as chaperones that redirect the activity of the cellular enzymes to new targets whose modification or destruction is required for successful virus infection. In addition, viral or cellular proteins may be rescued from proteasomal degradation by interfering with ubiquitination or with the degradation of ubiquitinated proteins by the proteasome. Finally, modulation of the efficiency and specificity of proteasomal processing is exploited by many viruses as a means of interfering with antigen presentation to protect the infected cells from immune attack. Examples of these viral strategies have been found in the life cycle of EBV (Figure 3.1).

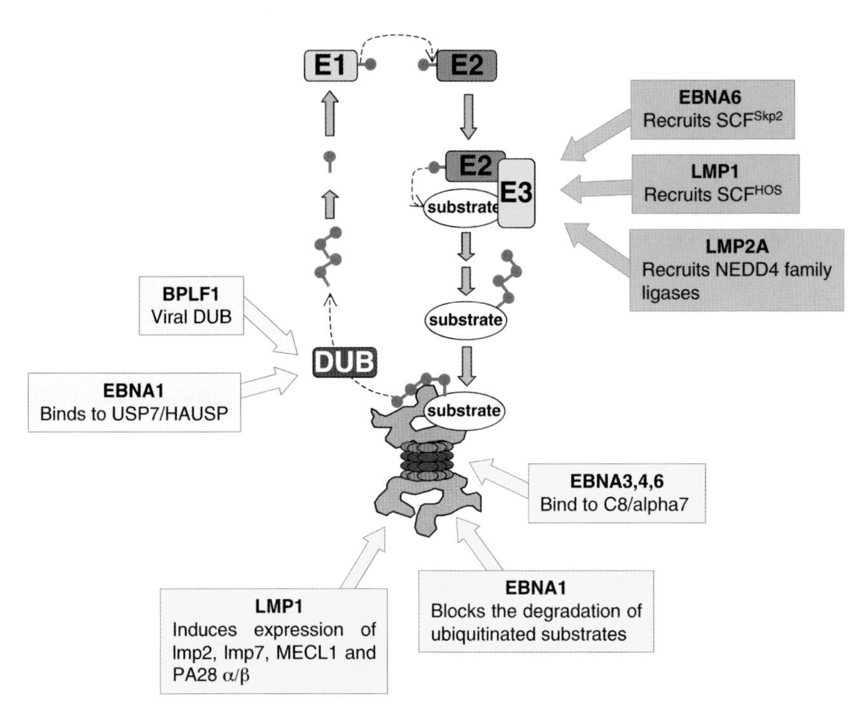

Fig. 3.1. EBV proteins that interfere with the ubiquitin–proteasome system. Proteins are targeted for proteasomal degradation by conjugation with ubiquitin, which requires the activity of a ubiquitin-activating enzyme (E1), a ubiquitin-conjugating enzyme (E2) and a ubiquitin ligase (E3). The polyubiquitinated substrate is then bound to the proteasome which unfolds the substrate and cleaves the protein into small peptides. Ubiquitin is recycled through the activity of deubiquitinating enzymes (DUB). Different components of the ubiquitin–proteasome system are targeted by EBV proteins.

3.3
The EBV Life Cycle

EBV is a largely non-pathogenic virus that establishes persistent infections in over 90% of the adults worldwide (reviewed in [7]). Primary infection usually occurs during early childhood and is generally asymptomatic while delay until adolescence or adulthood is often associated with a benign self-limiting lymphoproliferative disease known as glandular fever, or infectious mononucleosis (IM). Like other herpesviruses EBV establishes productive or latent infections in different cell types. Epithelial cells of the oropharynx are believed to be the major site of virus replication while B-lymphocytes sustain mainly non-productive infections and are the site of persistence in healthy carriers. Infection of B-lymphocytes *in vitro* results in the expression of a restricted set of viral genes that are collectively known as the "latent" gene to distinguish them from the "lytic" genes that characterize the productive virus cycle. The latent genes encode six nuclear and three membrane-associated proteins known as EBV nuclear antigens (EBNA)-1, -2, -3 (or 3A), -4 (or 3B), -5 (or LP) and -6 (or 3C) and latent membrane proteins (LMP)-1, -2A and -2B, respectively (reviewed in [2]). In addition, two untranslated RNAs, the EBER-1 and -2 involved in the regulation of interferon production [8], and other RNAs of unknown function have been detected in all the infected cells [8, 9]. Expression of the latent EBV proteins is associated with growth transformation *in vitro* and *in vivo*, which underscores their capacity to interfere with cellular pathways that regulate B-cell proliferation and differentiation.

Healthy EBV-infected individuals carry between 1 to 100 latently EBV-infected B lymphocytes per million cells in the peripheral blood [10–12]. The EBV genome is not integrated in the host-cell DNA and the infected B-cells carry multiple autonomously replicating virus episomes. This, and the establishment of latency in B-lymphocytes which cycle between resting and activated proliferative states, pose a special challenge to the life-long maintenance of the virus reservoir. In the case of EBV this problem has been solved by the development of multiple programs of viral gene expression that are adapted to different stages of B-cell activation/differentiation (Table 3.1). In the absence of effective immune surveillance, as observed *in vitro* or *in vivo* during primary infection and in severely immunosuppressed individuals, EBV-infected B-cells express a latency program, called Latency III, which includes all nine latent viral proteins. This program is associated with autonomous B-cell proliferation and is exemplified by the lymphoblastoid cell lines (LCLs) that can be established by *in vitro* EBV infection of B-cells from virtually all individuals. LCL cells resemble B-blasts which have been activated by encounter with the cognate antigen; they secrete immunoglobulins and express several activation markers and adhesion molecules. It is believed that this growth-inducing latency program is required to expand the pool of infected cells before the establishment of effective immunity, increasing thereby, the likelihood of access to the memory B-cell compartment. Indeed, proliferating EBV-infected cells are highly sensitive to innate and specific immune responses and are only found in the blood of IM patients before the establishment of specific immunity.

Table 3.1. EBV-latency programs and their expression in normal B-lymphocytes and EBV-associated malignancies.

Viral program	Expressed viral genes		B cell type	Disease
	Proteins	RNAs		
Latency I	(LMP2A EBNA1)	(EBERs; BARF0)	Memory cells[a]	Burkitt's lymphoma[b]
Latency II	EBNA1; LMP1; LMP2A; -2B	EBERs; BARF0; miRNAs	Centroblasts (germinal centers)	Hodgkin's disease, Peripheral T cell lymphoma, Nasal T/NK cell lymphoma, Nasopharyngeal carcinoma, Lympho-epithelioma (stomach, salivary glands)
Latency III	EBNA1-6; LMP1; LMP2A; -2B	EBERs; BARF0; miRNAs	Lymphoblasts	Infectious Mononucleosis; AIDS-related immunoblastic lymphoma; Post-transplant lymphoproliferative disease

a LMP-2A mRNA may be the only viral transcript detected in circulating memory cells. Complete silencing of the viral genome is likely to occur in some memory cells.

b LMP-2A is usually not expressed in BL cells while some tumors express EBNA-3, -4 and -6.

The virus-infected cells that circulate in the blood of healthy EBV carriers are non-proliferating memory B-lymphocytes where the viral genome is completely silenced or viral gene expression is restricted to the LMP2 membrane proteins either alone, or together with the nuclear antigen EBNA1. This viral program is called Latency I [10, 13]. Memory B-cells are both long-lived and poorly immunogenic and are therefore an ideal viral reservoir but, due to their continuous circulation though different body milieus, they are also exposed to new encounters with their cognate antigens. This could reactivate the latent virus and trigger the productive cycle, which yields infectious virus but is regularly accompanied by cell death. While continuous low levels of virus production and infection of new B-lymphocytes could potentially assure the persistence of the virus for the entire life of the infected host, the demonstration that the same virus strain can be isolated from healthy carriers over decades suggests that most if not all latently-infected cells never undergo lytic replication. As discussed below, this is achieved through the capacity of LMP2A to interfere with signaling through the B-cell receptor. An intermediate form of latency characterized by the expression of EBNA1, LMP1 and LMP2 (Latency II), has been identified in germinal center B-lymphocytes [14, 15]. LMP1 regulates both B-cell activation and apoptosis, which allows the survival, expansion and further differentiation into memory cells of infected B-lymphocytes which will reach the lymphoid follicles (reviewed in [16]).

Since very few EBV-infected cells are found in blood or lymphoid tissues of healthy carriers, our knowledge of viral gene expression in these cells rests exclusively on the detection of viral transcripts by highly sensitive PCRs as there is no evidence for protein expression. However, the expression of different combinations of latent viral proteins in different stages of B-cells activation/differentiation is strongly supported by studies of EBV-associated malignancies (Table 3.1). Thus, Latency III is expressed in the immunoblast-like cells of EBV carrying lymphoproliferative disorders that arise in severely immunosuppressed individuals, such as transplant recipients and AIDS patients [17], while Latency I is found in EBV carrying Burkitt's lymphoma (BL) whose cells are phenotypically similar to memory B lymphocytes [18]. In line with the germinal center cell origin of Hodgkin's Disease (HD) lymphomas, the EBV-positive forms of this tumor express Latency II (reviewed in [19]). This type of latency is also expressed in non-B-cell tumors of both hematopoietic and epithelial cell origin, including T cell lymphomas, NK cell lymphomas and hemophagocytic syndrome lymphomas, nasopharyngeal carcinoma (NPC) and lymphoepitheliomas originating from stomach, thymus and salivary glands (reviewed in [20, 21]). Each of the latency programs is likely to contribute in specific ways to the biology of the tumor in which it is expressed.

While studies on EBV-carrying cells of normal and tumor cell origin have yielded a wealth of information with regard to the mechanisms by which the virus manipulates the cellular environment during latency, our knowledge of other stages of the virus cycle is lagging behind due to the lack of an easily accessible *in vitro* model of lytic infection and difficulty in obtaining adequate amounts of infectious virus. This handicap is now being overcome with the help of recombinant DNA technologies, but major gaps still exist in our understanding of the early and late events of virus infection such as virus entry, uncoating and delivery of the viral DNA to the nucleus of the infected cells, the assembly of new virus and the release of infectious virus particles.

3.4 EBV and the Ubiquitin–Proteasome System

3.4.1 EBNA1

Because of its specific role in the virus life cycle, EBNA1 is the only viral protein expressed in all types of EBV-infected cells (Figure 3.2). Dimers of EBNA1 bind to the dyad symmetry and family of repeat sequences in the origin of latent plasmid replication (oriP) and coordinate the replication of the viral episomes with cellular DNA and their partitioning during cell division. EBNA1 is also a transcriptional regulator that acts on the two major latent promoters for EBNA transcription, Wp and Cp, and on its own latent promoter Qp (reviewed in [22]). Most of the identified functional domains of EBNA1, including a nuclear localization signal, dimerization and DNA binding domains, reside in the C-terminal half of the protein, while most of the N-terminal half is occupied by a Gly-Ala repeat (GAr) that varies in length between different EBV isolates [23].

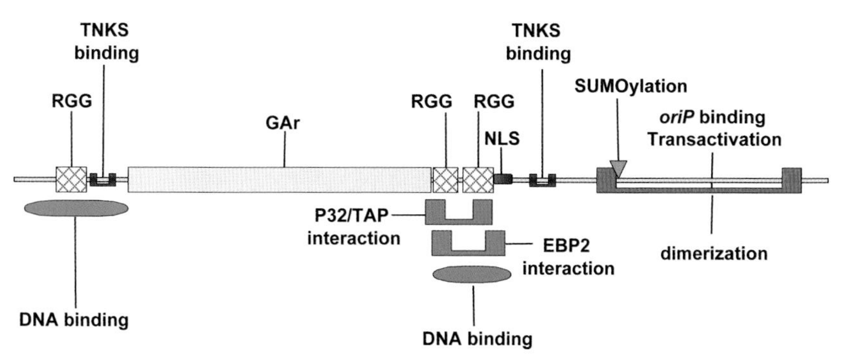

Fig. 3.2. Schematic representation of functional domains in EBNA1. EBNA1 is required for *oriP*-dependent virus replication and partitioning of the viral episome to replicating latently infected cells. These activities are regulated by interaction with tankyrase (TNKS) that also mediates poly-ADP ribose (PAR)-modification of EBNA1. Interaction of EBNA1 with EBNA1-binding protein 2 (EBP2) is required for tethering the viral episomes to the cellular mitotic chromosomes. The arginine–glycine repeat (RGG) domains are involved in binding to RNA. EBNA1 contains one nuclear localization signal (NLS) and one putative SUMOylation site. Interaction with p32/TAP is associated with the translocation of EBNA1 to the cytoplasm.

The regular or even exclusive expression of EBNA1 in EBV-associated malignancies makes it an ideal target for tumor-specific immune intervention, which has prompted an intensive search for specific T-cell responses that could be selectively boosted in cancer patients. This search was initially frustrated by the failure to identify EBV-specific CTLs capable of recognizing cells that express EBNA1 through recombinant vaccinia or adenovirus vectors (reviewed in [24]). This peculiar feature of EBNA1 was shown to be due to the presence of the GAr since removal of this domain resulted in accelerated protein turnover and efficient presentation of endogenous epitopes [25, 26], and reconstituted the capacity of EBNA1 to trigger specific rejection responses in a mouse tumor model [27]. Later studies have demonstrated that EBNA1-specific effectors exist but are either MHC class II restricted CD4 positive T-cells (reviewed in [28]), or MHC class I restricted CD8 positive T-cells that recognize epitopes derived from recombinant EBNA1 exogenously fed to antigen-presenting cells through an as yet poorly characterized cross-talk process between the endocytic – MHC class II restricted – and exocytic – MHC class I restricted – pathways of antigen processing [26]. More recent studies using highly sensitive methods capable of detecting small amounts of cytokines produced by individual effector cells have finally confirmed that MHC class I restricted epitopes can be generated from endogenously expressed EBNA1 [29–31], although processing appears to be less efficient than that for regular endogenous antigens. Several possible explanations can be envisaged to explain this escape from the

GAr-induced blockade of proteasomal processing. The EBNA1 epitopes may be produced from Defective Ribosomal Products (DRiPs) that, unlike the intact EBNA1, are targeted for proteasomal degradation [32]. Although attractive, this explanation does not account for the observation that the majority of the identified MHC Class I restricted epitopes in EBNA1 are located in the C-terminus of the protein, downstream of the GAr, and should therefore also be protected in the context of DRiPs. A more challenging possibility has been suggested by the recent finding that endogenous EBNA1 may gain access to the exocytic pathway of antigen processing by autophagy [33]. Thus, EBNA1 accumulates slowly in cytosolic autophagosomes after the inhibition of lysosome acidification while inhibition of autophagy decreases the presentation of MHC class II restricted epitopes to CD4-positive T-cell clones. It remains to be seen whether this pathway might also be involved in the generation of MHC class I restricted epitopes.

While the biological significance of the GAr in the context of EBV-specific immune surveillance awaits further clarification, elucidation of its mechanism of action is particularly interesting since this is the first example of a protein domain that blocks antigen presentation and this feature could be exploited in immunological and gene therapy settings. Using an *in-vitro* processing assay based on *in-vitro* translated substrates it was shown that the GAr is a specific inhibitor of ubiquitin–proteasome-dependent proteolysis [34] and acts as modular transferable element that can abrogate or severely inhibit the degradation of a variety of viral and cellular proteasome substrates. Several characteristics of this stabilization signal were resolved by using a set of chimeric reporters involving IκB [35], p53 [36], and green fluorescent protein (GFP)-based proteasome substrates [37]. The activity of the GAr was shown to be largely independent of its location in the target protein [34–37] and was not restricted by the type of ubiquitin ligase involved in substrate modification [35–37]. Fusions of the GAr to green fluorescent protein (GFP)-based reporters that are targeted for degradation with varying efficiency showed that the GAr counteracts the degradation signal in a length-dependent manner [37]. EBNA1 itself could also be targeted for ubiquitin-dependent proteolysis using a strong degradation signal, which resulted in the efficient presentation of EBNA1-derived CTL epitopes [38]. The only requirement for activity appears to be the presence of a sufficiently long stretch of alanines or similar small hybrophobic residues, preferably interspersed by one, two or three glycine residues which may act by increasing solubility [39]. This, together with the demonstration that ubiquitinated GAr containing IκB cannot form stable complexes with the proteasome [35], and that the repeat does not affect the interaction of ubiquitinated substrates with the S5a ubiquitin-binding subunit of the 19S cap [36, 40], suggests that the hydrophobic domain produced by the GAr may directly affect the interaction of the substrate with the proteasome. This possibility is supported by the demonstration that a synthetic GAr peptide could inhibit the degradation of biotinylated lysozyme *in vitro* [40]. Interestingly, GAr-containing chimeras were also shown to be protected from proteasomal degradation in yeast [41, 42], suggesting that the viral repeat targets a conserved step in proteasomal processing. By embedding the GAr within ornithine decarboxylase (ODC), a natural proteasome substrate that does not require ubiquitin conjugation, Zhang and Coffino demonstrated that the GAr acts as a stop signal for proteasome processing *in vitro*, resulting in partial proteolysis [42]. Introduction of the GAr into an ODC degradation domain-destabilized GFP, led to the accumulation of degradation products that still contain the repeat, suggesting that the GAr may interfere with the unfolding activity of the proteasome, which could halt degradation. This possibility was recently substantiated by the demonstration that the production of intermediates is influenced by the position of the GAr relative to a folded domain within the substrate. The spacing between the GAr and a downstream folded domain appears to be critical for intermediate production [43]. These findings support a model whereby positioning of the GAr domain within the ATPase ring reduces the efficiency of nucleotide hydrolysis and substrate unfolding. If this impairment takes place, insertion pauses and proteolysis are limited to the portion of the substrate that has already entered the catalytic chamber of the proteasome.

The finding that presentation of EBNA1 epitopes can occur in spite of the protective activity of the GAr, points to a non-immunological role for the effect of the repeat on EBNA1 stability. Indeed, using both affinity chromatography and TAP-tagging approaches it was recently shown that EBNA1 interacts with the ubiquitin-specific protease USP7 [44, 45], also known as herpes virus-associated ubiquitin-specific protease, HAUSP. This DUB was first identified by virtue of its interaction with the ICP0 protein of herpes simplex virus type 1 [46], a viral E3 ligase that is required for efficient initiation of the HSV-1 lytic cycle [47]. An EBNA1 mutant defective for USP7 binding exhibited the long half-life and lack of MHC class I presentation typical of wild-type EBNA1, indicating that USP7 is not directly involved in the regulation of EBNA1 turnover. However, disruption of USP7 binding enhanced the replication of an EBV *ori*P-containing plasmid. This may be due to a direct effect of USP7 on the ubiquitination of EBNA1 or of other cellular substrates that interact with, and regulate the activity of *oriP*. Indeed, tankirase-1 (TRF1), a negative regulator of telomere length that also interacts with *oriP* and binds to EBNA1 inhibiting *oriP*-dependent replication [48, 49], is a substrate for ubiquitin-dependent proteolysis [50]. Only the telomere-unbound form of TRF1 is ubiquitinated and degraded, suggesting that specific rescue of this protein bound to *oriP* could play an important role in the regulation of latent EBV replication. It is also possible that binding to EBNA1 may affect cellular functions that are normally regulated by USP7. Recent evidence indicates that USP7 is a key regulator of p53 and Mdm2 [51]. EBNA1 and p53 bind to the same pocket in USP7 but p53 makes less extensive contacts resulting in significantly lower binding affinity [45, 52]. Thus, EBNA1 could efficiently compete for p53 binding and prevent its deubiquitination, which would promote Mdm2-dependent degradation. Functional studies indicated that binding of EBNA1 to USP7 can protect cells from apoptosis by lowering the levels of p53 [52], providing a structural and conceptual framework for understanding how EBNA1 might contribute to the survival of EBV-infected cells.

3.4.2
EBNA6 (EBNA3C)

The progression of the cell cycle and stability of cell cycle checkpoint proteins is controlled by the ubiquitin–proteasome system. A critical regulator of cell cycle molecules is the Skp1/Cul1/F-box E3 ligase SCF^{Skp2} that mediates the polyubiquitination and degradation of E2F and several E2F transcriptional targets, including p27 and c-Myc. A link between the EBV nuclear antigen EBNA6 (EBNA3C) and SCF^{Skp2} was recently demonstrated, providing a new insight into the mechanism for cell cycle regulation by EBV (Figure 3.3). EBNA6 is one of three high molecular weight EBNAs encoded in the BamHI-E region of the EBV genome and was shown to be essential for B-cell immortalization [53]. Transfection of EBNA6 into EBV-negative cells of lymphoid and epithelial cell origin was shown to correlate with decreased Rb protein levels [54]. EBNA6 forms a stable complex with Rb in cells treated with inhibitors of the proteasome and interacts with Rb *in vitro* through a conserved motif within amino acids 140–149 that has been linked to the regulation of SCF^{Skp2}. Indeed, transfection of a dominant negative SCF^{Skp2} reduced the ability of EBNA6 to promote the degradation of Rb. SCF^{Skp2} has no detectable effect on Rb levels in the absence of EBNA6, suggesting that EBNA6 may specifically usurp this ligase to enhance Rb degradation. Capture of SCF^{Skp2} by EBNA6 may have additional effects on the regulation of cell cycle progression since EBNA6 also associates with the cyclin A/cdk2 complexes through a small region between amino acids 130 and 159 that shows high affinity for the conserved mammalian cyclin box in amino acids 206 to 226 of cyclin A [55]. Binding of EBNA6 to cyclin

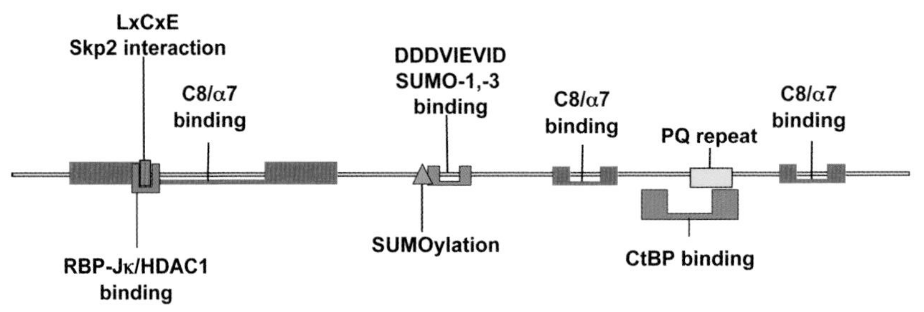

Fig. 3.3. Schematic representation of functional domains in EBNA6. EBNA6 (EBNA3C) is one of three nuclear antigens encoded in the BamHI E region of the viral genome and cooperates with EBNA2 in the regulation of viral and cellular genes. The RBP-Jκ, HDAC1 and CtBP binding sites, involved in transcription regulation, and the glutamine–proline repeats are indicated. EBNA6 interacts with SCF^{Skp2}, with SUMO-1 and SUMO-3 and contains one putative SUMOylation site. In addition, EBNA6 interacts with the C8/α7 subunit in the α-ring of the 20S proteasome.

A induces SCF^{Skp2}-dependent degradation of p27 and enhances the activity of the kinase [56].

In addition to its effect on cellular proteins, EBNA6 participates in the regulation of viral gene expression and cooperates with EBNA2 in transactivating the LMP1 promoter [57]. A region of the protein between amino acids 365 and 545, that was shown to be necessary and sufficient for LMP1 coactivation, interacts with SUMO-1 and SUMO-3 in yeast two-hybrid screens. This region is also required for localization of EBNA6 to PML bodies and for modification by SUMO-1, SUMO-2, and SUMO-3, but neither of these functions appears to be necessary for transcriptional coactivation [58]. Furthermore, coactivation was inhibited by mutation of a short sequence between amino acids 509 and 515 (DDDVIEVID) which bears close similarity to residues 84–90 of SUMO-1 (EEDVIV) and the resulting mutants lost the capacity to bind to SUMO-1 and SUMO-3. This conserved region is important for the binding of SUMO to UBC9 and ULP-1 [59] and a similar motif is found in the E1 enzymes of plants [60] and in SNF2 domain proteins that are involved in chromatin remodeling and transcriptional regulation [61], suggesting that it may mediate binding to SUMOylated substrates. It has been speculated that through this motif EBNA6 may bind to a repressor, such as HDAC-1, and inhibit its effects on another transcription factor(s) at EBNA2-regulated promoters.

A yeast two-hybrid screen using EBNA6 as bait revealed an interaction with the C8 (alpha7) subunit of the 20S proteasome [62]. The interaction was confirmed in glutathione S-transferase (GST) pull-down experiments that also revealed interaction between C8 and the two other members of the high molecular weight EBNA family: EBNA3 (EBNA3A) and EBNA4 (EBNA3B). Co-immunoprecipitation of the EBNA3 proteins with C8/alpha7 was also demonstrated after transfection of expression vectors into B cells. The interaction between these viral proteins and GST-C8 appears to be more robust than the interaction between C8 and the cyclin-dependent kinase inhibitor p21(WAF1/CIP1), which results in degradation of p21 by the 20S proteasome [63]. Consistent with their ability to bind directly to the 20S proteasome, the EBNAs were degraded *in vitro* using purified 20S proteasomes but the significance of this finding is unclear since the viral proteins have a relatively long half-life in EBV-infected cells. It remains to be seen whether these interactions serve any other purposes, such as for example, the targeting of specific cellular substrates to the proteasome.

3.4.3
LMP1

Expression of the Latency III program is potentially unfavorable for EBV since the uncontrolled proliferation of virus-infected cells could kill the host, as indeed happens in patients suffering from severe congenital or iatrogenic immune deficiencies. To counteract the hazard, viral proteins expressed in Latency III increase the immunogenicity of the infected cells by regulating the expression of adhesion and co-stimulatory molecules and enhance the activity of various components of

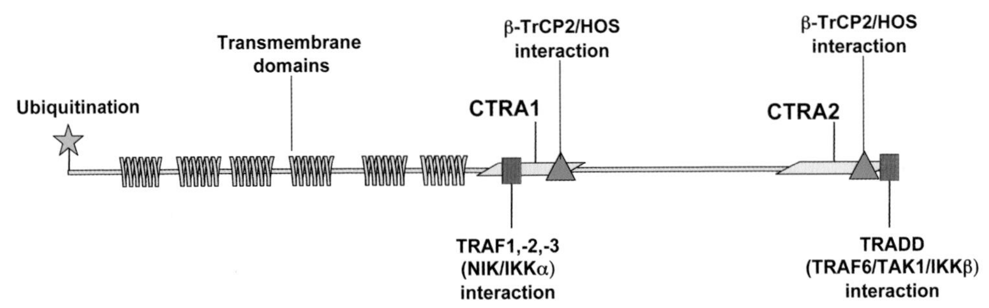

Fig. 3.4. Schematic representation of functional domains in LMP1. LMP1 is one of two EBV membrane proteins expressed in latently-infected cells and mediates NF-κB activation by acting as a constitutive receptor of the CD40/TNFR family. Two C-terminal activating regions (CTAR) are involved in NF-κB activation through the canonical (TRADD, TRAF6, TAK1, IKKβ) and alternative (TRAF2/NIK/IKKα) pathways. LMP1 is ubiquitinated at the N-terminus and contains two SCF^{HOS} binding sites.

the antigen presentation machinery, which renders the infected cells easily attacked by the immune responses. LMP1 plays a pivotal role in this regulation by interfering with a major effector of lymphoid cell activation and function, the NF-κB signaling pathway (Figure 3.4).

LMP1 is the only EBV protein with recognized oncogenic activity. Transfection in mouse or human fibroblasts and epithelial cells confers tumorigenicity in immunosuppressed animals [64–66] and LMP1 transgenic mice develop hyperproliferations and lymphomas [67, 68]. LMP1 contains a short N-terminal cytoplasmic domain followed by six membrane-spanning domains and a large cytoplasmic C-terminal domain that is involved in signaling (reviewed in [2]). Through the transmembrane domain LMP1 forms multimers that localize to lipid rafts together with LMP2A and a variety of cellular proteins involved in B-cell proliferation and function [69]. LMP1 acts as a constitutive receptor of the CD40/tumor necrosis factor-receptor (TNFR) family and induces expression of the NF-κB and JNK transcription factors [70, 71]. Similar to TNFR, LMP1 binds to TRAF1, TRAF2 and TRADD, and activates the NF-κB-inducing kinase (NIK) and the IkB kinases (IKKα and IKKβ) [72]. Activated IKK phosphorylates IκB, which leads to recognition by the WD40 domains of F-box proteins belonging to the β-TrCP family, including β-TrCP1 and β-TrCP2 (also termed FWD1/Fbw1a and HOS/Fbw1b). These F-box proteins interact with Skp1, Cullin1, and Roc1/Rbx1 to form the $SCF^{b\text{-}TrCP/HOS}$ E3 ubiquitin ligases (reviewed in [73]) that ubiquitinate IκB, which results in proteasomal degradation of the inhibitor and activation of NF-κB-dependent transcription [74]. Recent findings suggest that the regulation of NF-κB signaling by LMP1 may be more complex than originally thought. The LMP1 protein encoded by the prototype B95.8 strain of EBV was shown to interact with the HOS subunit of SCF^{HOS} via one canonical and one cryptic HOS recognition site [75]. Mutations of these sites abrogated HOS binding and increased the transforming activity of LMP1, which correlated with its increased ability to induce IκB degradation and NF-κB-mediated transcription without further activation of IKK. Furthermore, overexpression of HOS in cells expressing the B95.8 LMP1 enhanced the degradation of IκB and activation of NF-κB without significant effect on the stability of the viral protein. Thus, the B95.8 LMP1 appears to act as a pseudosubstrate for SCF^{HOS}, which by reducing the levels of endogenous HOS available to interact with phosphorylated IκB, may result in fine-tuning of LMP1-induced NF-κB signaling. Interestingly, the HOS interacting domains in LMP1 are mutated or deleted in the NPC-derived LMP1-Cao variant that exhibits enhanced tumorigenic properties in epithelial cells [64], providing an interesting clue to the possible role of LMP1 mutations in malignancies.

Several components of the antigen presentation pathway are also upregulated in LMP1-expressing cells including the transporters associated with antigen processing (TAPs, [76, 77] and some subunits of the proteasome [78], which results in enhanced enzymatic activity of the proteasome and altered cleavage specificity, thereby promoting the antigenicity of EBV-infected cells. In addition, LMP1 itself is a short-lived protein [79] and a substrate of ubiquitin-dependent proteolysis, which generates several epitopes for MHC class I restricted T-cell recognition of EBV-infected cells [80]. However, LMP1 also contains two epitopes with strong homology to an immunosuppressive domain found in a retroviral membrane protein, that strongly inhibits the activity of both cytotoxic T-lymphocytes and natural killer cells *in vitro* [81]. Thus, processing of LMP1 may suppress EBV-specific immune responses, as was shown to occur at the tumor site of EBV positive HD lymphomas [82]. The mechanism of ubiquitin-dependent processing of LMP1 is also quite puzzling since the single lysine residue in the C-terminal domain of the LMP1 encoded by the prototype B95.8 EBV strain, is often mutated in wild-type EBV isolates and, furthermore, is not required for ubiquitination. Minor modification of the N-terminus resulted in full stabilizations, confirming that this is the site for ubiquitin conjugation [83]. This unusual ubiquitination linkage has been described for only a few cellular substrates that either do not contain lysine residues or whose lysine residues may be inaccessible (reviewed in [84]). Interestingly, N-terminal ubiquitination is also involved in the degradation of the second EBV membrane protein, LMP2 [85]. The two EBV proteins co-localize in the cell membrane and it is therefore possible that the same ligase may be involved in their ubiquitination.

3.4.4 LMP2

Studies with EBV deletion mutants show that LMP2A and its N-terminal truncated variant LMP2B are not required for B-cell transformation [86] (Figure 3.5). Yet, the regular expression of these proteins in the most restricted forms of latency

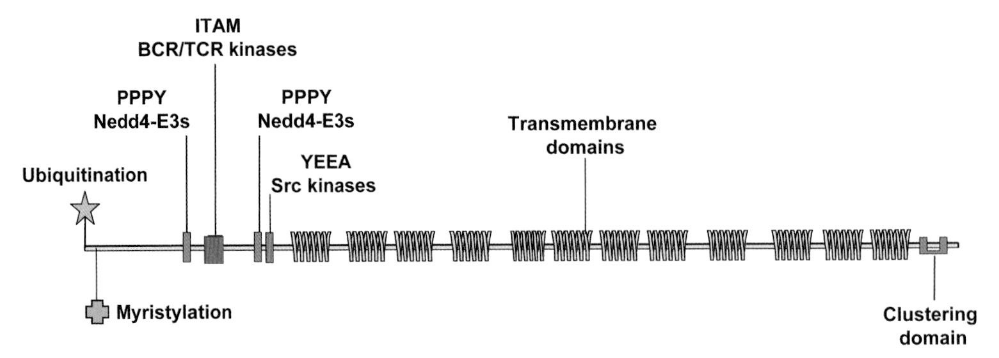

Fig. 3.5. Schematic representation of functional domains in LMP2. LMP2 is the second EBV-encoded membrane protein expressed in latently infected cells. It interferes with BCR/TCR signaling by acting as a scaffold for binding BCR/TCR-associated and Scr-family kinases and Nedd4-family ubiquitin ligases. LMP2 is ubiquitinated at the N-terminus.

suggests that they fulfill an important function in the biology of EBV infection. EBV persists in B-lymphocytes that are also the primary site of virus reactivation. The mechanisms that lead to the breakdown of latency are poorly understood but studies with tumor cell lines suggest that triggering the B-cell receptor (BCR) plays a critical role [87]. It has been proposed that LMP2A regulates this process through its ability to inhibit BRC signaling [88, 89] although recent studies question this scenario since LMP2A was shown to activate B-lymphocytes and induced antibody production and plasma cell differentiation in a transgenic mouse model [90]. The molecular mechanism by which LMP2 interferes with B-cell receptor signaling has been extensively studied. It has been shown that the N-terminus of LMP2A contains several ITAM motifs that mediate in the interaction with the SH2-domain of the BCR-associated tyrosine kinases Lyn and Syk, and two Pro-Pro-Pro-Pro-Tyr (PY) motifs that are known to interact with proteins containing WW modules [91]. In screening for interacting proteins two laboratories have independently identified members of the Nedd4 ubiquitin ligase family [92, 93]. These include AIP4/Itchy, WWP2/AIP2 and KIAA0439 which contain a C-terminal HECT domain. Binding of these E3s to LMP2A induces ubiquitination of Lyn and Syk and this correlates with the accelerated degradation of Lyn. Ubiquitination of Syk does not appear to affect its turnover although accelerated degradation of a small pool of activated Syk cannot be excluded. Other ligases are also involved in this signaling cascade. Phosphorylation of Syk causes constitutive activation of the Syk substrate SLP-65 (SH2 domain-containing leukocyte protein-65), which in turn induces the formation of a ternary complex composed of the ubiquitin ligase Cbl, C3G and the protooncogene CrkL (CT10 regulator of kinase-like enzyme) [94]. Cbl-b ubiquitinates Syk, which is important for its function as a negative regulator of BCR signaling [95]. LMP2A itself is specifically ubiquitinated by AIP4 and WWP2 at the N-terminus and this is required for modulation of BCR signaling. While further studies are needed to clarify the significance of LMP2A-induced ubiquitination of Lyn and Syk, these data confirm that the expression of scaffolds that recruit components of the ubiquitin–proteasome system is a common viral strategy for selective inactivation of cellular substrates.

LMP2A is also part of the Latency II program expressed in non-B cell tumors and its expression was shown to enhance the metastatic potential of NPC. The effect in T lymphocytes appears to be analogous to that observed in B cells since LMP2A was shown to interact with the T-cell receptor-associated tyrosine kinases Lck, Fyn and ZAP-70/Syk and stable expression of LMP2A downregulated T-cell receptor levels and signaling in the T cell line Jurkat [96]. Recruitment of IAP4 via the PPPPY motifs was required for LMP2A ubiquitination and regulated the stability of LMP2A and LMP2A–kinase complexes. In epithelial cells, LMP2A activates the phosphatidylinositol 3 -OH kinase/Akt and β-catenin signaling pathways [97, 98]. The ITAM was essential for the activation of Akt by LMP2A in human foreskin keratinocyte (HFK), while both the ITAM and PPPPY motifs contributed to LMP2A-mediated accumulation and nuclear translocation of β-catenin [99]. Furthermore, the PPPPY motifs were critical for LMP2A-mediated inhibition of epithelial cell differentiation in semi-solid methylcellulose medium. LMP2A was also shown to influence the adhesion and migration and invasiveness of epithelial cells, possibly by regulating the expression of the epithelial cell integrin $\alpha6\beta4$ [100]. An unexpected role of the Syk kinase in this phenotype was recently suggested by the finding that Tyr 74 and 85 in the ITAM of LMP2A are essential for both Syk activation and LMP2A-induced epithelial cell migration [101]. It remains to be seen whether IAP4, c-Cbl or other E3 ligases participate in this regulation.

3.4.5 BZLF1 (Zta) and BRLF1 (Rta)

The immediate early proteins of EBV, BZLF1 (Zta) and BRLF1 (Rta), mediate the switch from latent to productive infection by activating the transcription of early and late viral genes as well as a variety of cellular genes that are required for progression of the lytic virus cycle (reviewed in [102]). The molecular pathways leading to this switch are only partially understood but a common downstream event, for EBV as for other herpes viruses, is the dispersion of promyelocytic leukemia bodies (PML, ND10) in the nucleus of the host cell [103]. The PML protein is induced by interferon, involved in major histocompatibility complex class I presentation, and necessary for certain types of apoptosis [104]. Modification of PML by SUMO-1 is known to be required for the formation of PML bodies [105]. Both Zta [106] and Rta [107] are modified by SUMO-1 and Rta was shown to interact with the SUMO E2 conjugase, Ubc9 and E3 ligase, PIAS1, in yeast two-hybrid screens, GST-pull-down assays, co-immunoprecipitation, and confocal microscopy. Expression of Zta and Rta induces PML dispersion in EBV-positive cells but Zta alone is sufficient to produce this effect in EBV-negative cell lines. Similar to the corresponding proteins in HSV-1 (ICP0) and CMV (IE1), Zta reduces the amount of SUMO-1-modified PML, probably by competing with PML for limiting amounts of SUMO-1. Furthermore, the capacity of Rta to transactivate a reporter plasmid containing a specific responsive element was greatly increased by SUMOylation of Lys19 [107]. Thus, SUMO-1 modification of these viral immediate early proteins appears to promote lytic EBV replication by enhancing their transactivating activity as well as by modulating the function of cellular proteins that are targeted by SUMO-1 modification.

3.4.6 BPLF1

Recent reports describing the identification of genes encoding for proteins with DUB activity in adenovirus [108], SARS corona virus [109–111] and human herpes viruses [112], demonstrated that this step of the of the ubiquitin-dependent proteolytic machinery is also specifically targeted during viral infection. DUB activity was demonstrated in the N-terminal 500 residues of the largest tegument protein of herpes simplex virus 1 (HSV-1), UL36, as well as in the UL36 homologs of other herpesviruses, including the EBV-encoded BPLF1 protein [112]. These proteins are expressed late in the replication cycle of herpes viruses and are essential for the production of infectious virus particles [113]. UL36 has been implicated in a sequence-specific interaction with the viral genome [114] and studies carried out using a temperature-sensitive mutant suggest that it plays a role in the release of viral DNA from incoming nucleocapsids, viral DNA synthesis, and late gene expression [115]. The role of its DUB activity in these processes is not understood. The ubiquitination machinery regulates the membrane protein trafficking systems involved in the formation of caveolae and multivesicular bodies (MVBs) that are required for the entry and exit of some viruses from infected cells [116, 117]. At least two DUBs, AMSH and UBPY, regulate the activity of MVBs. Both DUBs interact with ESCRT proteins via STAM (signal transducing adaptor molecule) and show specificity for Lys63- and Lys48-linked polyubiquitin chains *in vitro*. AMSH (associated molecule with the SH3 domain of STAM) binds to clathrin and to mVps24/CHMP3, a component of ESCRT III complex, and is markedly stimulated by STAM, indicating that activation is coupled to association with the MVB-sorting machinery [118]. RNA-mediated knockdown of UBPY results in increased global levels of ubiquitinated protein and accumulation of ubiquitin on endosomes, while UBPY-depleted cells have more and larger MVBs [119]. Perhaps herpes viruses exploit the deubiquitinating activity encoded by UL36, and its homologs that are produced in the final stages of virus replication, to manipulate ubiquitin modifications during viral egress. Interestingly, the viral enzymes do not show strong homology with eukaryotic DUBs, making this new family of ubiquitin-specific proteases an attractive target for selective inhibition.

3.5 EBV-associated Malignancies

Specific genetic alterations contribute to the pathogenesis of EBV-associated malignancies. This concept is clearly illustrated by BL, a highly malignant B-cell tumor that is characterized by the regular presence of chromosomal translocations involving *c-myc* on chromosome 8 and one of the immunoglobulin heavy or light chain genes on chromosomes 14, 2 or 22 [120]. EBV-carrying BL is the most frequent childhood malignancy in endemic malaria areas of subtropical Africa and Papua New Guinea and occurs at a 100-fold higher frequency in AIDS patients. The role of EBV as a co-factor in the pathogenesis of this malignancy is supported by the observation that a genetically and phenotypically similar but EBV-negative variant of the tumor occurs worldwide at a much lower frequency. The EBV-positive tumors express only EBNA1 [18] or, in some cases, EBNA-1 and the high molecular weight EBNA-3, -4 and -6 [121]. The B-cell proliferation and activation-associated viral proteins, EBNA2 and LMP1 are not expressed and, as a result, EBV-positive BL cells do not express B-cell activation markers, adhesion or co-stimulatory molecules and grow as single cell suspensions rather than in clumps [18]. Also characteristic of the tumor cells are the low levels of MHC class I and selective loss of certain class I alleles, HLA A11 in particular [122], which may contribute to the poor immunogenicity of these cells [123]. In addition BL cells are deficient in their ability to present endogenous antigens [77], which correlates with the downregulation of the interferon- -inducible subunits of the proteasome and the peptide transporters TAP1 and TAP2 [76, 78]. Studies on *in vitro* EBV-transformed cell lines that carry inducible EBNA2 and *c-myc* genes and recapitulate the viral and cellular gene expression program of BLs, have shown that overexpression of c-Myc in cells expressing Latency I drives the BL cell phenotype [124].

Surprisingly, in spite of their apparent defect of proteasome function, BL cells were shown to be resistant to treatment with doses of proteasome inhibitors that are readily toxic for LCL cells [125]. Since the turnover of short- and long-lived protein proceeded virtually undisturbed in BL cells treated with the proteasome inhibitors, other proteases likely contribute to the regulation of protein turnover. Indeed, DUB activity and the serine protease tripeptidyl peptidase (TPP)-II, were shown to be overexpressed in BL cells and were upregulated on induction of c-Myc. The upregulation of DUBs may offer a selective advantage in cells with poor proteasome activity since the free polyubiquitin chains that may accumulate as a consequence of slow proteasomal processing are highly toxic. Furthermore, DUBs may also be involved in regulating the turnover of specific substrates that control the survival of these cells. Analysis of DUB expression and activity using a set of functional probes followed by mass spectrometry revealed consistent differences between BL and LCL cells. In particular, USP7/HAUSP was highly expressed in LCLs while UCH-L1 was regularly overexpressed in the tumor-derived cell line [126]. UCH-L1, also known as PGP9.5, is an abundant neuronal protein and mutations at or around the putative catalytic site have been associated with familial

Parkinson's disease [127] and with other neurodegenerative disorders characterized by the formation of protein aggregates including spinocerebellar ataxia [128] and Huntington's disease [129]. In addition, high levels of UCH-L1 have been detected in a variety of human malignancies including neuroblastoma [130], colon carcinoma [131], non small-cell-lung carcinoma [132, 133], pancreatic carcinoma [134], prostate and breast carcinomas [135, 136] and renal carcinoma [137] and appear to be associated with the more malignant and invasive forms of these tumors. The physiological targets of UCH-L1 are unknown. *In vitro*, UCH-L1 acts as a C-terminal hydrolase for ubiquityl esters and amides but has no activity with regard to larger ubiquitin conjugates [138]. In addition, recent evidence suggests that, depending on its level of expression and capacity to form homodimers, UCH-L1 may act as an E3 ligase [139]. Interestingly, UCH-L1 was shown to interact with Ubc9, RanBPM and Jab1 in a yeast two-hybrid screen [140]. Ubc9 is the E2 for SUMO and RanBPM is a regulator of the small G protein Ran [141] while Jab1 is a Jun activation domain-binding protein that can bind to p27(Kip1) and is involved in the cytoplasmic transport of p27(Kip1) for its degradation [142]. Jab1 is also a component of the COP9 signalosome [143] and acts as a Cullin de-Neddylase [144]. UCH-L1 co-localizes with Jab1 in the perinuclear cytoplasm of contact-inhibited cells while, under serum-re-stimulation, nuclear translocation of both UCH-L1 and Jab1 coincides with a reduced level of p27(Kip1) in the nucleus. Therefore, UCH-L1 may contribute to p27(Kip1) degradation via its interaction and nuclear translocation with Jab1.

Recent findings suggest an unexpected role for TPPII in the phenotype of BL cells. TPPII is an evolutionarily conserved cytosolic serine protease of the subtilisin family that removes tripeptides from the free N-terminus of oligopeptides [145–149]. The 138-kDa subunit of TPPII forms large oligomeric complexes of 5–9 MDa with a twisted ribbon structure that are detected in the cytoplasm and associate with the plasma membrane of most cell types [145, 146, 150, 151]. It has been proposed that TPPII may act downstream of the proteasome to accelerate the production of free amino acids from longer precursors by generating tripeptide intermediates that are easily degraded by other cellular exopeptidases [152]. In addition, TPPII appears to play a more specific function in the processing of certain cellular substrates as it is the main cholecystokinin-inactivating enzyme in rat brain [151] and regulates apoptotic responses by promoting the maturation of procaspase-1 in macrophages infected with the enteropathogenic bacterium *Shigella flexneri* [153]. Recent evidence suggests that TPPII may have a role in antigen processing [154, 155], probably through its capacity to cleave long peptides fragments produced by the proteasome [154, 156]. The observations that TPPII is overexpressed in cells adapted to grow in the presence of lethal concentrations of proteasome inhibitors [150, 157, 158], rescues transfected cells from acute intoxication with these inhibitors [150, 152, 157] and increases the survival and tumorigenicity of proteasome inhibitor adapted cells in a mouse lymphoma model [159], suggest a possible role for this enzyme in tumor progression. Indeed, TPPII overexpression correlates with accelerated proliferation and accumulation of centrosome and chromosome aberrations in BL as well as in transfected 293HEK cells, whereas functional knockdown of TPPII by shRNA resulted in growth retardation and the generation of polynucleated cells that failed to complete cell division [160]. Furthermore, TPPII overexpressing cells evaded mitotic arrest induced by spindle poisons and displayed high levels of polyploidy despite the constitutively high expression of major components of the spindle checkpoints [161]. This was accompanied by upregulation of inhibitors of apoptosis (IAPs) and resistance to p53-induced apoptosis, suggesting that TPPII may allow the transit through mitosis and the survival of cells with severe mitotic spindle damage. Collectively, these findings suggest that TPPII may participate in the regulation of critical events that control the homeostasis of cell division. In particular, the accumulation of centrosome abnormalities and multipolar spindles together with the capacity to overcome spindle checkpoints, point to a possible involvement of TPPII in the early phases of mitosis, at or around the time of centrosome duplication.

3.6
Concluding Remarks

The unraveling of the different strategies by which EBV exploits and manipulates ubiquitin/proteasome-dependent proteolysis, together with a better understanding of how these manipulations assist the virus during productive infection, latency and malignant transformation, will provide new strategies for meeting the challenge of this potentially oncogenic viral infection. Interference with the interaction between LMP2A and BCR/TCR-associated tyrosine kinases may release the blockade of productive infection and promote the elimination of virus-infected cells either directly or indirectly, through the exposure of numerous highly immunogenic epitopes that could target the infected cells for destruction by CTLs. Induction of EBNA1 processing may also sensitize the infected cells to CTL-mediated rejection since EBNA1-specific precursors are present in EBV carriers. It is likely that the endocytic, exocytic and autophagic pathways of antigen processing play different roles in different cell types or pathologic conditions and a better understanding of the interplay between these modes of protein processing will be required to develop effective strategies for enhancing the degradation of the Gar-containing protein. Our current detailed understanding of the interaction between EBNA1 and USP7/HAUSP based on structure determination, may also provide a conceptual basis for the rational design of inhibitors that will selectively affect the capacity of the viral protein to interfere with cellar functions that control proliferation and apoptosis. Finally, it seems reasonable to assume that, as clearly demonstrated for other virus infections, modulation of components of the ubiquitin–proteasome system will interfere with EBV assembly and maturation. Changes in the pattern of antibody responses to viral antigens associated with the productive cycle and increased virus load are observed in EBV-associated malignancies and play an important diagnostic and prognostic role in NPC, pointing to a possible role of virus production in the pathogenesis of this tumor [162]. Thus, interference with the late stages of virus replication by modulation of the ubiquitin–proteasome system may provide a new means by which to influence the natural history of at least some EBV-associated malignancies.

Acknowledgments

I wish to thank all the colleagues who have contributed to this work and in particular Ramachrishna Sompallae for help with the illustrations for this chapter.

This research was supported by grants from the Swedish Cancer Society, the Swedish Research Council and the Karolinska Institute, Stockholm, Sweden.

References

1 Glickman, M.H. and Ciechanover, A. (2002) The ubiquitin-proteasome proteolytic pathway: destruction for the sake of construction. *Physiol Rev* **82**, 373–428.

2 Kieff, E. (1996), in *Fields Virology*, 3rd. edn, Vol. 2, (eds B.N. Fields, D.M. Knipe, P.M. Howley et al.), Lippincott, Raven Publishers, Philadelphia, pp. 2343–2396.

3 Benaroudj, N., Tarcsa, E., Cascio, P. and Goldberg, A.L. (2001) The unfolding of substrates and ubiquitin-independent protein degradation by proteasomes. *Biochimie* **83**, 311–318.

4 Fang, S. and Weissman, A.M. (2004) A field guide to ubiquitylation. *Cell Mol Life Sci* **61**, 1546–1561.

5 Nijman, S.M., Luna-Vargas, M.P., Velds, A., Brummelkamp, T.R., Dirac, A.M., Sixma, T.K. and Bernards, R. (2005) A genomic and functional inventory of deubiquitinating enzymes. *Cell* **123**, 773–786.

6 Kerscher, O., Felberbaum, R. and Hochstrasser, M. (2006) Modification of proteins by ubiquitin and ubiquitin-like proteins. *Annu Rev Cell Dev Biol* **22**, 159–180.

7 Rickinson, A.B. and Kieff, E. (1996), in *Virology* (eds B.N. Fields, D.M. Knipe, P.M. Howley et al.), Lippincott, Raven Publishers, Philadelphia, pp. 2397–446.

8 Samanta, M., Iwakiri, D., Kanda, T., Imaizumi, T. and Takada, K. (2006) EB virus-encoded RNAs are recognized by RIG-I and activate signaling to induce type I IFN. *EMBO J* **25**, 4207–4214.

9 Cai, X., Schafer, A., Lu, S., Bilello, J.P., Desrosiers, R.C., Edwards, R., Raab-Traub, N. and Cullen, B.R. (2006) Epstein-Barr virus microRNAs are evolutionarily conserved and differentially expressed. *PLoS Pathog* **2**, e23.

10 Chen, F., Zou, J.-Z., Di Renzo, L., Winberg, G., Hu, L.-F., Klein, E., Klein, G. and Ernberg, I. (1995) A subpopulation of normal B cells latently infected with Epstein-Barr virus resembles Burkitt lymphoma cells in expressing EBNA-1 but not EBNA-2 or LMP1. *J Virol* **69**, 3752–3758.

11 Babcock, G.J., Decker, L.L., Freeman, R.B. and Thorley-Lawson, D.A. (1999) Epstein-Barr virus-infected resting memory B cells, not proliferating lymphoblasts, accumulate in the peripheral blood of immunosuppressed patients. *J Exp Med* **190**, 567–576.

12 Babcock, G.J., Decker, L.L., Volk, M. and Thorley-Lawson, D.A. (1998) EBV persistence in memory B cells *in vivo*. *Immunity* **9**, 395–404.

13 Miyashita, E.M., Yang, B., Babcock, G.J. and Thorley-Lawson, D.A. (1997) Identification of the site of Epstein-Barr virus persistence in vivo as a resting B cell. *J Virol* **71**, 4882–4891.

14 Babcock, G.J. and Thorley-Lawson, D.A. (2000) Tonsillar memory B cells, latently infected with Epstein-Barr virus, express

the restricted pattern of latent genes previously found only in Epstein-Barr virus-associated tumors. *Proc Natl Acad Sci USA* **97**, 12250–12255.

15 Babcock, G.J., Hochberg, D. and Thorley-Lawson, A.D. (2000) The expression pattern of Epstein-Barr virus latent genes *in vivo* is dependent upon the differentiation stage of the infected B cell. *Immunity* **13**, 497–506.

16 Thorley-Lawson, D.A. (2001) Epstein-Barr virus: exploiting the immune system. *Nat Rev Immunol* **1**, 75–82.

17 Thomas, J., Allday, M. and Crawford, D. (1991) Epstein-Barr virus-associated lymphoproliferative disorders in immunocompromised individuals. *Adv Cancer Res* **57**, 329–380.

18 Rowe, D.T., Rowe, M., Evan, G.I., Wallace, L., Farrell, P. and Rickinson, A.B. (1986) Restricted expression of EBV latent genes and T-lymphocyte-detected membrane antigen in Burkitt's lymphoma cells. *EMBO J* **5**, 2599–2607.

19 Niedobitek, G., Baumann, I., Brabletz, T., Lisner, R., Winkelmann, C., Helm, G. and Kirchner, T. (2000) Hodgkin's disease and peripheral T-cell lymphoma: composite lymphoma with evidence of Epstein-Barr virus infection. *J Pathol* **191**, 394–399.

20 Knecht, H., Berger, C., Rothenberger, S., Odermatt, B.F. and Brousset, P. (2001) The role of Epstein-Barr virus in neoplastic transformation. *Oncology* **60**, 289–302.

21 Dolcetti, R. and Masucci, M.G. (2003) Epstein-Barr virus: induction and control of cell transformation. *J Cell Physiol* **196**, 207–218.

22 Leight, E.R. and Sugden, B. (2000) EBNA-1: a protein pivotal to latent infection by Epstein-Barr virus. *Rev Med Virol* **10**, 83–100.

23 Falk, K., Gratama, J.W., Rowe, M., Zou, J.Z., Khanim, F., Young, L.S., Oosterveer, M.A.P. and Ernberg, I. (1995) The role of repetitive DNA sequences in the size variation of Epstein-Barr virus (EBV) nuclear antigens, and the identification of different EBV isolates using RFLP and PCR analysis. *J Gen Virol* **76**, 779–790.

24 Rickinson, A.B., Murray, R.J., Brooks, J., Griffin, H., Moss, D.J. and Masucci, M.G. (1992) T cell recognition of Epstein-Barr virus associated lymphomas. *Cancer Surv* **13**, 53–80.

25 Levitskaya, J., Coram, M., Levitsky, V., Imreh, S., Steigerwald-Mullen, P.M., Klein, G., Kurilla, M.G. and Masucci, M.G. (1995) Inhibition of antigen processing by the internal repeat region of the Epstein-Barr virus nuclear antigen-1. *Nature* **375**, 685–688.

26 Blake, N., Lee, S., Redchenko, I., Thomas, W., Steven, N., Leese, A., Steigerwald-Mullen, P., Kurilla, M.G., Frappier, L. and Rickinson, A. (1997) Human CD8+ T cell responses to EBV EBNA1: HLA class I presentation of the (Gly-Ala)-containing protein requires exogenous processing. *Immunity* **7**, 791–802.

27 Mukherjee, S., Trivedi, P., Dorfman, D.M., Klein, G. and Townsend, A. (1998) Murine cytotoxic T lymphocytes recognize an epitope in an EBNA-1 fragment, but fail to lyse EBNA-1-expressing mouse cells. *J Exp Med* **187**, 445–450.

28 Paludan, C. and Munz, C. (2003) CD4+ T cell responses in the immune control against latent infection by Epstein-Barr virus. *Curr Mol Med* **3**, 341–347.

29 Tellam, J., Connolly, G., Green, K.J., Miles, J.J., Moss, D.J., Burrows, S.R. and Khanna, R. (2004) Endogenous presentation of CD8+ T cell epitopes from Epstein-Barr virus-encoded nuclear antigen 1. *J Exp Med* **199**, 1421–1431.

30 Lee, S.P., Brooks, J.M., Al-Jarrah, H., Thomas, W.A., Haigh, T.A., Taylor, G.S., Humme, S., Schepers, A., Hammerschmidt, W., Yates, J.L., Rickinson, A.B. and Blake, N.W. (2004) CD8 T cell recognition of endogenously expressed Epstein-Barr virus nuclear antigen 1. *J Exp Med* **199**, 1409–1420.

31 Voo, K.S., Fu, T., Wang, H.Y., Tellam, J., Heslop, H.E., Brenner, M.K., Rooney, C.M. and Wang, R.F. (2004) Evidence for the presentation of major histocompatibility complex class I-restricted Epstein-Barr virus nuclear antigen 1 peptides to CD8+ T lymphocytes. *J Exp Med* **199**, 459–470.

32 Yin, Y., Manoury, B. and Fahraeus, R. (2003) Self-inhibition of synthesis and antigen presentation by Epstein-Barr virus-encoded EBNA1. *Science* **301**, 1371–1374.

33 Paludan, C., Schmid, D., Landthaler, M., Vockerodt, M., Kube, D., Tuschl, T. and Munz, C. (2005) Endogenous MHC class II processing of a viral nuclear antigen after autophagy. *Science* **307**, 593–596.

34 Levitskaya, J., Sharipo, A., Leonchiks, A., Ciechanover, A. and Masucci, M.G. (1997) Inhibition of ubiquitin/proteasome-dependent protein degradation by the Gly-Ala repeat domain of the Epstein-Barr virus nuclear antigen 1. *Proc Natl Acad Sci USA* **94**, 12616–12621.

35 Sharipo, A., Imreh, M., Leonchiks, A., Imreh, S. and Masucci, M.G. (1998) A minimal glycine-alanine repeat prevents the interaction of ubiquitinated I kappaB alpha with the proteasome: a new mechanism for selective inhibition of proteolysis. *Nature Med* **4**, 939–944.

36 Heessen, S., Leonchiks, A., Issaeva, N., Sharipo, A., Selivanova, G., Masucci, M. G. and Dantuma, N.P. (2002) Functional p53 chimeras containing the Epstein-Barr virus Gly-Ala repeat are protected from Mdm2- and HPV-E6-induced proteolysis. *Proc Natl Acad Sci USA* **99**, 1532–1537.

37 Dantuma, N., Heesseen, S., Lindsten, K., Jellne, M. and Masucci, M.G. (2000) Inhibition of proteasomal degradation by the Gly-Ala repeat of Epstein-Barr virus is influenced by the length of the repeat and the strength of the degradation signal. *Proc Natl Acad Sci USA* **97**, 8381–8385.

38 Tellam, J., Sherritt, M., Thomson, S., Tellam, R., Moss, D.J., Burrows, S.R., Wiertz, E. and Khanna, R. (2001) Targeting of EBNA1 for rapid intracellular degradation overrides the inhibitory effects of the Gly-Ala repeat domain and restores CD8+ T cell recognition. *J Biol Chem* **276**, 33353–33360.

39 Sharipo, A., Imreh, M., Leonchiks, A., Brändén, C.I. and Masucci, M.G. (2001) *Cis*-acting inhibition of proteasomal degradation by viral repeats: impact of length and amino acid composition. *FEBS Lett* **499**, 137–142.

40 Leonchiks, A., Stavropoulou, V., Sharipo, A. and Masucci, M.G. (2002) Inhibition of ubiquitin-dependent proteolysis by a synthetic glycine-alanine repeat peptide that mimics an inhibitory viral sequence. *FEBS Lett* **522**, 93–98.

41 Heessen, S., Dantuma, N.P., Tessarz, P., Jellne, M. and Masucci, M.G. (2003) Inhibition of ubiquitin/proteasome-dependent proteolysis in *Saccharomyces cerevisiae* by a Gly-Ala repeat. *FEBS Lett* **555**, 397–404.

42 Zhang, M. and Coffino, P. (2004) Repeat sequence of Epstein-Barr virus-encoded nuclear antigen 1 protein interrupts proteasome substrate processing. *J Biol Chem* **279**, 8635–8641.

43 Hoyt, M.A., Zich, J., Takeuchi, J., Zhang, M., Govaerts, C. and Coffino, P. (2006) Glycine-alanine repeats impair proper substrate unfolding by the proteasome. *EMBO J* **25**, 1720–1729.

44 Holowaty, M.N., Zeghouf, M., Wu, H., Tellam, J., Athanasopoulos, V., Greenblatt, J. and Frappier, L. (2003) Protein profiling with Epstein-Barr nuclear antigen-1 reveals an interaction with the herpesvirus-associated ubiquitin-specific protease HAUSP/USP7. *J Biol Chem* **278**, 29987–29994.

45 Holowaty, M.N., Sheng, Y., Nguyen, T., Arrowsmith, C. and Frappier, L. (2003) Protein interaction domains of the ubiquitin-specific protease, USP7/HAUSP. *J Biol Chem* **278**, 47753–47761.

46 Everett, R.D., Meredith, M., Orr, A., Cross, A., Kathorai, M. and Parkinson, J. (1997) A novel ubiquitin-specific protease is dynamically associated with the PML nuclear domain and binds to a herpesvirus regulatory protein. *EMBO J* **16**, 1519–1530.

47 Boutell, C., Dasis, S. and Everett, R.D. (2002) Herpes simplex virus type 1 immediate-early protein ICP0 and its isolated RING finger domain act as ubiquitin E3 ligases *in vitro*. *J Virol* **76**, 841–850.

48 Deng, Z., Atanasiu, C., Zhao, K., Marmorstein, R., Sbodio, J.I., Chi, N.W. and Lieberman, P.M. (2005) Inhibition of Epstein-Barr virus OriP function by tankyrase, a telomere-associated poly-ADP ribose polymerase that binds and modifies EBNA1. *J Virol* **79**, 4640–4650.

49 Deng, Z., Lezina, L., Chen, C.J., Shtivelband, S., So, W. and Lieberman, P.M. (2002) Telomeric proteins regulate episomal maintenance of Epstein-Barr virus origin of plasmid replication. *Mol Cell* **9**, 493–503.

50 Chang, W., Dynek, J.N. and Smith, S. (2003) TRF1 is degraded by ubiquitin-mediated proteolysis after release from telomeres. *Genes Dev* **17**, 1328–1333.

51 Brooks, C.L. and Gu, W. (2004) Dynamics in the p53-Mdm2 ubiquitination pathway. *Cell Cycle* **3**, 895–899.

52 Saridakis, V., Sheng, Y., Sarkari, F., Holowaty, M.N., Shire, K., Nguyen, T., Zhang, R.G., Liao, J., Lee, W., Edwards, A.M., Arrowsmith, C.H. and Frappier, L. (2005) Structure of the p53 binding domain of HAUSP/USP7 bound to Epstein-Barr nuclear antigen 1 implications for EBV-mediated immortalization. *Mol Cell* **18**, 25–36.

53 Tomkinson, B., Robertson, E. and Kieff, E. (1993) Epstein-Barr virus nuclear proteins EBNA-3A and EBNA-3C are essential for B-lymphocyte growth transformation. *J Virol* **67**, 2014–2025.

54 Knight, J.S., Sharma, N. and Robertson, E.S. (2005) Epstein-Barr virus latent antigen 3C can mediate the degradation of the retinoblastoma protein through an SCF cellular ubiquitin ligase. *Proc Natl Acad Sci USA* **102**, 18562–18566.

55 Knight, J.S. and Robertson, E.S. (2004) Epstein-Barr virus nuclear antigen 3C regulates cyclin A/p27 complexes and enhances cyclin A-dependent kinase activity. *J Virol* **78**, 1981–1991.

56 Knight, J.S., Sharma, N. and Robertson, E.S. (2005) SCFSkp2 complex targeted by Epstein-Barr virus essential nuclear antigen. *Mol Cell Biol* **25**, 1749–1763.

57 Lin, J., Johannsen, E., Robertson, E. and Kieff, E. (2002) Epstein-Barr virus nuclear antigen 3C putative repression domain mediates coactivation of the LMP1 promoter with EBNA-2. *J Virol* **76**, 232–242.

58 Rosendorff, A., Illanes, D., David, G., Lin, J., Kieff, E. and Johannsen, E. (2004) EBNA3C coactivation with EBNA2 requires a SUMO homology domain. *J Virol* **78**, 367–377.

59 Liu, Q., Jin, C., Liao, X., Shen, Z., Chen, D.J. and Chen, Y. (1999) The binding interface between an E2 (UBC9) and a ubiquitin homologue (UBL1). *J Biol Chem* **274**, 16979–16987.

60 Hatfield, P.M., Gosink, M.M., Carpenter, T.B. and Vierstra, R.D. (1997) The ubiquitin-activating enzyme (E1) gene family in Arabidopsis thaliana. *Plant J* **11**, 213–226.

61 Johnston, H., Kneer, J., Chackalaparampil, I., Yaciuk, P. and Chrivia, J. (1999) Identification of a novel SNF2/SWI2 protein family member, SRCAP, which interacts with CREB-binding protein. *J Biol Chem* **274**, 16370–16376.

62 Touitou, R., O'nions, J., Heaney, J. and Allday, M.J. (2005) Epstein-Barr virus EBNA3 proteins bind to the C8/alpha7 subunit of the 20S proteasome and are degraded by 20S proteasomes *in vitro*, but are very stable in latently infected B cells. *J Gen Virol* **86**, 1269–1277.

63 Touitou, R., Richardson, J., Bose, S., Nakanishi, M., Rivett, J. and Allday, M.J. (2001) A degradation signal located in the C-terminus of p21WAF1/CIP1 is a binding site for the C8 alpha-subunit of the 20S proteasome. *EMBO J* **20**, 2367–2375.

64 Hu, L.F., Chen, F., Zheng, X., Ernberg, I., Cao, S.L., Christensson, B., Klein, G. and Winberg, G. (1993) Clonability and tumorigenicity of human epithelial cells expressing the EBV encoded membrane protein LMP1. *Oncogene* **8**, 1575–1583.

65 Yang, X., Sham, J.S., Ng, M.H., Tsao, S.W., Zhang, D., Lowe, S.W. and Cao, L. (2000) LMP1 of Epstein-Barr virus induces proliferation of primary mouse embryonic fibroblasts and cooperatively transforms the cells with a p16-insensitive CDK4 oncogene. *J Virol* **74**, 883–891.

66 Takanashi, M., Li, J., Shirakata, M., Mori, S. and Hirai, K. (1999) Tumorigenicity of mouse BALB/c 3T3 fibroblast cells which express Epstein-Barr virus-encoded LMP1 and show normal growth phenotypes in vitro is correlated with loss of transforming growth factor-beta 1-mediated growth inhibition. *Arch Virol* **144**, 241–257.

67 Kulwichit, W., Edwards, R.H., Davenport, E.M., Baskar, J.F., Godfrey, V. and Raab-Traub, N. (1998) Expression of the Epstein-Barr virus latent membrane protein 1 induces B cell lymphoma in transgenic mice. *Proc Natl Acad Sci USA* **95**, 11963–11968.

68 Curran, J.A., Laverty, F.S., Campbell, D., Macdiarmid, J. and Wilson, J.B. (2001) Epstein-Barr virus encoded latent membrane protein-1 induces epithelial cell proliferation and sensitizes transgenic mice to chemical carcinogenesis. *Cancer Res* **61**, 6730–6738.

69 Rothenberger, S., Rousseaux, M., Knecht, H., Bender, F.C., Legler, D.F. and Bron, C. (2002) Association of the Epstein-Barr virus latent membrane protein 1 with lipid rafts is mediated through its N-terminal region. *Cell Mol Life Sci* **59**, 171–180.

70 Mosialos, G., Birkenbach, M., Yalamanchili, R., Vanarsdale, T., Ware, C. and Kieff, E. (1995) The Epstein-Barr virus transforming protein LMP1 engages signaling proteins for the tumor necrosis factor receptor family. *Cell* **80**, 389–399.

71 Kieser, A., Kaiser, C. and Hammerschmidt, W. (1999) LMP1 signal transduction differs substantially from TNF receptor 1 signaling in the molecular functions of TRADD and TRAF2. *EMBO J* **18**, 2511–2521.

72 Sylla, B.S., Hung, S.C., Davidson, D.M., Hatzivassiliou, E., Malinin, N.L., Wallach, D., Gilmore, T.D., Kieff, E. and Mosialos, G. (1998) Epstein-Barr virus-transforming protein latent infection membrane protein 1 activates transcription factor NF-kappaB through a pathway that includes the NF-kappaB-inducing kinase and the IkappaB kinases IKKalpha and IKKbeta. *Proc Natl Acad Sci USA* **95**, 10106–10111.

73 Deshaies, R.J. (1999) SCF and Cullin/Ring H2-based ubiquitin ligases. *Annu Rev Cell Dev Biol* **15**, 435–467.

74 Fuchs, S.Y., Chen, A., Xiong, Y., Pan, Z.Q. and Ronai, Z. (1999) HOS, a human homolog of Slimb, forms an SCF complex with Skp1 and Cullin1 and targets the phosphorylation-dependent degradation of IkappaB and beta-catenin. *Oncogene* **18**, 2039–2046.

75 Tang, W., Pavlish, O.A., Spiegelman, V.S., Parkhitko, A.A. and Fuchs, S.Y. (2003) Interaction of Epstein-Barr virus latent membrane protein 1 with SCFHOS/beta-TrCP E3 ubiquitin ligase regulates extent of NF-kappaB activation. *J Biol Chem* **278**, 48942–48949.

76 Rowe, M., Khanna, R., Jacob, C.A., Argaet, V., Kelly, A., Powis, S., Belich, M., Croom-Carter, D., Lee, S., Burrows, S.R. et al. (1995) Restoration of endogenous antigen processing in Burkitt's lymphoma cells by Epstein-Barr virus latent membrane protein-1: coordinate up-regulation of peptide transporters and HLA-class I antigen expression. *Eur J Immunol* **25**, 1374–1384.

77 Frisan, T., Zhang, Q.J., Levitskaya, J., Coram, M., Kurilla, M.G. and Masucci, M.G. (1996) Defective presentation of MHC class I-restricted cytotoxic T-cell epitopes in Burkitt's lymphoma cells. *Int J Cancer* **68**, 251–258.

78 Frisan, T., Levitsky, V., Polack, A. and Masucci, M.G. (1998) Phenotype-dependent differences in proteasome subunit composition and cleavage specificity in B cell lines. *J Immunol* **160**, 3281–3289.

79 Moorthy, R. and Thorley-Lawson, D.A. (1990) Processing of the Epstein-Barr virus-encoded latent membrane protein p63/LMP. *J Virol* **64**, 829–837.

80 Gottschalk, S., Edwards, O.L., Sili, U., Huls, M.H., Goltsova, T., Davis, A.R., Heslop, H.E. and Rooney, C.M. (2003) Generating CTLs against the subdominant Epstein-Barr virus LMP1 antigen for the adoptive immunotherapy of EBV-associated malignancies. *Blood* **101**, 1905–1912.

81 Dukers, D.F., Meij, P., Vervoort, M.B., Vos, W., Scheper, R.J., Meijer, C.J., Bloemena, E. and Middeldorp, J.M. (2000) Direct immunosuppressive effects of EBV-encoded latent membrane protein 1. *J Immunol* **165**, 663–670.

82 Frisan, T., Sjoberg, J., Dolcetti, R., Boiocchi, M., De Re, V., Carbone, A., Brautbar, C., Battat, S., Biberfeld, P., Eckman, M. et al. (1995) Local suppression of Epstein-Barr virus (EBV)-specific cytotoxicity in biopsies of EBV-positive Hodgkin's disease. *Blood* **86**, 1493–1501.

83 Aviel, S., Winberg, G., Masucci, M.G. and Ciechanover, A. (2000) Degradation of Epstein-Barr virus latent membrane protein-1 (LMP1) by the ubiquitin-proteasome pathway: targeting via ubiquitination of the N-terminal residue. *J Biol Chem* **275**, 23491–23499.

84 Ciechanover, A. and Ben-Saadon, R. (2004) N-terminal ubiquitination: more protein substrates join in. *Trends Cell Biol* **14**, 103–106.

85 Ikeda, M., Ikeda, A. and Longnecker, R. (2002) Lysine-independent ubiquitination of Epstein-Barr virus LMP2A. *Virology* **300**, 153–159.

86 Longnecker, R., Miller, C.L., Miao, X.Q., Tomkinson, B. and Kieff, E. (1993) The last seven transmembrane and carboxy-terminal cytoplasmic domains of Epstein-Barr virus latent membrane protein 2 (LMP2) are dispensable for lymphocyte infection and growth transformation in vitro. *J Virol* **67**, 2006–2013.

87 Rowe, M., Lear, A., Croom-Carter, D., Davies, A. and Rickinson, A. (1992) Three pathways of Epstein-Barr virus gene activation from EBNA-1 positive latency in B lymphocytes. *J Virol* **66**, 122–131.

88 Miller, C.L., Lee, J.H., Kieff, E. and Longnecker, R. (1994) An integral membrane protein (LMP2) blocks reactivation of Epstein-Barr virus from latency following surface immunoglobulin crosslinking. *Proc Natl Acad Sci USA* **91**, 772–776.

89 Miller, C.L., Lee, J.H., Kieff, E., Burkhardt, A.L., Bolen, J.B. and Longnecker, R. (1994) Epstein-Barr virus protein LMP2A regulates reactivation from latency by negatively regulating tyrosine kinases involved in sIg-mediated signal transduction. *Infect Agents Dis* **3**, 128–136.

90 Swanson-Mungerson, M., Bultema, R. and Longnecker, R. (2006) Epstein-Barr virus LMP2A enhances B-cell responses *in vivo* and *in vitro*. *J Virol* **80**, 6764–6770.

91 Staub, O., Dho, S., Henry, P., Correa, J., Ishikawa, T., Mcglade, J. and Rotin, D. (1996) WW domains of Nedd4 bind to the proline-rich PY motifs in the epithelial Na+ channel deleted in Liddle's syndrome. *EMBO J* **15**, 2371–2380.

92 Ikeda, M., Ikeda, A., Longan, L.C. and Longnecker, R. (2000) The Epstein-Barr virus latent membrane protein 2A PY motif recruits WW domain-containing ubiquitin-protein ligases. *Virology* **268**, 178–191.

93 Winberg, G., Matskova, L., Chen, F., Plant, P., Rotin, D., Gish, G., Ingham, R., Ernberg, I. and Pawson, T. (2000) Latent membrane protein 2A of Epstein-Barr virus binds WW domain E3 protein-ubiquitin ligases that ubiquitinate B-cell tyrosine kinases. *Mol Cell Biol* **20**, 8526–8535.

94 Engels, N., Merchant, M., Pappu, R., Chan, A.C., Longnecker, R. and Wienands, J. (2001) Epstein-Barr virus latent membrane protein 2A (LMP2A) employs the SLP-65 signaling module. *J Exp Med* **194**, 255–264.

95 Sohn, H.W., Gu, H. and Pierce, S.K. (2003) Cbl-b negatively regulates B cell antigen receptor signaling in mature B cells through ubiquitination of the tyrosine kinase Syk. *J Exp Med* **197**, 1511–1524.

96 Ingham, R.J., Raaijmakers, J., Lim, C.S., Mbamalu, G., Gish, G., Chen, F., Matskova, L., Ernberg, I., Winberg, G. and Pawson, T. (2005) The Epstein-Barr virus protein, latent membrane protein 2A, co-opts tyrosine kinases used by the

T cell receptor. *J Biol Chem* **280**, 34133–34142.

97 Morrison, J.A., Klingelhutz, A.J. and Raab-Traub, N. (2003) Epstein-Barr virus latent membrane protein 2A activates beta-catenin signaling in epithelial cells. *J Virol* **77**, 12276–12284.

98 Scholle, F., Bendt, K.M. and Raab-Traub, N. (2000) Epstein-Barr virus LMP2A transforms epithelial cells, inhibits cell differentiation, and activates Akt. *J Virol* **74**, 10681–10689.

99 Morrison, J.A. and Raab-Traub, N. (2005) Roles of the ITAM and PY motifs of Epstein-Barr virus latent membrane protein 2A in the inhibition of epithelial cell differentiation and activation of {beta}-catenin signaling. *J Virol* **79**, 2375–2382.

100 Pegtel, D.M., Subramanian, A., Sheen, T.S., Tsai, C.H., Golub, T.R. and Thorley-Lawson, D.A. (2005) Epstein-Barr-virus-encoded LMP2A induces primary epithelial cell migration and invasion: possible role in nasopharyngeal carcinoma metastasis. *J Virol* **79**, 15430–15442.

101 Lu, J., Lin, W.H., Chen, S.Y., Longnecker, R., Tsai, S.C., Chen, C.L. and Tsai, C.H. (2006) Syk tyrosine kinase mediates Epstein-Barr virus latent membrane protein 2A-induced cell migration in epithelial cells. *J Biol Chem* **281**, 8806–8814.

102 Amon, W. and Farrell, P.J. (2005) Reactivation of Epstein-Barr virus from latency. *Rev Med Virol* **15**, 149–156.

103 Moller, A. and Schmitz, M.L. (2003) Viruses as hijackers of PML nuclear bodies. *Arch Immunol Ther Exp (Warsz)* **51**, 295–300.

104 Zimber, A., Nguyen, Q.D. and Gespach, C. (2004) Nuclear bodies and compartments: functional roles and cellular signalling in health and disease. *Cell Signal* **16**, 1085–1104.

105 Duprez, E., Saurin, A.J., Desterro, J.M., Lallemand-Breitenbach, V., Howe, K., Boddy, M.N., Solomon, E., De The, H., Hay, R.T. and Freemont, P.S. (1999) SUMO-1 modification of the acute promyelocytic leukaemia protein PML: implications for nuclear localisation. *J Cell Sci* **112**(Pt 3), 381–393.

106 Adamson, A.L. and Kenney, S. (2001) Epstein-Barr virus immediate-early protein BZLF1 is SUMO-1 modified and disrupts promyelocytic leukemia bodies. *J Virol* **75**, 2388–2399.

107 Chang, L.K., Lee, Y.H., Cheng, T.S., Hong, Y.R., Lu, P.J., Wang, J.J., Wang, W.H., Kuo, C.W., Li, S.S. and Liu, S.T. (2004) Post-translational modification of Rta of Epstein-Barr virus by SUMO-1. *J Biol Chem* **279**, 38803–38812.

108 Balakirev, M.Y., Jaquinod, M., Haas, A.L. and Chroboczek, J. (2002) Deubiquitinating function of adenovirus proteinase. *J Virol* **76**, 6323–6331.

109 Ratia, K., Saikatendu, K.S., Santarsiero, B.D., Barretto, N., Baker, S.C., Stevens, R.C. and Mesecar, A.D. (2006) Severe acute respiratory syndrome coronavirus papain-like protease: structure of a viral deubiquitinating enzyme. *Proc Natl Acad Sci USA* **103**, 5717–5722.

110 Lindner, H.A., Fotouhi-Ardakani, N., Lytvyn, V., Lachance, P., Sulea, T. and Menard, R. (2005) The papain-like protease from the severe acute respiratory syndrome coronavirus is a deubiquitinating enzyme. *J Virol* **79**, 15199–15208.

111 Barretto, N., Jukneliene, D., Ratia, K., Chen, Z., Mesecar, A.D. and Baker, S.C. (2005) The papain-like protease of severe acute respiratory syndrome coronavirus has deubiquitinating activity. *J Virol* **79**, 15189–15198.

112 Kattenhorn, L.M., Korbel, G.A., Kessler, B.M., Spooner, E. and Ploegh, H.L. (2005) A deubiquitinating enzyme encoded by HSV-1 belongs to a family of cysteine proteases that is conserved across the family Herpesviridae. *Mol Cell* **19**, 547–557.

113 Knipe, D.M., Batterson, W., Nosal, C., Roizman, B. and Buchan, A. (1981) Molecular genetics of herpes simplex virus. VI. Characterization of a temperature-sensitive mutant defective in the expression of all early viral gene products. *J Virol* **38**, 539–547.

114 Chou, J. and Roizman, B. (1989) Characterization of DNA sequence-common and sequence-specific proteins binding to cis-acting sites for cleavage of the terminal a sequence of the herpes simplex virus 1 genome. *J Virol* **63**, 1059–1068.

115 Batterson, W., Furlong, D. and Roizman, B. (1983) Molecular genetics of herpes simplex virus. VIII. Further characterization of a temperature-sensitive mutant defective in release of viral DNA and in other stages of the viral reproductive cycle. *J Virol* **45**, 397–407.

116 Tagawa, A., Mezzacasa, A., Hayer, A., Longatti, A., Pelkmans, L. and Helenius, A. (2005) Assembly and trafficking of caveolar domains in the cell: caveolae as stable, cargo-triggered, vesicular transporters. *J Cell Biol* **170**, 769–779.

117 Von Schwedler, U.K., Stuchell, M., Muller, B., Ward, D.M., Chung, H.Y., Morita, E., Wang, H.E., Davis, T., He, G.P., Cimbora, D.M., Scott, A., Krausslich, H.G., Kaplan, J., Morham, S.G. and Sundquist, W.I. (2003) The protein network of HIV budding. *Cell* **114**, 701–713.

118 Bowers, K., Piper, S.C., Edeling, M.A., Gray, S.R., Owen, D.J., Lehner, P.J. and Luzio, J.P. (2006) Degradation of endocytosed epidermal growth factor and virally ubiquitinated major histocompatibility complex class I is independent of mammalian ESCRTII. *J Biol Chem* **281**, 5094–5105.

119 Row, P.E., Prior, I.A., Mccullough, J., Clague, M.J. and Urbe, S. (2006) The ubiquitin isopeptidase UBPY regulates endosomal ubiquitin dynamics and is essential for receptor down-regulation. *J Biol Chem* **281**, 12618–12624.

120 Klein, G. (1994) Role of EBV and Ig/myc translocation in Burkitt lymphoma. *Antibiot Chemother* **46**, 110–116.

121 Kelly, G., Bell, A. and Rickinson, A. (2002) Epstein-Barr virus-associated Burkitt lymphomagenesis selects for downregulation of the nuclear antigen EBNA2. *Nature Med* **8**, 1098–1104.

122 Andersson, M.L., Stam, N.J., Klein, G., Ploegh, H.L. and Masucci, M.G. (1991) Aberrant expression of HLA class-I antigens in Burkitt lymphoma cells. *Int J Cancer* **47**, 544–550.

123 Avila-Carino, J., Torsteinsdottir, S., Ehlin-Henriksson, B., Lenoir, G., Klein, G., Klein, E. and Masucci, M.G. (1987) Paired Epstein-Barr virus (EBV)-negative and EBV converted Burkitt's lymphoma lines. Stimulatory capacity in allogeneic mixed lymphocyte cultures. *Int J Cancer* **40**, 691–97.

124 Staege, M.S., Lee, S.P., Frisan, T., Mautner, J., Scholz, S., Pajic, A., Rickinson, A.B., Masucci, M.G., Polack, A. and Bornkamm, G.W. (2002) MYC overexpression imposes a nonimmunogenic phenotype on Epstein-Barr virus-infected B cells. *Proc Natl Acad Sci USA* **99**, 4550–4555.

125 Gavioli, R., Frisan, T., Vertuani, S., Bornkamm, G.W. and Masucci, M.G. (2001) c-myc overexpression activates alternative pathways for intracellular proteolysis in lymphoma cells. *Nature Cell Biol* **3**, 283–288.

126 Borodovsky, A., Ovaa, H., Kolli, N., Gan-Erdene, T., Wilkinson, K.D., Ploegh, H.L. and Kessler, B.M. (2002) Chemistry-based functional proteomics reveals novel members of the deubiquitinating enzyme family. *Chem Biol* **9**, 1149–1159.

127 Lansbury, P.T. Jr and Brice, A. (2002) Genetics of Parkinson's disease and biochemical studies of implicated gene products. *Curr Opin Cell Biol* **14**, 653–660.

128 Fernandez-Funez, P., Nino-Rosales, M.L., De Gouyon, B., She, W.C., Luchak, J.M., Martinez, P., Turiegano, E., Benito, J., Capovilla, M., Skinner, P.J., Mccall, A., Canal, I., Orr, H.T., Zoghbi, H.Y. and Botas, J. (2000) Identification of genes that modify ataxin-1-induced neurodegeneration. *Nature* **408**, 101–106.

129 Naze, P., Vuillaume, I., Destee, A., Pasquier, F. and Sablonniere, B. (2002) Mutation analysis and association studies of the ubiquitin carboxy-terminal hydrolase L1 gene in Huntington's disease. *Neurosci Lett* **328**, 1–4.

130 Yanagisawa, T.Y., Sasahara, Y., Fujie, H., Ohashi, Y., Minegishi, M., Itano, M., Morita, S., Tsuchiya, S., Hayashi, Y., Ohi, R. and Konno, T. (1998) Detection of the PGP9.5 and tyrosine hydroxylase mRNAs for minimal residual neuroblastoma cells in bone marrow and peripheral blood. *Tohoku J Exp Med* **184**, 229–240.

131 Yamazaki, T., Hibi, K., Takase, T., Tezel, E., Nakayama, H., Kasai, Y., Ito, K., Akiyama, S., Nagasaka, T. and Nakao, A. (2002) PGP9.5 as a marker for invasive colorectal cancer. *Clin Cancer Res* **8**, 192–195.

132 Sasaki, H., Yukiue, H., Moriyama, S., Kobayashi, Y., Nakashima, Y., Kaji, M., Fukai, I., Kiriyama, M., Yamakawa, Y. and Fujii, Y. (2001) Expression of the protein gene product 9.5, PGP9.5, is correlated with T-status in non-small cell lung cancer. *Jpn J Clin Oncol* **31**, 532–535.

133 Brichory, F., Beer, D., Le Naour, F., Giordano, T. and Hanash, S. (2001) Proteomics-based identification of protein gene product 9.5 as a tumor antigen that induces a humoral immune response in lung cancer. *Cancer Res* **61**, 7908–7912.

134 Tezel, E., Hibi, K., Nagasaka, T. and Nakao, A. (2000) PGP9.5 as a prognostic factor in pancreatic cancer. *Clin Cancer Res* **6**, 4764–4767.

135 Aumuller, G., Renneberg, H., Leonhardt, M., Lilja, H. and Abrahamsson, P.A. (1999) Localization of protein gene product 9.5 immunoreactivity in derivatives of the human Wolffian duct and in prostate cancer. *Prostate* **38**, 261–267.

136 Schumacher, U., Mitchell, B.S. and Kaiserling, E. (1994) The neuronal marker protein gene product 9.5 (PGP 9.5) is phenotypically expressed in human breast epithelium, in milk, and in benign and malignant breast tumors. *DNA Cell Biol* **13**, 839–843.

137 D'andrea, V., Malinovsky, L., Berni, A., Biancari, F., Biassoni, L., Di Matteo, F. M., Corbellini, L., Falvo, L., Santoni, F., Spyrou, M. and De Antoni, E. (1997) The immunolocalization of PGP 9.5 in normal human kidney and renal cell carcinoma. *G Chir* **18**, 521–524.

138 Larsen, C.N., Krantz, B.A. and Wilkinson, K.D. (1998) Substrate specificity of deubiquitinating enzymes: ubiquitin C-terminal hydrolases. *Biochemistry* **37**, 3358–3368.

139 Liu, Y., Fallon, L., Lashuel, H.A., Liu, Z. and Lansbury, P.T. Jr. (2002) The UCH-L1 gene encodes two opposing enzymatic activities that affect alpha-synuclein degradation and Parkinson's disease susceptibility. *Cell* **111**, 209–218.

140 Caballero, O.L., Resto, V., Patturajan, M., Meerzaman, D., Guo, M.Z., Engles, J., Yochem, R., Ratovitski, E., Sidransky, D. and Jen, J. (2002) Interaction and colocalization of PGP9.5 with JAB1 and p27(Kip1). *Oncogene* **21**, 3003–3010.

141 Nishimoto, T. (1999) A new role of ran GTPase. *Biochem Biophys Res Commun* **262**, 571–574.

142 Tomoda, K., Kubota, Y. and Kato, J. (1999) Degradation of the cyclin-dependent-kinase inhibitor p27Kip1 is instigated by Jab1. *Nature* **398**, 160–165.

143 Chamovitz, D.A. and Segal, D. (2001) JAB1/CSN5 and the COP9 signalosome. A complex situation. *EMBO Rep* **2**, 96–101.

144 Cope, G.A., Suh, G.S., Aravind, L., Schwarz, S.E., Zipursky, S.L., Koonin, E.V. and Deshaies, R.J. (2002) Role of predicted metalloprotease motif of Jab1/Csn5 in cleavage of Nedd8 from Cul1. *Science* **298**, 608–611.

145 Balow, R.M., Ragnarsson, U. and Zetterqvist, O. (1983) Tripeptidyl aminopeptidase in the extralysosomal fraction of rat liver. *J Biol Chem* **258**, 11622–11628.

146 Balow, R.M., Tomkinson, B., Ragnarsson, U. and Zetterqvist, O. (1986) Purification, substrate specificity, and classification of tripeptidyl peptidase II. *J Biol Chem* **261**, 2409–2417.

147 Macpherson, E., Tomkinson, B., Balow, R.M., Hoglund, S. and Zetterqvist, O. (1987) Supramolecular structure of tripeptidyl peptidase II from human erythrocytes as studied by electron microscopy, and its correlation to enzyme activity. *Biochem J* **248**, 259–263.

148 Balow, R.M. and Eriksson, I. (1987) Tripeptidyl peptidase II in haemolysates and liver homogenates of various species. *Biochem J* **241**, 75–80.

149 Tomkinson, B., Hansen, M. and Cheung, W.F. (1997) Structure-function studies of recombinant murine tripeptidyl-peptidase II: the extra domain which is subject to alternative splicing is involved in complex formation. *FEBS Lett* **405**, 277–280.

150 Geier, E., Pfeifer, G., Wilm, M., Lucchiari-Hartz, M., Baumeister, W., Eichmann, K. and Niedermann, G. (1999) A giant protease with potential to substitute for some functions of the proteasome. *Science* **283**, 978–981.

151 Rose, C., Vargas, F., Bourgeat, P. and Schwartz, J.C. (1996) A radioimmunoassay for the tripeptide Gly-Trp-Met, a major metabolite of endogenous cholecystokinin in brain. *Neuropeptides* **30**, 231–235.

152 Wang, E.W., Kessler, B.M., Borodovsky, A., Cravatt, B.F., Bogyo, M., Ploegh, H.L. and Glas, R. (2000) Integration of the ubiquitin-proteasome pathway with a cytosolic oligopeptidase activity. *Proc Natl Acad Sci USA* **97**, 9990–9995.

153 Hilbi, H., Puro, R.J. and Zychlinsky, A. (2000) Tripeptidyl peptidase II promotes maturation of caspase-1 in Shigella flexneri-induced macrophage apoptosis. *Infect Immun* **68**, 5502–5508.

154 York, I.A., Bhutani, N., Zendzian, S., Goldberg, A.L. and Rock, K.L. (2006) Tripeptidyl peptidase II is the major peptidase needed to trim long antigenic precursors, but is not required for most MHC class I antigen presentation. *J Immunol* **177**, 1434–1443.

155 Seifert, U., Maranon, C., Shmueli, A., Desoutter, J.F., Wesoloski, L., Janek, K., Henklein, P., Diescher, S., Andrieu, M., De La Salle, H., Weinschenk, T., Schild, H., Laderach, D., Galy, A., Haas, G., Kloetzel, P.M., Reiss, Y. and Hosmalin, A. (2003) An essential role for tripeptidyl peptidase in the generation of an MHC class I epitope. *Nat Immunol* **4**, 375–379.

156 Reits, E., Neijssen, J., Herberts, C., Benckhuijsen, W., Janssen, L., Drijfhout, J.W. and Neefjes, J. (2004) A major role for TPPII in trimming proteasomal degradation products for MHC class I antigen presentation. *Immunity* **20**, 495–506.

157 Glas, R., Bogyo, M., Mcmaster, J.S., Gaczynska, M. and Ploegh, H.L. (1998) A proteolytic system that compensates for loss of proteasome function. *Nature* **392**, 618–622.

158 Deng, L., Wang, C., Spencer, E., Yang, L., Braun, A., You, J., Slaughter, C., Pickart, C. and Chen, Z.J. (2000) Activation of the IkappaB kinase complex by TRAF6 requires a dimeric ubiquitin-conjugating enzyme complex and a unique polyubiquitin chain. *Cell* **103**, 351–361.

159 Hong, X., Lei, L. and Glas, R. (2003) Tumors acquire inhibitor of apoptosis protein (IAP)-mediated apoptosis resistance through altered specificity of cytosolic proteolysis. *J Exp Med* **197**, 1731–1743.

160 Stavropoulou, V., Xie, J., Henriksson, M., Tomkinson, B., Imreh, S. and Masucci, M.G. (2005) Mitotic infidelity and centrosome duplication errors in cells overexpressing tripeptidyl-peptidase II. *Cancer Res* **65**, 1361–1368.

161 Stavropoulou, V., Vasquez, V., Cereser, B., Freda, E. and Masucci, M.G. (2006) TPPII promotes genetic instability by allowing the escape from apoptosis of cells with activated mitotic checkpoints. *Biochem Biophys Res Commun* **346**, 415–425.

162 Fan, H., Nicholls, J., Chua, D., Chan, K.H., Sham, J., Lee, S. and Gulley, M.L. (2004) Laboratory markers of tumor burden in nasopharyngeal carcinoma: a comparison of viral load and serologic tests for Epstein-Barr virus. *Int J Cancer* **112**, 1036–1041.

4
HECT Ubiquitin-protein Ligases in Human Disease

Martin Scheffner and Olivier Staub

4.1
Introduction

It is commonly accepted that the specific recognition of substrate proteins of the ubiquitin-conjugation system is mainly mediated by the action of E3 ubiquitin-protein ligases [1–3]. It is, therefore, not surprising that E3s constitute the largest class of enzymes known to be involved in ubiquitination, with the human genome encoding more than 500 putative E3s or E3 complexes. Deregulation of the activity of a still increasing number of E3s has been associated with the development of human disease including cancer, cardiovascular, immunological, and neurological disorders. For example, mutations in the *brca1* gene that interferes with the E3 activity of the respective protein product, have been linked to hereditary breast cancer in a certain percentage of cases [4, 5]. Similarly, amplification of the *hdm2* gene resulting in increased levels of Hdm2, which acts as an E3 ligase for the tumor suppressor p53, has been linked to a certain percentage of soft tissue sarcomas [6, 7]. Thus, both inappropriate inactivation ("loss of function", e.g. BRCA1) and inappropriate activation ("gain of function", e.g. Hdm2) of E3s can have pathophysiological consequences. Based on the presence of distinct amino acid sequence motifs, proteins with E3 activity can be roughly grouped into three classes: HECT E3s, RING-finger E3s, and U-box E3s [8, 9]. The basic mechanisms by which E3s facilitate ubiquitination of substrate proteins.

4.2
Definition of HECT E3s

Functional and structural studies indicate that in a simplified view, all known E3s have a modular structure consisting of at least two functional domains. The RING-finger domain, the U-box, or the HECT domain of the respective E3 mediates the interaction with its cognate ubiquitin-conjugating enzyme E2 [1–3]. The other domain is required for specific interaction with the respective target protein

Protein Degradation, Vol. 4: The Ubiquitin-Proteasome System and Disease.
Edited by R. J. Mayer, A. Ciechanover, M. Rechsteiner

ISBN: 978-3-527-31436-2

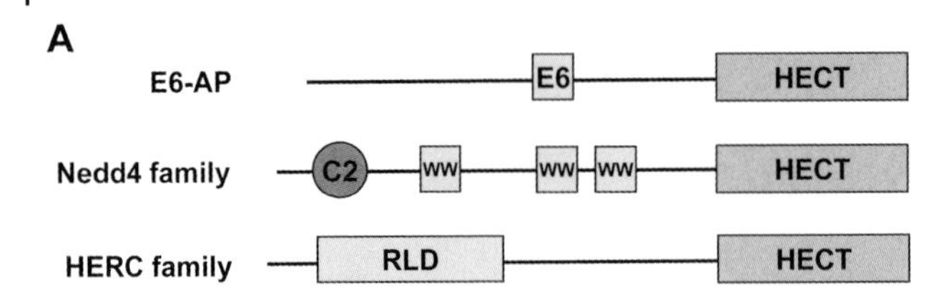

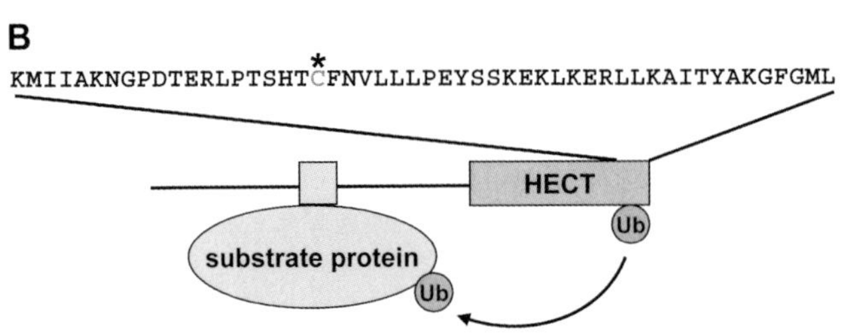

Fig. 4.1. The family of HECT ubiquitin-protein ligases. (A) Schematic representation of the three subfamilies of HECT E3s. All members of the HECT family of E3s are characterized by the C-terminal HECT domain. The HECT domain consists of approximately 350 amino acid residues and represents the catalytic domain of HECT E3s. Nedd4 family members are characterized by an N-terminal C2 domain and the presence of several WW domains (as representative the schematic structure of Smurf2 is shown; see also Figure 4.3). The HERC family comprises six members, which are characterized by the presence of one or several RLD domains (as representative the structure of HERC5 is shown schematically). The schematic structure of E6-AP, the founding member of the HECT family, is shown as representative of the third subfamily. Members of this subfamily are characterized by the notion that, apart from the HECT domain, they do not share any significant similarities in amino acid sequence. (B) HECT E3s have the ability to form thioester complexes with ubiquitin in the presence of E1 and their cognate E2s and are assumed to directly catalyze the covalent attachment of ubiquitin to lysine residues of substrate proteins. The sequence of the C-terminal 52 amino acid residues of E6-AP are shown with the cysteine residue involved in thioester complex formation and with ubiquitin marked with an asterisk.

and, thus, determines the substrate specificity of the respective E3. In the case of HECT E3s, both domains are generally displayed on a single polypeptide chain (Figure 4.1), although auxiliary factors may affect or alter the substrate specificity of a given HECT E3.

Members of the HECT E3 family are large proteins ranging in size from approximately 80 kDa to more than 500 kDa and have been found in all eukaryotic organisms examined. They are characterized by a C-terminal region of about 350 amino acids in length that shows significant similarity to the C terminus of E6-AP (see below) and, therefore, has been termed the "HECT domain" (*H*omologous to *E*6-AP *C* *T*erminus) [8, 9]. Functional characterization of E6-AP revealed that, in the presence of its cognate E2s, the HECT domain forms thioester complexes with ubiquitin. Furthermore, a conserved cysteine residue within the HECT domain is required for both ubiquitin thioester complex formation and E3 ligase activity [8, 9]. This indicates that, in contrast to RING-finger and U-box E3s, the HECT domain does not only function as a binding site for E2s but, in addition, plays a catalytic role in the final attachment of ubiquitin to a substrate protein (Figure 4.1).

Similarly to RING-finger and U-box E3s, a common nomenclature that would unambiguously identify a given protein as member of the HECT E3 family has not been introduced. In the following, we will therefore refer to individual members of the HECT E3 family with their respective trivial names that are most commonly used in the literature.

4.3 Human HECT E3s and their Role in Disease

Database analyses indicate that the human genome encodes 28 different HECT proteins (K. Hoffmann and H. Scheel, personal communication). Based on the presence of distinct amino acid sequence motifs, human HECT E3s can be classified into three subfamilies: HECT E3s with RCC1-like domains (RLDs) termed HERC E3s (*HE*CT and *RC*C1-like domain), HECT E3s with WW domains (Nedd4/Nedd4-like proteins), and HECT E3s that neither contain RLDs nor WW domains (Figure 4.1). RLDs and WW domains represent known protein–protein interaction domains and, thus, provide some information about potential interaction partners of the respective E3s.

RLDs were originally described as interaction sites for small GTP binding proteins [10, 11]. Indeed, HERC1 was shown to bind to and act as guanine nucleotide exchange factor for ARF1 [12]. However, no evidence has been provided that ARF1 represents a ubiquitination substrate for HERC1. Thus, the interaction with ARF1 may not be related to the E3 function of HERC1 suggesting that, at least in some cases, HECT E3s are multifunctional proteins. Recent evidence has suggested that HERC1 interacts with TSC2, a GTPase-activating protein of the Rheb GTPase, and targets it for degradation [13]. Interestingly, TSC2 negatively affects the mTOR pathway and has been associated with tuberous sclerosis complex, an inherited disease characterized by hamartoma formation in various organs. However, if and how HERC1 is involved in the development of this disease remains to be elucidated.

Although the RLD may represent an interaction motif for small GTP-binding proteins, it is also involved in the interaction with other proteins. We have recently obtained evidence that HERC2 interacts with E6-AP via RLD2 (Figure 4.1) (U. Kogel, S. Glockzin and M. Scheffner, unpublished data) indicating that, similar to RING-finger E3s, HECT E3s have the potential to form hetero-oligomeric E3 complexes. Apart from HERC1, none of the members of the HERC subgroup of HECT E3s has been etiologically associated with human disease so far. However, mutations in the *herc2* gene have been linked to pathophysiological phenotypes in mice [14, 15] and *herc2* expression may be affected in some Angelman syndrome patients (see below). Furthermore, HERC5 has recently been reported to act as an E3 ligase for ISG15, a ubiquitin-like protein that is expressed when cells are stimulated with interferon [16, 17]. However, the physiological significance of this observation is presently unclear and will not be discussed further.

The WW domain represents a highly conserved protein domain that binds to proline-rich regions of interacting proteins [18–20]. Based on the actual sequence of the proline-rich region bound, WW domains have been classified into four groups, with group I domains having a preference for PPXY motifs, group II/III domains for poly-P regions, and group IV for P motifs containing phosphorylated S or T [19, 20]. The best characterized members of the WW domain subgroup of HECT E3s are Nedd4, Itchy, and Smurf. Since these proteins have been associated with human disease or processes with pathophysiological potential, these proteins will be discussed below in more detail. As indicated above, the third subgroup of HECT E3s is characterized by the notion that its members do not share any known protein–protein interaction motifs with other HECT E3s. Members of this subgroup comprise E6-AP and MULE/ARF-BP1/HectH9, both of which have been associated with human cancer.

4.4 E6-AP

E6-AP was the first human E3 that was identified at the amino acid sequence level and represents the founding member of the HECT family of E3s. E6-AP (*E6-Associated Protein*) was originally isolated as a protein that binds to the E6 oncoprotein of cancer-associated human papillomaviruses (HPVs) and, in complex with E6, targets p53 for ubiquitination and proteasome-mediated degradation [21, 22]. Subsequently, it was shown that loss of the E3 activity of E6-AP is associated with the development of a hereditary neurological disorder, the Angelman syndrome [23–25]. Thus, E6-AP represents a prime example of the hypothesis that deregulation of components of the ubiquitin-conjugation system contributes to human disease: inappropriate activation of E6-AP contributes to carcinogenesis ("gain of function"), while inactivation results in a neurological disease ("loss of function") (Figure 4.2).

4.4.1 E6-AP and Cervical Cancer (Cancer of the Uterine Cervix)

Cervical cancer represents the second most frequent cancer in women worldwide with approximately 400 000 new cases each year. It is commonly accepted that infection with certain HPV types represents the most significant risk factor for the development of cervical cancer (for review see [26]). Approximately 30 HPV types have been associated with lesions of the anogenital tract and these HPVs

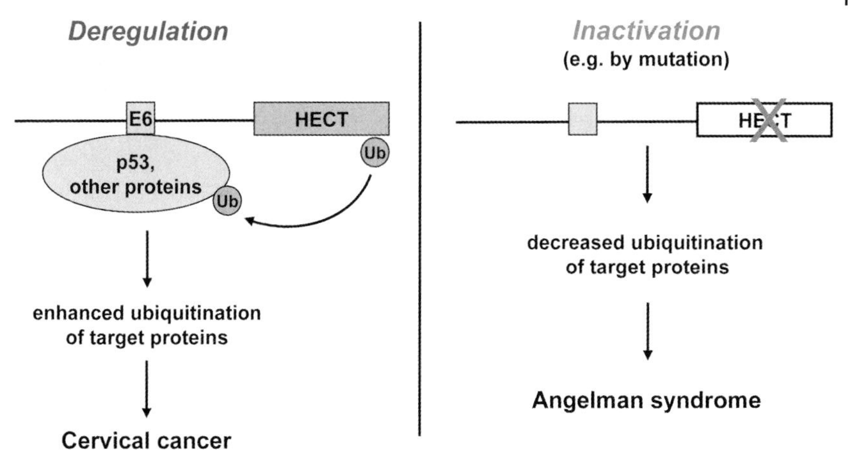

Fig. 4.2. Deregulation of E6-AP activity contributes to the development of human disease. For details, see text.

can be classified into "low risk" and "high risk" types based on their association with malignant lesions. While low risk HPVs are generally associated with benign lesions including condyloma accuminata, women that are infected with high risk HPVs have a significantly increased risk of developing cervical cancer. High risk HPVs encode two major oncoproteins termed E6 and E7 and the respective genes are the only viral genes that are generally retained and expressed in cervical cancer tissues. Furthermore, a number of studies have shown that continuous expression of both E6 and E7 is required for the viability of HPV-positive cancer cells.

As indicated above, it is well established that E6 utilizes E6-AP to target p53 for degradation and, thus, inactivation. Similarly, E7 targets the retinoblastoma susceptibility gene product pRB and the pRB-related proteins p107 and p130 for proteasome-mediated degradation [27–30]. A potential reason for inactivating p53 and pRB family members via degradation rather than by other means is provided by the notion that in HPV-positive cells, E6 and E7 are maintained at rather low levels. Thus, a catalytic mechanism (assuming that one E6 or E7 molecule can target more than one p53 or pRB molecule for degradation) would assure that p53 and pRB family members are efficiently inactivated even if E6 and E7 are expressed at lower amounts in infected cells than their respective target proteins [30].

What is the physiological significance of E6/E6-AP-induced degradation of p53? In contrast to many other tumor types (approximately 40% of all human tumors harbor a mutated p53 gene) the p53 gene is very rarely mutated in cervical carcinomas [26, 30]. Thus, E6/E6-AP-induced degradation of p53 can be considered as functionally equivalent to inactivation of p53 by mutation of the p53 gene, although the situation in HPV-positive cancers may be somewhat more complicated (for detailed discussion of this issue, see [30–32]). Furthermore, there is good evidence to indicate that in HPV-negative cells, E6-AP plays no, or only a minor, role in p53 degradation [33–36]. In fact, a number of RING-finger E3s including Hdm2, Pirh2, COP1, and HECTH9/ARF-BP1/MULE (see below) have been reported to mediate p53 ubiquitination and degradation in HPV-negative cells [37–42]. Furthermore, genetic experiments in mice clearly indicate that Mdm2, the mouse homolog of Hdm2, is a major antagonist of p53 [43]. Taken together, the published data indicate, that the E6 oncoprotein utilizes E6-AP to target p53 for degradation under conditions where the normal pathways for p53 degradation are not functional. Indeed, it was reported that, for as yet unknown reasons, the Hdm2-dependent pathway of p53 degradation is inactive in cervical carcinoma cell lines [44]. In support of this notion, interference with E6-AP expression by antisense RNA-based approaches or by RNA interference results in the accumulation of p53 and activation of its transcriptional and growth-suppressive properties [36, 45].

It is commonly accepted that the ability of high risk E6s to target p53 for degradation contributes to virus-induced cellular transformation. However, it is also clear that the E6 protein has oncogenic activities that are independent of p53. Numerous cellular proteins including the PDZ domain-containing proteins hDlg, MAGI-1, MUPP-1, and hScrib have been reported to interact with high risk E6 proteins [46–50] and experiments with transgenic mice indicate that the ability of E6 to interact with PDZ domain-containing proteins contributes to its oncogenic potential [51, 52]. Furthermore, it has recently been reported that the E6/E6-AP complex targets NFX1-91 for ubiquitination and degradation [53]. NFX1-91 acts as a transcriptional repressor of the gene encoding hTERT, the rate-limiting and catalytic subunit of telomerase. Thus, this activity of the E6/E6-AP complex may critically contribute to the immortalizing activities of E6.

Unlike the interaction with p53, binding of E6 to PDZ domain proteins is independent of E6-AP [46]. Interestingly, E6 targets hScrib for degradation in an E6-AP-dependent manner, whereas hDlg, MAGI-1, and MUPP-1 have been reported to be targeted for degradation by E6 in an E6-AP-independent manner [54–56]. Thus, it was speculated that E6 may interact with E3 ligases other than E6-AP. However, by using E6-AP-deficient cells we have recently obtained strong evidence that, within cells, E6-mediated degradation of hDlg and MAGI-1 is dependent on the presence of E6-AP (P. Kuballa, K. Matentzoglu, and M. Scheffner, unpublished data). We are therefore proposing that all the known proteolytic properties of E6 depend on E6-AP. Finally, it should be mentioned that E6-AP may not only be utilized by E6 but may represent a direct target for E6, since binding of E6 targets E6-AP for self-ubiquitination and degradation [57]. Although the physiological significance of this observation is presently unclear, it can be speculated that an E6-induced decrease of intracellular E6-AP levels should have profound effects on the stability of E6-independent substrates of E6-AP. However, whether this is indeed the case, remains to be determined. Thus, to fully understand the role of E6-AP in cervical carcinogenesis, it is important to identify and characterize the cellular pathways that are affected by E6-AP in normal (i.e. HPV-negative) cells.

4.4.2
E6-AP and Angelman Syndrome

The "Angelman syndrome" (AS) was first described in 1965 by the pediatrician Harry Angelman [58]. AS is characterized by mental retardation, movement or balance disorder, characteristic abnormal behaviors, and severe limitations in speech and language. It is a genetic disorder with an incidence of approximately 1 in 10000 to 1 in 40000 and has been linked to chromosome 15q11-13 known as the Prader–Willi/Angelman region [59, 60]. This region is known to contain a bipartite imprinting center and, accordingly, contains maternally and paternally imprinted genes. The Prader-Willi syndrome and AS represent two clinically distinct disorders with Prader–Willi syndrome resulting from paternal genetic deficiency and AS from maternal genetic deficiency [61]. Interestingly, studies in mice have shown that the ube3a gene encoding E6-AP is biallelically expressed in all somatic cells with the exception of Purkinje cells (cerebellum), hippocampal neurons and mitral cells of the olfactory bulb, in which the paternally-derived ube3a gene is silenced [62]. Indeed, all of the genetic abnormalities associated with AS affect expression of the maternal ube3a gene and/or the ubiquitin ligase activity of E6-AP. Development of AS appears to be the result of several genetic mechanisms with deletion of the 15q11-13 region of the maternal chromosome accounting for approximately 70% of cases. The Prader–Willi/Angelman region is approximately 4 Mb in size and is bounded by duplicons of the *herc2* gene that may predispose to chromosomal rearrangements. Other mechanisms include uniparental paternal disomy, defects in imprinting, and single point mutations in the ube3a gene [59–61, 63]. In this context, it should be noted that E6-AP affects the activity of nuclear hormone receptors and this property of E6-AP does not appear to be related to its ubiquitin ligase function [64, 65]. However, since the relevance of this property of E6-AP for the development of AS or cervical cancer is unclear (e.g. this property is not affected in those E6-AP mutants that are derived from AS patients with point mutations in the ube3a gene), it will not be discussed further.

To understand why loss of the ubiquitin ligase activity of E6-AP results in the development of AS, it is essential to identify the cellular pathways that involve E6-AP. An obvious possibility in this regard is the identification and characterization of proteins that serve as ubiquitination substrates of E6-AP. Several E6-independent substrates of E6-AP have been reported, including HHR23A and HHR23B (the human homologs of *S. cerevisiae* RAD23), Blk (a member of the Src-family of non-receptor tyrosine kinases), Bak (a human pro-apoptotic protein), and Mcm7 (which is involved in DNA replication) [66–69]. However, whether deregulated degradation of these proteins is involved in the pathogenesis of AS patients is presently unclear. Similarly, it should be noted that in E6-AP null mice, cytoplasmic levels of p53 are significantly increased in postmitotic neurons [70]. Although it is possible that this suggests that E6-AP may play a more prominent role in p53 degradation in certain tissues or at certain stages of cellular differentiation, it seems likely that the observed increase in p53 levels is an indirect rather than a direct effect of the loss of E6-AP expression. Taken together, although the function of E6-AP in cervical cancer is at least in part understood, the pathways that involve E6-AP in the absence of the HPV E6 oncoproteins and that are deregulated in AS patients remain to be identified.

4.5 HECTH9

In 1995, a HECT protein termed UREB1 (*U*pstream *R*egulatory *E*lement *B*inding protein *1*) was reported to interfere with p53-mediated transcriptional transactivation [71]. However, the physiological significance of this interaction remained unclear and, subsequently, it was found that the UREB1 protein studied represented a significantly N-terminally-truncated version of the actual full-length protein. Recently, several groups have reported on the identification of substrate proteins of full-length UREB1 including histones, p53, the anti-apoptotic protein Mcl-1, and the proto-oncoprotein c-Myc and the respective authors have referred to UREB1 using various names including E3histone, ARF-BP1 (ARF-Binding Protein 1), MULE (Mcl-1 Ubiquitin Ligase E3), and HECTH9 [42, 72–74] (in the following discussion we will refer to UREB1/E3histone/ARF-BP1/MULE/HectH9 simply as HectH9 [75]). HECTH9 consists of 4374 amino acid residues and, in addition to the HECT domain, contains three domains known to serve as protein–protein interaction sites, namely a BH3 domain, a WWE domain, and a UBA domain (UBA domains mediate interaction with ubiquitin). The BH3 domain is required for the interaction of HECTH9 with Mcl-1 [73], whereas the interactions sites for p53 and c-Myc have not as yet been mapped in detail [42, 74]. Interestingly, while HECTH9-mediated ubiquitination targets p53 and Mcl-1 for proteasome-mediated degradation, c-Myc is not targeted for degradation by HECTH9. HECTH9 modifies c-Myc with ubiquitin chains that are linked via lysine residue 63 (K63) of ubiquitin [42, 73, 74] and K63-linked ubiquitin chains are known to serve non-proteolytic roles [1–3]. Indeed, HECTH9-mediated ubiquitination of c-Myc appears to be required for transactivation of multiple target genes of c-Myc and induction of cell proliferation. The notion that HECTH9 acts as a positive effector of cell proliferation is supported by the observations that HECTH9 plays an important role in p53 degradation, that its activity is negatively regulated by the human tumor suppressor p14ARF (or p19ARF in mice), and that RNAi-mediated downregulation of HECTH9 expression interferes with the growth of p53 null cells. The latter observation indicates that HECTH9 has p53-independent pro-proliferative properties [74]. However, whether these p53-independent properties are related to HECTH9-mediated activation of c-Myc and whether p14ARF affects c-Myc activation, remains to be determined. Finally, it should be noted that the observation that HECTH9 targets the anti-apoptotic protein Mcl-1 for degradation cannot be readily associated with its pro-proliferative properties. Nonetheless, the findings that the HECTH9 gene is overexpressed in various human tumors and that HECTH9 expression is required for proliferation of at least a subset of tumor cells, suggests that HECTH9 represents an attractive target for the development of molecular strategies in the treatment of cancer [76].

4.6 HECT E3s with WW Domains

HECT E3s with WW domains share a common structure including an N-terminal calcium-dependent phospholipid binding C2 domain [77], two to four WW domains [18, 20], and the HECT domain (for recent reviews on Nedd4/Nedd4-like proteins see [78–80]). There are nine members of Nedd4-like HECT E3s encoded in the human or mouse genome (Figure 4.3), and orthologs are found in all eukaryotic organisms including fly, worm, and yeast [78]. Extensive alternative splicing of some, possibly all family members, contributes to the diversity of the protein family [81–85]. Nedd4/Nedd4-like proteins are involved in a plethora of functions, including endocytosis, trafficking, degradation of membrane proteins, control of cell growth, and virus budding [78] and therefore play a role in a number of pathologies including hypertension, cancer, and immune diseases.

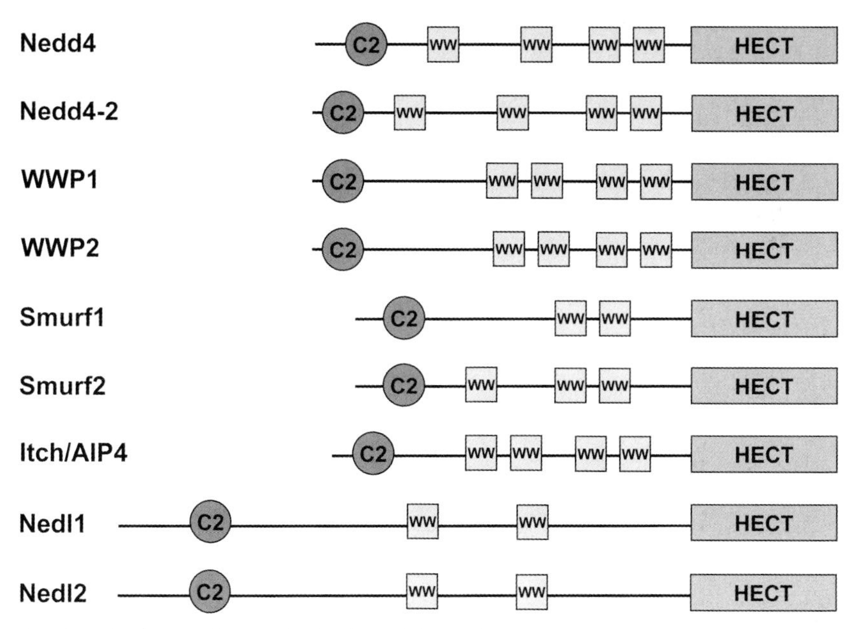

Fig. 4.3. The family of human Nedd4/Nedd4-like proteins. Schematic view of the family members showing the C2 domain, WW domains, and the HECT domain.

4.6.1 Nedd4/Nedd4-2

Nedd4 (Neuronal precursor cell Expressed Developmentally Downregulated 4) is the founding member of the Nedd4/Nedd4-like family of proteins and was originally identified in a subtractive screen using mRNAs derived from different stages during brain development [86]. The gene encoding Nedd4 (also referred to as Nedd4-1) is located on chromosome 15q22 and is ubiquitously expressed in all tissues [87–89]. It was originally described as a protein containing three WW domains and a C-terminal HECT domain [90], but it is now evident that there are multiple alternative transcripts encoding different forms of Nedd4 which either have or do not have the C2 domain and varying numbers of WW domains. Nedd4 has a close relative, Nedd4-2, encoded by a different gene at chromosomal location 18q21. As is the case for Nedd4, Nedd4-2 is ubiquitously expressed, with abundant expression in the kidney, heart, and liver [89, 91, 92]. It has been suggested that both proteins play a role in a hereditary form of hypertension, Liddle's syndrome.

4.6.1.1 Nedd4/Nedd4-2 and Liddle's Syndrome

Liddle's syndrome is a rare genetic form of hypertension first described in 1963 [93, 94]; it is characterized by early onset of severe hypertension, hypokalemia, metabolic alkalosis, and low circulating levels of renin and aldosterone. Patients are treated with a low Na^+ diet and administration of triamterenes, which are specific inhibitors of the epithelial Na^+ channel (ENaC). All the available data suggest that there is a defect in the Na^+-reabsorbing epithelia of the distal nephron. Such cells have on their apical side (facing the urinary compartment) ENaC, which allows entry of Na^+ into the cell, and on the basolateral side the Na^+,K^+-ATPase, which extrudes Na^+ out of the cell into the blood compartment. This Na^+ transport is highly regulated by the renin–angiotensin–aldosterone system. In support of a defect in this transport system, mutations in the genes encoding ENaC subunits were found to be at the origin of Liddle's syndrome [95, 96]. ENaC is a transmembrane protein complex that is composed of three homologous subunits which assemble in a tetrameric structure ($2\alpha1\beta1\gamma$); each subunit contains two transmembrane domains, one extracellular loop and two short cytosolic N and C termini. Of relevance are the short proline-rich regions in the C termini of each subunit, which are referred to as PY-motifs (consensus: L/P-P-X-Y-X-X-ø, where ø is a hydrophobic amino acid [97]). These PY-motifs serve as binding sites for the WW domains of Nedd4 and/or Nedd4-2, which ubiquitinate ENaC on the N termini of the ENaC subunits [88, 89, 91, 92, 97–101]. Such ubiquitination leads to the rapid internalization of ENaC and rapid degradation in the endosomal/lysosomal system, as described for many membrane proteins [102–106]. In Liddle's syndrome, a PY-motif is mutated or deleted either in the β- or in the γ-subunit [95, 96, 107], leading to impaired interaction with Nedd4 or Nedd4-2 and consequently reduced ubiquitination and internalization, resulting in the accumulation of channels at the plasma membrane (Figure 4.4).

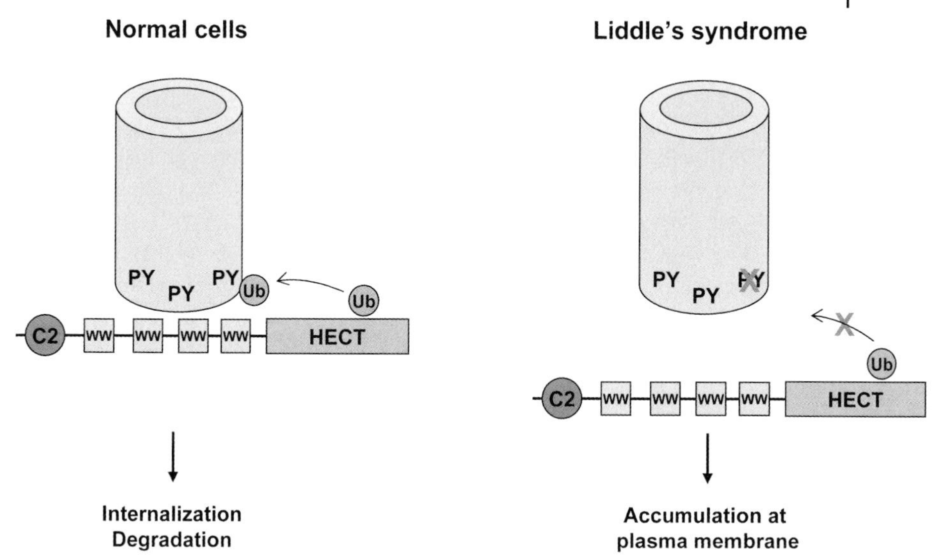

Fig. 4.4. Defective regulation of ENaC by Nedd4-2 in Liddle's syndrome. In normal cells, Nedd4 or Nedd4-2 interacts via the WW domain/PY-motif interaction with the ENaC complex. In Liddle's syndrome, this interaction is impaired, causing reduced ubiquitination, internalization of ENaC, and consequently accumulation of the channel at the plasma membrane.

Which Nedd4/Nedd4-like protein is physiologically controlling ENaC? The majority of evidence indicates Nedd4-2 to be the relevant regulator in the kidney: (1) functional and biochemical evidence from heterologous expression experiments indicates that Nedd4-2 is a more potent regulator of ENaC than Nedd4-1 and can be co-immunoprecipitated with ENaC from intact cells [89, 91, 98, 100]. (2) RNA interference experiments suggest that Nedd4-2, but not Nedd4-1, is involved in the negative control of transepithelial Na^+ transport in epithelial cells [108]. (3) Nedd4-2 is expressed together with ENaC in the principal cells of the distal nephron and its expression is controlled by dietary Na^+. A high-Na^+ diet leads to high Nedd4-2 expression (and inversely to low ENaC expression), whereas a low-Na^+ diet produces the opposite effect, suggesting that Nedd4-2 is itself regulated in the physiology of Na^+ reabsorption [109]. (4) Nedd4-2 interacts with and becomes phosphorylated by the aldosterone-induced "serum and glucocorticoid-induced kinase" Sgk1, a key regulator of Na^+ transport. Phosphorylation of Nedd4-2 creates binding sites for 14-3-3 proteins, which interfere with the ENaC/Nedd4-2 interaction, causing, as in Liddle's syndrome, an increase in the number of channels at the cell surface [98, 100, 104, 105, 110–112]. (5) Polymorphisms in the Nedd4-2 gene are linked to hypertension [113–116]. However, definite proof that Nedd4-2 controls Na^+ transport in the distal nephron will require its inactivation and analysis in a transgenic mouse KO model, or the genetic linkage of the Nedd4-2 locus to Liddle's syndrome or other forms of hypertension.

4.6.1.2 Nedd4 and Retrovirus Budding

Nedd4/Nedd4-like proteins have attracted considerable attention following the finding that they play a role in budding of many viruses. Viral budding is a process that largely takes advantage of the cellular mechanisms that control targeting of proteins for lysosomal degradation via the multiple vesicle body (MVB) pathway (for reviews, see [102, 117]). Essentially, late domains on viral gag proteins recruit components of the MVB machinery (e.g. Tsg101) and promote the budding of the virus. Late domains may contain either a P(S/T)AP, a PY motif, or a tyrosine-based sorting motif $YP_x(n)L$. P(S/T)AP has been shown to interact with the UEV domain of Tsg101, allowing recruitment of ESCRT-I and, subsequently, ESCRT-II and ESCRT-III complexes. These complexes usually induce membrane invagination and vesicle formation towards the interior of the late endosome. In the case of viruses, this interaction allows the virus to bud into the lumen of late endosomes, and eventually to leave the cell via exocytosis, or alternatively, to assemble the ESCRT complexes at the plasma membrane and subsequently bud from there. The P(S/T)AP and the PY-motif have been demonstrated to be functionally exchangeable.

Although it has been shown that the viral PY-motifs bind to Nedd4 family members, it is not clear how this promotes particle release. Ubiquitylation of Gag may recruit the MVB machinery, for example via binding to the Tsg101 protein or other ubiquitin-binding proteins of the ESCRT pathway. Alternatively, it has been shown that Nedd4 and Tsg101 can interact with each other, which may be a way to recruit Tsg101 to the Gag protein and induce virus budding [118]. Furthermore, studies using HTLV-1 mutants, which have either the PY motif or the P(S/T)AP motif mutated, showed that mutation of the PY-motif leads to the accumulation of the virus at the plasma membrane, whereas mutations of the P(S/T)AP motif leads to accumulation in endosomes, suggesting that HTLV-1 first interacts at the plasma membrane with Nedd4 and later with Tsg101 in endosomes [119].

4.6.2 Itch and the Immune Response

Mouse Itch, the ortholog of human AIP4, is a member of the Nedd4/Nedd4-like HECT E3 family and was identified as the protein encoded by the itchy locus of chromosome 2 in non-agouti lethal 18H mice [120, 121]. In these mice, Itch is inactivated by gene inversion, simultaneously with promoter inactivation of the agouti gene, leading to a darker colored coat [120]. Depending on the genetic background, the mice present two different, but related phenotypes: with the JC/Ct background, they display an inflammatory disease of the large intestine, whereas with the C57L/6J background they present a fatal disease involving changes in the lung, spleen, lymph nodes, skin, ear, thymus, and stomach. In each organ, the phenotype suggests hyperactivation of processes typical of chronic inflammation.

Moreover, these mice, also referred to as *itchy* mice, are characterized by skin and ear scarring, caused by constant scratching when older than 16 weeks. Larger spleens and lymph nodes, likely due to hyperproliferation of lymphocytes, are also observed. This phenotype points to a critical role for Itch in the negative control of the immune system. Indeed, defects in T-cell development and function are characteristic in Itch null mice. When pathogens invade the host, an adaptive immune response is elicited through the generation of specific T cells. Antigenic stimulation drives naïve $CD4^+$ T cells into Th1 and Th2 cells which can be distinguished by their cytokine profiles and functions [122, 123]. Th1 cells produce IL-2 and IFN-γ, which are effective in counteracting viral infections and other intracellular pathogens, whereas Th2 cells produce IL-4 and IL-5, which are involved in the elimination of extracellular helminthic pathogens. Itch plays a role in Th2 differentiation as evidenced in older Itch–/– mice. In this context, T cells can be chronically activated and display increased proliferation in response to stimulation with anti-CD3 or anti-CD3 plus anti-CD28. This stimulation is accompanied by increased production of IL-4 and IL-5 and a biased differentiation of $CD4^+$ cells into Th2 cells [124]. Consistently, the Th2-dependent serum concentrations of IgG1 and IgE in itchy mice are increased. During Itch-mediated T-cell differentiation, Itch binds to either JunB or c-jun, transcription factors which are involved in the gene regulation of Th2 cytokines such as IL-4. Itch binds via its WW domains to the PPXY motif in Jun-B/c-jun and promotes Jun-B/c-jun ubiquitination and consequently their proteasomal degradation (Figure 4.5) [124].

Itch may also play a role in immune self tolerance. Although an efficient immune response is important for protection of the host, the immune system also has mechanisms to prevent excessive damage to normal cells and tissues, a process known as self tolerance. Defects in Th1 responses can result in autoimmune diseases such as type 1 diabetes or rheumatoid arthritis, whereas excessive Th2 cell activation can lead to asthmatic or allergic symptoms. There are a number of

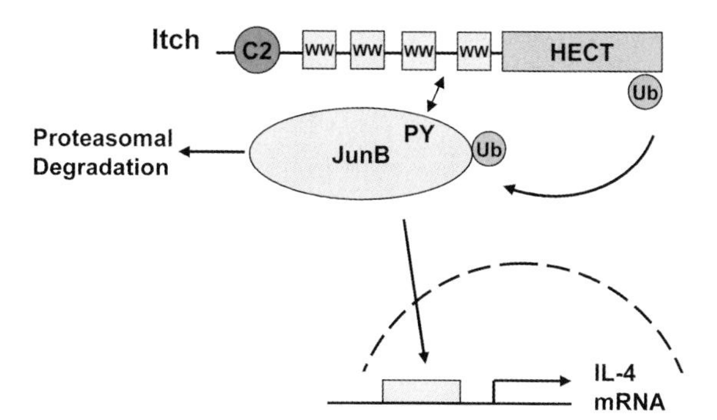

Fig. 4.5. Itch-dependent regulation of c-jun/JunB. Itch interacts via the WW domain/PY-motif with c-jun or JunB, thereby ubiquitinating/targeting c-jun for proteasomal degradation. In Itch null mice, increased c-jun activity leads to increased expression of IL-4.

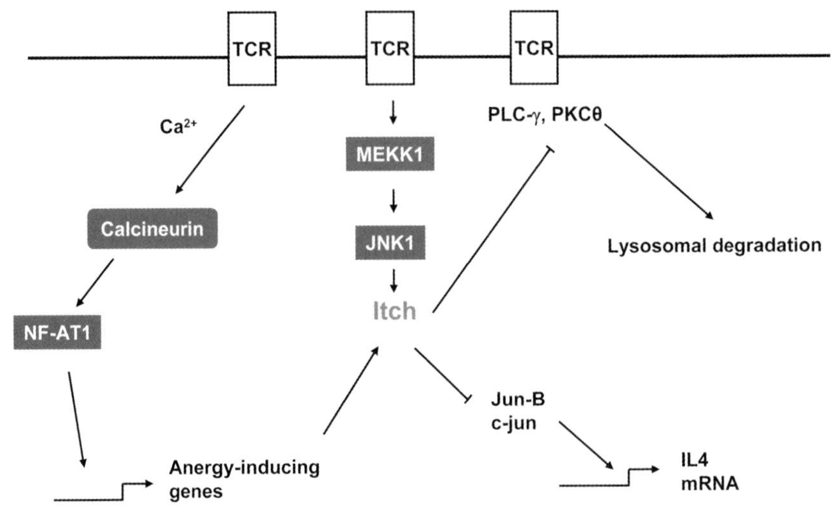

Fig. 4.6. Role of Itch in T-cell anergy. Engagement of the T-cell receptor without co-stimulation leads to mobilization of intracellular Ca^{2+} and activation of calcineurin. Calcineurin dephosphorylates and activates NF-AT1, promoting the transcription of the genes involved in anergy, including Itch. Itch becomes phosphorylated via the MEKK1/JNK1 pathway and ubiquitinates c-jun/JunB, PLC-γ, and PKCθ, which are then targeted for proteasomal (c-jun/Junb) and lysosomal (PLC-γ, PKCθ) degradation, respectively.

mechanisms involved in the induction of peripheral immune tolerance, including peripheral deletion of autoreactive T cells, generation of Tregs, and T-cell anergy. [125, 126]. Of particular interest is T-cell anergy, which is a state of unresponsiveness in T cells that is achieved when the T-cell receptor (TCR) is engaged without co-stimulation of accessory molecules such as the CD86 receptor [127]. There are multiple mechanisms that seem to be relevant in T-cell anergy and some of them involve Itch (Figure 4.6). In the absence of co-factors, TCR engagement is sufficient to stimulate mobilization of intracellular free calcium ions, causing activation of calcineurin, a calcium-sensitive phosphatase. Calcineurin dephosphorylates the NF-AT1 transcription factor which becomes activated and stimulates the transcription of putative "anergy genes", whose products are responsible for keeping the T cell in an anergic state [128]. Among a number of signaling proteins and other ubiquitin-protein ligases (Grail, Cbl-B), Itch is induced under anergic conditions [129]. Upon stimulation, Itch and its close relative Nedd4 become membrane associated, most likely via their C2 domains, resulting in binding and subsequent mono-ubiquitination of PKC- and PLC-γ1, proteins which are important in TCR signaling. The mono-ubiquitinated PKC- and PLC-γ1 are channeled via the ESCRT pathway (including Tsg101, which is also upregulated in calcium induced T-cell anergy) into lysosomes, where they are degraded [129].

The activity of Itch can be regulated by phosphorylation, either negatively or positively. Tyrosine phosphorylation by the Fyn kinase causes a negative effect on

Itch activity [130], whereas TCR-mediated signaling via the mitogen and extracellular kinase 1 (MEKK1) and JNK1 [131] has a positive effect. Consistent with this, phosphorylation of Itch by MEKK1 and JNK1 stimulates JunB and c-jun turnover. T cells derived from MEKK1 *KD* mice that express a catalytically inactive form of MEKK1, show reduced JNK activation following engagement of the TCR and the CD28 auxiliary receptor. Moreover, both peripheral T cells and thymocytes hyperproliferate in response to stimulation with antibodies to CD3 and CD28, and within 4 h of receptor engagement express larger amounts of IL-4, consistent with the requirement of phosphorylated Itch to negatively control Th2 activation. The MEKK1–JNK–Itch pathway is important in T-cell anergy and Th2 tolerance and airway inflammation [132]. Both *in-vitro* and *in-vivo* assays show that Th2 cells that either express an inactive MEKK1 mutant or are devoid of JNK1 or Itch cannot become immune tolerized, whereas Th1 tolerance is not affected. This breakdown of Th2 tolerization leads to airway inflammation [132].

4.6.3 **Smurfs**

Smurf1 and Smurf2 are two other, closely related members of the Nedd4/Nedd4-like proteins family that have been extensively studied in recent years and are particularly important in a number of pathophysiological conditions, including cancer and bone homeostasis.

Smurf1 (*Sm*ad *u*biquitination *r*egulatory *f*actor 1) was originally identified as an interacting partner of Smad1, a protein involved in TGF-β signaling [133]. The TGF-β superfamily of cytokines is involved in a plethora of biological functions, including embryonic development, regulation of cell growth and differentiation, and apoptosis. Moreover, mutations in TGF-β pathway components are associated with a number of human diseases, including cancer and osteoporosis. There are roughly 40 members of the TGF-β ligands and they can be divided into two groups, the TGF/activin/nodals and the bone morphogenetic proteins (BMPs), each group stimulating different signaling pathways. TGF-β ligands bind to Ser/Thr kinase receptors. These receptors are composed of type I and type II class receptors, which heterodimerize on ligand binding. The type II receptor phosphorylates the type I receptors, which become activated and stimulate Smads, which are involved in the TGF-β or the BMP pathways. The type I receptor phosphorylates serine residues on the receptor-regulated Smads (or R-Smads), Smad-1, -2, -3, -5, and -8. Smad-1, -5, and -8 are involved in the BMP pathway, whereas Smad-2 and -3 have a signaling function in the TGF-β pathway. Once phosphorylated, R-Smads complex with the common co-SMAD, Smad-4, and translocate to the nucleus, where they interact with DNA-binding proteins and regulate transcription. In addition to R-Smads and co-Smads, there are so-called inhibitory or I-Smads (Smad-6 and -7) that are important for the negative regulation of TGF-β and BMP pathways. They inhibit TGF-β and BMP signaling by competing with R-Smads for association with type I receptors or by targeting receptors for ubiquitin-mediated degradation. Ubiquitination, via Smurf1 or -2, is an alternative negative

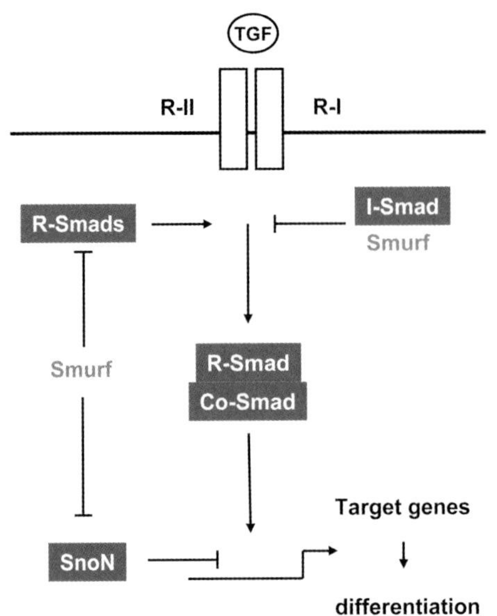

Fig. 4.7. Role of Smurf in the TGF-β pathway. TGF-β ligand stimulates heterodimerization of type I and type II Ser/Thr kinase receptors, leading to the phosphorylation of the type I receptor by the type II receptor. This recruits receptor-regulated Smads (R-Smads), which become phosphorylated. Upon phosphorylation, R-Smads interact with the common Smad, Smad4, and the complex translocates to the nucleus, where it interacts with co-factors and stimulates transcription of the genes involved in differentiation. The pathway is negatively regulated by inhibitory Smads (I-Smads), by SnoN, and by Smurfs. Smurfs can interact with and ubiquitinate R-Smads and can be recruited by I-Smads to the receptor where they induce receptor ubiquitination and internalization. Moreover, Smurfs are also involved in SnoN ubiquitination.

mechanism by which TGF-β and BMP signaling is regulated. Thus Smurf1/2 can act at different levels in the signaling pathway (Figure 4.7).

As mentioned above, Smurf1 has been identified as a ubiquitin-protein ligase that interacts with Smad1 via WW domain/PY-motif interaction. By doing so, it ubiquitinates and suppresses the steady state levels of the BMP signaling proteins Smad-1 and Smad-5 (but not Smad-2 or -4) [133]. Smurf2 associates with Smad-1, -2, and -3, giving rise to a decrease in the steady state level of Smad1 and, to a lesser extent, of Smad2 [134–136]. The interaction between Smurf2 and Smad2 is dramatically increased when Smad2 becomes activated by the TGF-β pathway, indicating that Smurf2 plays a particular role in the ligand-dependent regulation of the pathway. Destruction of activated Smad2 occurs in the nucleus via proteasomal degradation [137]. Similarly, activated Smad3 is also targeted for proteasomal degradation but in this case, it is ubiquitinated by the SCF/Roc1 E3 ligase complex [138].

Smurfs can also use adapters for the regulation of the TGF-β pathway. Smad2/Smurf2 complexes are formed after TGF-β stimulation and Smad2 then recruits Smurf2 to the transcriptional co-repressor SnoN [136]. As SnoN is an inhibitor of the TGF-β transcriptional response, its degradation may thereby facilitate TGFβ signaling. A different mechanism involves Smad7, which is able to interact with Smurf2 in the nucleus. This interaction causes export of Smurf2 from the nucleus and interaction of the complex with the TGF-β receptor. In concert with the E2 enzyme UbcH7 [139], Smad7 becomes ubiquitinated and triggers the internalization and degradation of itself and the receptor [140]. Interestingly, this action involves the lipid raft/caveolar internalization pathway; alternatively the receptor can also internalize via the classical clathrin-dependent pathway, thereby maintaining TGF-β signaling [141]. In a similar manner to Smurf2, Smurf1 is also recruited by Smad7 to the TGF-β receptor [142]. For nuclear export, it uses the protein CRM-1 [143], which is an importin β-related nuclear transport receptor and physically interacts with a nuclear export signal in the HECT domain of Smurf1 [142, 144]. Furthermore, both Smad6 and Smad7 can also translocate Smurf1 to the BMP receptors and induce ubiquitination of theses receptors [145].

4.6.3.1 Smurfs and Cancer

The TGF-β pathway generally has an inhibitory effect on growth, both *in vitro* and *in vivo*. As described above, steady-state expression levels of the proteins in this pathway are tightly regulated to ensure proper function. It can therefore be expected that improper steady state levels are frequently associated with cancer [146]. Although the majority of Smad mutations associated with cancer are loss of function mutations [147, 148], there are a number of others that affect the steady state level [146]. Oncogenic missense mutations in Smad4 (L43S, G65V, R100T, or P130S) or in Smad 2 (R133C, or a nonsense mutation at position 515, Q407R, L369R) lead to increased ubiquitination and proteasomal degradation as compared to the normal proteins [149–154]. Hence, certain oncogenic Smad mutations destabilize the protein and thereby inactivate TGF-β regulation. It is not yet known which ubiquitin-protein ligase(s) is/are involved, but they may very well be different from Smurf1 or Smurf2. It seems conceivable that the above-mentioned mutations may cause misfolding of the proteins and, thus, the misfolded proteins may be recognized by E3 ligases involved in quality control. However, there is also evidence for the involvement of Smurf ligases in certain types of cancer. It has been found that Smurf2 is highly overexpressed in esophageal squamous carcinoma and is associated with a poor prognosis of this disease [155].

In addition to its tumor suppressive effect, TGF-β can also display, under certain circumstances such as in late-stage tumors, a tumor-promoting effect [156]. This can be attributed to its ability to promote malignant progress, invasiveness, and metastasis. Hence a decrease in Smurf levels causing an increase in TGF-β signaling may stimulate late-stage tumors. RNF11, a Ring-H2 protein that is highly expressed in invasive breast cancer [157], was recently shown to interact with Smurf2 and to target Smurf2 for ubiquitin–proteasome-mediated degradation. Furthermore, it has been found that RNF11 can interfere with Smurf2-mediated ubiquitination of the TGFβ receptor. By blocking Smurf2 activity, RNF11 may thus enhance TGF-β signaling and its tumor-promoting activity [157].

4.6.3.2 Smurfs and Bone Homeostasis

The bone mass is balanced between constant resorption and new formation. These properties are assured by specialized cells, the osteoclasts which are responsible for resorption, and the osteoblasts which produce bone [158]. Osteoblasts develop from mesenchymal progenitors and require the osteoblast-specific transcription factors RunX2 and Osterix. These proteins are essential for the differentiated osteoblasts to synthesize alkaline phosphatases, type 1 collagen, osteocalcin, and bone sialoprotein and to deposit them in the bone extracellular matrix. Osteoblast differentiation and their bone-forming activities are subject to control by members of the TGF-β/BMP superfamily (Figure 4.8). BMPs promote osteoblast differentiation and bone ECM deposition, whereas TGF-β interferes with this process by inhibiting Runx2 and osteocalcin activity. Because both TGF-β and BMPs take advantage of the Smads, it is not surprising that the Smurfs, especially Smurf1, also play a role in osteoblast function. Overexpressing Smurf1 in osteoblast precursor cells interferes with BMP-induced osteoblast differentiation [159, 160], whereas RNA interference-mediated downregulation of Smurf1 expression or expression of catalytically inactive Smurf1 enhances osteoblast differentiation [159, 161]. Thus, overexpression or suppression of Smurf1 controls the expression of Smad proteins, particularly of Smad1 and/or Smad5, thereby enhancing or interfering with the BMP signaling pathway [159–161]. Alternatively, Smurf1 is also capable of binding to Runx2, partly via interaction with the WW domain/PY motifs and additionally via Smad6, this interaction then destabilizes Runx2 and inhibits osteo-

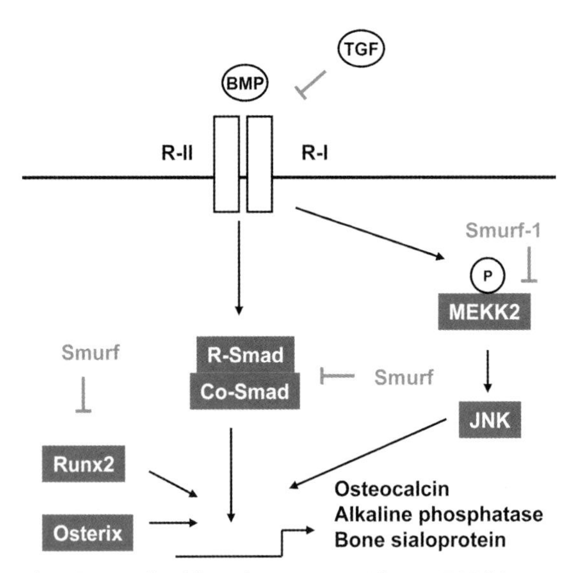

Fig. 4.8. Smurf and bone homeostasis. Whereas TGF-β has a negative effect on Runx2 and osteocalcin activity, BMP stimulates these processes via the Smad pathway. Smurfs can interfere with BMP signaling by ubiquitinating R-Smads, Runx2, and phosphorylated MEKK2.

blast differentiation [160–162]. Transgenic mice which overexpress Smurf1 display significantly reduced bone formation and this provides further support for the suggestion that Smurf1 plays a role in bone homeostasis [160].

The strongest argument for a role in bone homeostasis comes from Smurf1–/– mice [163]. These mice are viable and develop normally, become fertile and have a similar life expectancy as wild-type mice. However, starting from approximately 4 months of age, they show an increase in bone mass caused by enhanced osteoblast activity [163]. Because there is no developmental defect in bone formation, it is likely that Smurf1 does not play an essential role in early osteoblast formation, but rather is important in bone-forming activities in mature osteoblasts. Surprisingly, neither BMP signaling nor Runx2 activity seem to be affected by the inactivation of Smurf1. Rather it is JNK, which has previously been shown to play a role in osteoblast function, which is constantly phosphorylated in Smurf–/– mice, whereas phosphorylation of JNK in normal mice requires stimulation by BMP. Consequential to this phosphorylation, downstream transcription activity is enhanced in Smurf–/– cells. Smurf1 does not interact with JNK, but with MEKK2, an upstream activator of JNK. Autophosphorylation of MEKK2 appears to be indispensable for its interaction with Smurf1 and for its ubiquitin-dependent degradation. Expression of constitutively active MEKK2 or JNK, or inactive MEKK2 in osteoblasts demonstrates that these kinases regulate osteoblast activity.

The lack of effect of Smurf1 inactivation on the BMP pathway may be explained by the compensatory action of Smurf2. Indeed, Smurf1–/– mice show an increase in Smurf2 expression, and, double KO mice for Smurf1 and Smurf2 are embryonic lethal, supporting the idea that Smurf1 and 2 have overlapping functions. It remains to be shown how MEKK2 becomes activated in the control of osteoblast activity. The involvement of Smurf1 in human pathologies with dysregulated bone homeostasis (such as osteoporosis) remains to be demonstrated. However, the fact that Smurf1 inactivation has no effect on the maintenance of skeletal integrity may be useful for developing therapeutic strategies for treating age-related bone loss such as osteoporosis.

4.7
Concluding Remarks

As described above, deregulation of the activity of certain HECT E3s is intrinsically involved in the development of distinct human diseases. Thus, an important question is whether HECT E3s or – in those cases where E3 activity is lost during disease development via mutation – their respective substrate proteins or proteins that regulate the interaction of HECT E3s with their substrate proteins (e.g. Sgk1 kinase which negatively affects Nedd4-2/substrate interaction) represent potential targets in the development of treatments for the respective disease. Unfortunately, the physiologically relevant substrate proteins of most HECT E3s are not yet known and their identification may be hampered by the notion that a given protein is not only recognized as a substrate by a single E3 but by several

E3s. Nonetheless, identification of the substrate proteins of E6-AP, for example, should provide significant insight into the mechanisms by which loss of the E3 activity of E6-AP results in the development of the Angelman syndrome. In cases in which inappropriate activation of HECT E3s contributes to disease (e.g. overexpression of the respective HECT E3 or of auxiliary factors affecting the substrate specificity of the E3), there may be direct interference with the E3 activity of the respective HECT E3 or interference with the interaction between the E3 and particular substrate proteins, for example by small molecules. Indeed, downregulation of E6-AP expression has been reported to selectively activate the p53 tumor suppressor pathway in HPV-positive cells. Although this (inactivation of E6-AP) may not appear to be an attractive approach for the treatment of cervical cancer (since the viability of HPV-positive cells depends on the continuous presence of E6 and E7, molecular strategies to target these viral oncoproteins represent the approaches of choice), this observation indicates that HECT E3s can serve as potential targets for the development of molecular therapeutic approaches. Finally, it should be noted, that ubiquitination is a reversible process by the action of deubiquitinating enzymes. Since deubiquitinating enzymes are proteolytic enzymes with a spatially-defined catalytic center, they may more easily be made tractable by small molecules than by HECT E3s. It will, therefore, be particularly important to understand which of the E3s/deubiquitinating enzymes are involved in regulating the stability of distinct target proteins.

References

1 Pickart, C.M. (2001) Mechanisms underlying ubiquitylation. *Annu Rev Biochem* **70**, 503–533.

2 Glickman, M.H. and Ciechanover, A. (2002) The ubiquitin-proteasome proteolytic pathway: destruction for the sake of construction. *Physiol Rev* **82**, 373–428.

3 Kerscher, O., Felberbaum, R. and Hochstrasser, M. (2006) Modification of Proteins by Ubiquitin and Ubiquitin-Like Proteins. *Annu Rev Cell Dev Biol* **22**, 159–180.

4 Hashizume, R., Fukuda, M., Maeda, I., Nishikawa, H., Oyake, D., Yabuki, Y., Ogata, H. and Ohta, T. (2001) The RING heterodimer BRCA1-BARD1 is a ubiquitin ligase inactivated by a breast cancer-derived mutation. *J Biol Chem* **276**, 14537–14540.

5 Starita, L.M. and Parvin, J.D. (2006) Substrates of the BRCA1-dependent ubiquitin ligase. *Cancer Biol Ther* **5**, 137–141.

6 Oliner, J.D., Kinzler, K.W., Meltzer, P.S., George, D.L. and Vogelstein, B. (1992) Amplification of a gene encoding a p53-associated protein in human sarcomas. *Nature* **358**, 80–83.

7 Leach, F.S., Tokino, T., Meltzer, P., Burrell, M., Oliner, J.D., Smith, S., Hill, D.E., Sidransky, D., Kinzler, K.W. and Vogelstein, B. (1993) p53 mutation and MDM2 amplification in human soft tissue sarcomas. *Cancer Res* **53**, 2231–2234.

8 Huibregtse, J.M., Scheffner, M., Beaudenon, S. and Howley, P.M. (1995) A family of proteins structurally and functionally related to the E6-AP ubiquitin-protein ligase. *Proc Natl Acad Sci USA* **92**, 2563–2567.

9 Scheffner. M., Nuber, U. and Huibregtse, J.M. (1995) Protein ubiquitination

involving an E1-E2-E3 enzyme ubiquitin thioester cascade. *Nature* **373**, 81–83.
10 Bischoff, F.R. and Ponstingl, H. (1991) Catalysis of guanine nucleotide exchange on Ran by the mitotic regulator RCC1. *Nature* **354**, 80–82.
11 Renault, L., Nassar, N., Vetter, I., Becker, J., Klebe, C., Roth, M. and Wittinghofer, A. (1998) The 1.7 A crystal structure of the regulator of chromosome condensation (RCC1) reveals a seven-bladed propeller. *Nature* **392**, 97–101.
12 Rosa, J.L., Casaroli-Marano, R.P., Buckler, A.J., Vilaro, S. and Barbacid, M. (1996) p619, a giant protein related to the chromosome condensation regulator RCC1, stimulates guanine nucleotide exchange on ARF1 and Rab proteins. *EMBO J* **15**, 4262–4273.
13 Chong-Kopera, H., Inoki, K., Li, Y., Zhu, T., Garcia-Gonzalo F.R., Rosa J.L. and Guan K.L. (2006) TSC1 stabilizes TSC2 by inhibiting the interaction between TSC2 and the HERC1 ubiquitin ligase. *J Biol Chem* **281**, 8313–8316.
14 Lehman, A.L., Nakatsu, Y., Ching, A., Bronson, R.T., Oakey, R.J., Keiper-Hrynko, N., Finger, J.N., Durham-Pierre, D., Horton, D.B., Newton, J.M., Lyon, M.F. and Brilliant, M.H. (1998) A very large protein with diverse functional motifs is deficient in rjs (runty, jerky, sterile) mice. *Proc Natl Acad Sci USA* **95**, 9436–9441.
15 Rinchik, E.M., Carpenter, D.A. and Handel, M.A. (1995) Pleiotropy in microdeletion syndromes: neurologic and spermatogenic abnormalities in mice homozygous for the p6H deletion are likely due to dysfunction of a single gene. *Proc Natl Acad Sci USA* **92**, 6394–6398.
16 Dastur, A., Beaudenon, S., Kelley, M., Krug, R.M. and Huibregtse, J.M. (2006) Herc5, an interferon-induced HECT E3 enzyme, is required for conjugation of ISG15 in human cells. *J Biol Chem* **281**, 4334–4338.
17 Wong, J.J., Pung, Y.F., Sze, N.S. and Chin, K.C. (2006) HERC5 is an IFN-induced HECT-type E3 protein ligase that mediates type I IFN-induced ISGylation of protein targets. *Proc Natl Acad Sci USA* **103**, 10735–10740.
18 Staub, O. and Rotin, D. (1996) WW domains. *Structure* **4**, 495–499.
19 Sudol, M. and Hunter, T. (2000) NeW wrinkles for an old domain. *Cell* **103**, 1001–1004.
20 Macias, M.J., Wiesner, S. and Sudol, M. (2002) WW and SH3 domains, two different scaffolds to recognize proline-rich ligands. *FEBS Lett* **513**, 30–37.
21 Huibregtse, J.M., Scheffner, M. and Howley, P.M. (1993) Cloning and expression of the cDNA for E6-AP, a protein that mediates the interaction of the human papillomavirus E6 oncoprotein with p53. *Mol Cell Biol* **13**, 775–784.
22 Scheffner, M., Huibregtse, J.M., Vierstra, R.D. and Howley, P.M. (1993) The HPV-16 E6 and E6-AP complex functions as a ubiquitin-protein ligase in the ubiquitylation of p53. *Cell* **75**, 495–505.
23 Kishino, T., Lalande, M. and Wagstaff, J. (1997) UBE3A/E6-AP mutations cause Angelman syndrome. *Nat Genet* **15**, 70–73.
24 Matsuura, T., Sutcliffe, J.S., Fang, P., Galjaard, R.J., Jiang, Y.H., Benton, C.S., Rommens, J.M. and Beaudet, A.L. (1997) *De novo* truncating mutations in E6-AP ubiquitin-protein ligase gene (UBE3A) in Angelman syndrome. *Nat Genet* **15**, 74–77.
25 Yamamoto, Y., Huibregtse, J.M. and Howley, P.M. (1997) The human E6-AP gene (UBE3A) encodes three potential protein isoforms generated by differential splicing. *Genomics* **41**, 263–266.
26 Zur Hausen, H. (2002) Papillomaviruses and cancer: from basic studies to clinical application. *Nat Rev Cancer* **2**, 342–350.
27 Boyer, S.N., Wazer, D.E. and Band, V. (1996) E7 protein of human papilloma virus-16 induces degradation of retinoblastoma protein through the ubiquitin-proteasome pathway. *Cancer Res* **56**, 4620–4624.
28 Gonzalez, S.L., Stremlau, M., He, X., Basile, J.R. and Munger, K. (2001) Degradation of the retinoblastoma tumor suppressor by the human papillomavirus

type 16 E7 oncoprotein is important for functional inactivation and is separable from proteasomal degradation of E7. *J Virol* **75**, 7583–7591.

29 Zhang, B., Chen, W. and Roman, A. (2006) The E7 proteins of low- and high-risk human papillomaviruses share the ability to target the pRB family member p130 for degradation. *Proc Natl Acad Sci USA* **103**, 437–442.

30 Scheffner, M. and Whitaker, N.J. (2003) Human papillomavirus-induced carcinogenesis and the ubiquitin-proteasome system. *Semin Cancer Biol* **13**, 59–67.

31 Butz, K., Shahabeddin, L., Geisen, C., Spitkovsky, D., Ullmann, A. and Hoppe-Seyler, F. (1995) Functional p53 protein in human papillomavirus-positive cancer cells. *Oncogene* **10**, 927–936.

32 Butz, K., Geisen, C., Ullmann, A., Spitkovsky, D. and Hoppe-Seyler, F. (1996) Cellular responses of HPV-positive cancer cells to genotoxic anti-cancer agents: repression of E6/E7-oncogene expression and induction of apoptosis. *Int J Cancer* **68**, 506–513.

33 Beer-Romero, P., Glass, S. and Rolfe, M. (1997) Antisense targeting of E6AP elevates p53 in HPV-infected cells but not in normal cells. *Oncogene* **14**, 595–602.

34 Talis, A.L., Huibregtse, J.M. and Howley, P.M. (1998) The role of E6AP in the regulation of p53 protein levels in human papillomavirus (HPV)-positive and HPV-negative cells. *J Biol Chem* **273**, 6439–6445.

35 Kim, Y., Cairns, M.J., Marouga, R. and Sun, L.Q. (2003) E6AP gene suppression and characterization with *in vitro* selected hammerhead ribozymes. *Cancer Gene Ther* **10**, 707–716.

36 Hengstermann, A., D'silva, M.A., Kuballa, P., Butz, K., Hoppe-Seyler, F. and Scheffner, M. (2005) Growth suppression induced by downregulation of E6-AP expression in human papillomavirus-positive cancer cell lines depends on p53. *J Virol* **79**, 9296–9300.

37 Haupt, Y., Maya, R., Kazaz, A. and Oren, M. (1997) Mdm2 promotes the rapid degradation of p53. *Nature* **387**, 296–299.

38 Kubbutat, M.H., Jones, S.N. and Vousden, K.H. (1997) Regulation of p53 stability by Mdm2. *Nature* **387**, 299–303.

39 Honda, R. and Yasuda, H. (2000) Activity of MDM2, a ubiquitin ligase, toward p53 or itself is dependent on the RING finger domain of the ligase. *Oncogene* **19**, 1473–1476.

40 Leng, R.P., Lin, Y., Ma, W., Wu, H., Lemmers, B., Chung, S., Parant, J.M., Lozano, G., Hakem, R. and Benchimol, S. (2003) Pirh2, a p53-induced ubiquitin-protein ligase, promotes p53 degradation. *Cell* **112**, 779–791.

41 Dornan, D., Wertz, I., Shimizu, H., Arnott, D., Frantz, G.D., Dowd, P., O'Rourke, K., Koeppen, H. and Dixit, V. M. (2004) The ubiquitin ligase COP1 is a critical negative regulator of p53. *Nature* **429**, 86–92.

42 Chen, D., Kon, N., Li, M., Zhang, W., Qin, J. and Gu, W. (2005) ARF-BP1/Mule is a critical mediator of the ARF tumor suppressor. *Cell* **121**, 1071–1083.

43 Montes De Oca Luna, R., Wagner, D.S. and Lozano, G. (1995) Rescue of early embryonic lethality in mdm2-deficient mice by deletion of p53. *Nature* **378**, 203–206.

44 Hengstermann, A., Linares, L.K., Ciechanover, A., Whitaker, N.J. and Scheffner, M. (2001) Complete switch from Mdm2 to human papillomavirus E6-mediated degradation of p53 in cervical cancer cells. *Proc Natl Acad Sci USA* **98**, 1218–1223.

45 Kelley, M.L., Keiger, K.E., Lee, C.J. and Huibregtse, J.M. (2005) The global transcriptional effects of the human papillomavirus E6 protein in cervical carcinoma cell lines are mediated by the E6AP ubiquitin ligase. *J Virol* **79**, 3737–3747.

46 Mantovani, F. and Banks, L. (2001) The human papillomavirus E6 protein and its contribution to malignant progression. *Oncogene* **20**, 7874–7887.

47 Lee, S.S., Weiss, R.S. and Javier, R.T. (1997) Binding of human virus oncoproteins to hDlg/SAP97, a mammalian homolog of the Drosophila discs large tumor suppressor protein. *Proc Natl Acad Sci USA* **94**, 6670–6775.

48 Lee, S.S., Glaunsinger, B., Mantovani, F., Banks, L. and Javier, R.T. (2000) Multi-PDZ domain protein MUPP1 is a cellular target for both adenovirus E4-ORF1 and high-risk papillomavirus type 18 E6 oncoproteins. *J Virol* **74**, 9680–9693.

49 Glaunsinger, B.A., Lee, S.S., Thomas, M., Banks, L. and Javier, R. (2000) Interactions of the PDZ-protein MAGI-1 with adenovirus E4-ORF1 and high-risk papillomavirus E6 oncoproteins. *Oncogene* **19**, 5270–5280.

50 Nakagawa, S. and Huibregtse, J.M. (2000) Human scribble (Vartul) is targeted for ubiquitin-mediated degradation by the high-risk papillomavirus E6 proteins and the E6AP ubiquitin-protein ligase. *Mol Cell Biol* **20**, 8244–8253.

51 Nguyen, M.L., Nguyen, M.M., Lee, D., Griep, A.E. and Lambert, P.F. (2003) The PDZ ligand domain of the human papillomavirus type 16 E6 protein is required for E6's induction of epithelial hyperplasia in vivo. *J Virol* **77**, 6957–6964.

52 Simonson, S.J., Difilippantonio, M.J. and Lambert, P.F. (2005) Two distinct activities contribute to human papillomavirus 16 E6's oncogenic potential. *Cancer Res* **65**, 8266–8273.

53 Gewin, L., Myers, H., Kiyono, T. and Galloway, D.A. (2004) Identification of a novel telomerase repressor that interacts with the human papillomavirus type-16 E6/E6-AP complex. *Genes Dev* **18**, 2269–2282.

54 Pim, D., Thomas, M., Javier, R., Gardiol, D. and Banks, L. (2000) HPV E6 targeted degradation of the discs large protein: evidence for the involvement of a novel ubiquitin ligase. *Oncogene* **19**, 719–725.

55 Pim, D., Thomas, M. and Banks, L. (2002) Chimaeric HPV E6 proteins allow dissection of the proteolytic pathways regulating different E6 cellular target proteins. *Oncogene* **21**, 8140–8148.

56 Grm, H.S. and Banks, L. (2004) Degradation of hDlg and MAGIs by human papillomavirus E6 is E6-AP-independent. *J Gen Virol* **85**, 2815–2819.

57 Kao, W.H., Beaudenon, S.L., Talis, A.L., Huibregtse, J.M. and Howley, P.M. (2000) Human papillomavirus type 16 E6 induces self-ubiquitination of the E6AP ubiquitin-protein ligase. *J Virol* **74**, 6408–6417.

58 Angelman, H. (1965) "Puppet children". A report of three cases. *Dev Med Child Neurol* **7**, 681–688.

59 Jiang, Y., Lev-Lehman, E., Bressler, J., Tsai, T.F. and Beaudet, A.L. (1999) Genetics of Angelman syndrome. *Am J Hum Genet* **65**, 1–6.

60 Clayton-Smith, J. and Laan, L. (2003) Angelman syndrome: a review of the clinical and genetic aspects. *J Med Genet* **40**, 87–95.

61 Nicholls, R.D. and Knepper, J.L. (2001) Genome organization, function, and imprinting in Prader–Willi and Angelman syndromes. *Annu Rev Genomics Hum Genet* **2**, 153–175.

62 Albrecht, U., Sutcliffe, J.S., Cattanach, B.M., Beechey, C.V., Armstrong, D., Eichele, G. and Beaudet, A.L. (1997) Imprinted expression of the murine Angelman syndrome gene, Ube3a, in hippocampal and Purkinje neurons. *Nat Genet* **17**, 75–78.

63 Fang, P., Lev-Lehman, E., Tsai, T.F., Matsuura, T., Benton, C.S., Sutcliffe, J.S., Christian, S.L., Kubota, T., Halley, D.J., Meijers-Heijboer, H., Langlois, S., Graham, J.M. Jr., Beuten, J., Willems, P.J., Ledbetter, D.H. and Beaudet, A.L. (1999) The spectrum of mutations in UBE3A causing Angelman syndrome. *Hum Mol Genet* **8**, 129–135.

64 Nawaz, Z., Lonard, D.M., Smith, C.L., Lev-Lehman, E., Tsai, S.Y., Tsai, M.J. and O'Malley, B.W. (1999) The Angelman syndrome-associated protein, E6-AP, is a coactivator for the nuclear hormone receptor superfamily. *Mol Cell Biol* **19**, 1182–1189.

65 Smith, C.L., Devera, D.G., Lamb, D.J., Nawaz, Z., Jiang, Y.H., Beaudet, A.L. and O'Malley, B.W. (2002) Genetic ablation of the steroid receptor coactivator-ubiquitin ligase, E6-AP, results in tissue-selective steroid hormone resistance and defects in reproduction. *Mol Cell Biol* **22**, 525–535.

66 Kumar, S., Talis, A.L. and Howley, P.M. (1999) Identification of HHR23A as a substrate for E6-associated protein-mediated ubiquitination. *J Biol Chem* **274**, 18785–18792.

67 Oda, H., Kumar, S. and Howley, P.M. (1999) Regulation of the Src family tyrosine kinase Blk through E6AP-mediated ubiquitination. *Proc Natl Acad Sci USA* **96**, 9557–9562.

68 Thomas, M. and Banks, L. (1998) Inhibition of Bak-induced apoptosis by HPV-18 E6. *Oncogene* **17**, 2943–2954.

69 Kuhne, C. and Banks, L. (1998) E3-ubiquitin ligase/E6-AP links multicopy maintenance protein 7 to the ubiquitination pathway by a novel motif, the L2G box. *J Biol Chem* **273**, 34302–34309.

70 Jiang, Y.H., Armstrong, D., Albrecht, U., Atkins, C.M., Noebels, J.L., Eichele, G., Sweatt, J.D. and Beaudet, A.L. (1998) Mutation of the Angelman ubiquitin ligase in mice causes increased cytoplasmic p53 and deficits of contextual learning and long-term potentiation. *Neuron* **21**, 799–811.

71 Gu, J., Dubner, R., Fornace, A.J. Jr. and Iadarola, M.J. (1995) UREB1, a tyrosine phosphorylated nuclear protein, inhibits p53 transactivation. *Oncogene* **11**, 2175–2178.

72 Liu, Z., Oughtred, R. and Wing, S.S. (2005) Characterization of E3Histone, a novel testis ubiquitin protein ligase which ubiquitinates histones. *Mol Cell Biol* **25**, 2819–2831.

73 Zhong, Q., Gao, W., Du, F. and Wang, X. (2005) Mule/ARF-BP1, a BH3-only E3 ubiquitin ligase, catalyzes the polyubiquitination of Mcl-1 and regulates apoptosis. *Cell* **121**, 1085–1095.

74 Adhikary, S., Marinoni, F., Hock, A., Hulleman, E., Popov, N., Beier, R., Bernard, S., Quarto, M., Capra, M., Goettig, S., Kogel, U., Scheffner, M., Helin, K. and Eilers, M. (2005) The ubiquitin ligase HectH9 regulates transcriptional activation by Myc and is essential for tumor cell proliferation. *Cell* **123**, 409–421.

75 Schwarz, S.E., Rosa, J.L. and Scheffner, M. (1998) Characterization of human hect domain family members and their interaction with UbcH5 and UbcH7. *J Biol Chem* **273**, 12148–12154.

76 Chen, D., Brooks, C.L. and Gu, W. (2006) ARF-BP1 as a potential therapeutic target. *Br J Cancer* **94**, 1555–1558.

77 Rizo, J. and Südhof, T.C. (1998) C2-domains, structure and function of a universal Ca^{2+}-binding domain. *J Biol Chem* **273**, 15879–15882.

78 Shearwin-Whyatt, L., Dalton, H.E., Foot, N. and Kumar, S. (2006) Regulation of functional diversity within the Nedd4 family by accessory and adaptor proteins. *Bioessays* **28**, 617–628.

79 Ingham, R.J., Gish, G. and Pawson, T. (2004) The Nedd4 family of E3 ubiquitin ligases: functional diversity within a common modular architecture. *Oncogene* **23**, 1972–1984.

80 Rotin, D. (2000) Regulation of the epithelial sodium channel (ENaC) by accessory proteins. *Curr Opin Nephrol Hypertens* **9**, 529–534.

81 Flasza, M., Gorman, P., Roylance, R., Canfield, A.E. and Baron, M. (2002) Alternative Splicing Determines the Domain Structure of WWP1, a Nedd4 Family Protein. *Biochem. Biophys Res Commun* **290**, 431–437.

82 Chen, H., Ross, C.A., Wang, N., Huo, Y., Mackinnon, D.F., Potash, J.B., Simpson, S.G., Mcmahon, F.J., Depaulo Jr., J.R. and Mcinnis, M.G. (2001) NEDD4L on human chromosome 18q21 has multiple forms of transcripts and is a homologue of the mouse Nedd4-2 gene. *Eur J Hum Genet* **9**, 922–930.

83 Dunn, D.M., Ishigami, T., Pankow, J., Von Niederhausern, A., Alder, J., Hunt, S.C., Leppert, M.F., Lalouel, J.M. and Weiss, R.B. (2002) Common variant of human NEDD4L activates a cryptic splice site to form a frameshifted transcript. *J Hum Genet* **47**, 665–676.

84 Itani, O.A., Stokes, J.B. and Thomas, C.P. (2005) Nedd4-2 isoforms differentially associate with ENaC and regulate its activity. *Am J Physiol Renal Physiol* **289**, F334–F346.

85 Itani, O.A., Campbell, J.R., Herrero, J., Snyder, P.M. and Thomas, C.P. (2003) Alternative promoter usage and differential splicing leads to human Nedd4-2 isoforms with a C2 domain and with variable number of WW domains. *Am J Physiol* **285**, F916–F929.

86 Kumar, S., Tomooka, Y. and Noda, M. (1992) Identification of a set of genes with developmentally down-regulated expression in the mouse brain. *Biochem Biophys Res Commun* **185**, 1155–1161.

87 Kumar, S., Harvey, K.F., Kinoshita, M., Copeland, N.G., Noda, M. and Jenkins, N.A. (1997) cDNA cloning, expression analysis, and mapping of the mouse Nedd4 gene. *Genomics* **40**, 435–443.

88 Staub, O., Dho, S., Henry, P.C., Correa, J., Ishikawa, T., Mcglade, J. and Rotin, D. (1996) WW domains of Nedd4 bind to the proline-rich PY motifs in the epithelial Na^+ channel deleted in Liddle's syndrome. *EMBO J* **15**, 2371–2380.

89 Kamynina, E., Tauxe, C. and Staub, O. (2001) Differential characteristics of two human Nedd4 proteins with respect to epithelial Na^+ channel regulation. *Am J Physiol* **281**, F469–F477.

90 André, B. and Springael, J.-Y. (1994) WWP, a new amino acid motif present in single or multiple copies in various proteins including dystrophin and the SH3-binding Yes-associated protein YAP65. *Biochem Biophys Res Commun* **205**, 1201–1205.

91 Kamynina, E., Debonneville, C., Bens, M., Vandewalle, A. and Staub, O. (2001) A novel mouse Nedd4 protein suppresses the activity of the epithelial Na^+ channel. *FASEB J* **15**, 204–214.

92 Harvey, K.F., Dinudom, A., Cook, D.I. and Kumar, S. (2001) The Nedd4-like Protein KIAA0439 Is a Potential Regulator of the Epithelial Sodium Channel. *J Biol Chem* **276**, 8597–8601.

93 Liddle, G.W., Bledsoe, T. and Coppage, W.S. Jr. (1963) A familial renal disorder simulating primary aldosteronism but with negligible aldosterone secretion. *Trans Assoc Am Physicians* **76**, 199–213.

94 Botero-Velez, M., Curtis, J.J. and Warnock, D.G. (1994) Brief report: Liddles's syndrome revisited. *N Engl J Med* **330**, 178–181.

95 Shimkets, R.A., Warnock, D.G., Bositis, C.M., Nelson-Williams, C., Hansson, J. H., Schambelan, M., Gill, J.R., Ulick, S., Milora, R.V., Findling, J.W., et al. (1994) Liddle's syndrome: heritable human hypertension caused by mutations in the β subunit of the epithelial sodium channel. *Cell* **79**, 407–414.

96 Hansson, J.H., Nelson-Williams, C., Suzuki, H., Schild, L., Shimkets, R.A., Lu, Y., Canessa, C.M., Iwasaki, T., Rossier, B.C. and Lifton, R.P. (1995) Hypertension caused by a truncated epithelial sodium channel gamma subunit: Genetic heterogeneity of Liddle syndrome. *Nat Genet* **11**, 76–82.

97 Kanelis, V., Rotin, D. and Forman-Kay, J.D. (2001) Solution structure of a Nedd4 WW domain-ENaC peptide complex. *Nat Struct Biol* **8**, 1–6.

98 Debonneville, C., Flores, S.Y., Kamynina, E., Plant, P.J., Tauxe, C., Thomas, M.A., Munster, C., Chraibi, A., Pratt, J.H., Horisberger, J.D., et al. (2001) Phosphorylation of Nedd4-2 by Sgk1 regulates epithelial Na(+) channel cell surface expression. *EMBO J* **20**, 7052–7059.

99 Harvey, K.F., Dinudom, A., Komwatana, P., Jolliffe, C.N., Day, M.L., Parasivam, G., Cook, D.I. and Kumar, S. (1999) All three WW domains of murine Nedd4 are involved in the regulation of epithelial sodium channels by intracellular Na^+. *J Biol Chem* **274**, 12525–12530.

100 Snyder, P.M., Olson, D.R. and Thomas, B.C. (2002) Serum and glucocorticoid-regulated kinase modulates Nedd4-2-mediated inhibition of the epithelial Na^+ channel. *J Biol Chem* **277**, 5–8.

101 Asher, C., Chigaev, A. and Garty, H. (2001) Characterization of interactions between Nedd4 and beta and gammaENaC using Surface Plasmon Resonance. *Biochem Biophys Res Commun* **286**, 1228–1231.

102 Staub, O. and Rotin, D. (2006) Role of ubiquitylation in cellular membrane transport. *Physiol Rev* **86**, 669–707.

103 Zhou, R. and Snyder, P.M. (2005) Nedd4-2 phosphorylation induces serum and glucocorticoid-regulated kinase (SGK) ubiquitination and degradation. *J Biol Chem* **280**, 4518–4523.

104 Bhalla, V., Daidie, D., Li, H., Pao, A.C., Lagrange, L.P., Wang, J., Vandewalle, A., StockAnd, J.D., Staub, O. and Pearce, D. (2005) Serum- and glucocorticoid-regulated kinase 1 regulates ubiquitin ligase neural precursor cell-expressed, developmentally down-regulated protein 4-2 by inducing interaction with 14-3-3. *Mol Endocrinol* **19**, 3073–3084.

105 Ichimura, T., Yamamura, H., Sasamoto, K., Tominaga, Y., Taoka, M., Kakiuchi, K., Shinkawa, T., Takahashi, N., Shimada, S. and Isobe, T. (2005) 14-3-3 proteins modulate the expression of epithelial Na+ channels by phosphorylation-dependent interaction with Nedd4-2 ubiquitin ligase. *J Biol Chem* **280**, 13187–13194.

106 Malik, B., Schlanger, L., Al-Khalili, O., Bao, H.F., Yue, G., Price, S.R., Mitch, W.E. and Eaton, D.C. (2001) Enac degradation in A6 cells by the ubiquitin-proteosome proteolytic pathway. *J Biol Chem* **276**, 12903–12910.

107 Hansson, J.H., Schild, L., Lu, Y., Wilson, T.A., Gautschi, I., Shimkets, R.A., Nelson-Williams, C., Rossier, B.C. and Lifton, R.P. (1995) A de novo missense mutation of the β subunit of the epithelial sodium channel causes hypertension and Liddle syndrome, identifying a proline-rich segment critical for regulation of channel activity. *Proc Natl Acad Sci USA* **25**, 11495–11499.

108 Snyder, P.M., Steines, J.C. and Olson, D.R. (2004) Relative contribution of Nedd4 and Nedd4-2 to ENaC regulation in epithelia determined by RNA interference. *J Biol Chem* **279**, 5042–5046.

109 Loffing-Cueni, D., Flores, S.Y., Sauter, D., Daidie, D., Siegrist, N., Meneton, P., Staub, O. and Loffing, J. (2006) Dietary sodium intake regulates the ubiquitin-protein ligase nedd4-2 in the renal collecting system. *J Am Soc Nephrol* **17**, 1264–1274.

110 Loffing, J., Flores, S.Y. and Staub, O. (2006) Sgk kinases and their role in epithelial transport. *Annu Rev Physiol* **68**, 461–490.

111 Flores, S.Y., Loffing-Cueni, D., Kamynina, E., Daidie, D., Gerbex, C., Chabanel, S., Dudler, J., Loffing, J. and Staub, O. (2005) Aldosterone-induced serum and glucocorticoid-induced kinase 1 expression is accompanied by Nedd4-2 phosphorylation and Increased Na+ transport in cortical collecting duct cells. *J Am Soc Nephrol* **16**, 2279–2287.

112 Liang, X., Peters, K.W., Butterworth, M.B. and Frizzell, R.A. (2006) 14-3-3 isoforms are induced by aldosterone and participate in its regulation of epithelial sodium channels. *J Biol Chem* **281**, 16323–16332.

113 Fava, C., Von Wowern, F., Berglund, G., Carlson, J., Hedblad, B., Rosberg, L., Burri, P., Almgren, P. and Melander, O. (2006) 24-h ambulatory blood pressure is linked to chromosome 18q21-22 and genetic variation of NEDD4L associates with cross-sectional and longitudinal blood pressure in Swedes. *Kidney Int* **70**, 562–569.

114 Russo, C.J., Melista, E., Cui, J., Destefano, A.L., Bakris, G.L., Manolis, A. J., Gavras, H. and Baldwin, C.T. (2005) Association of NEDD4L ubiquitin ligase with essential hypertension. *Hypertension* **46**, 488–491.

115 Pankow, J.S., Dunn, D.M., Hunt, S.C., Leppert, M.F., Miller, M.B., Rao, D.C., Heiss, G., Oberman, A., Lalouel, J.M. and Weiss, R.B. (2005) Further evidence of a quantitative trait locus on chromosome 18 influencing postural change in systolic blood pressure: the Hypertension Genetic Epidemiology Network (HyperGEN) Study. *Am J Hypertens* **18**, 672–678.

116 Fouladkou, F., Alikhani-Koopaei, R., Vogt, B., Flores, S.Y., Malbert-Colas, L., Lecomte, M.C., Loffing, J., Frey, F.J., Frey, B.M. and Staub, O. (2004) A naturally occurring human Nedd4-2 variant displays impaired ENaC regulation in *Xenopus laevis* oocytes. *Am J Physiol Renal Physiol* **287**, F550–F561.

117 Demirov, D.G. and Freed, E.O. (2004) Retrovirus budding. *Virus Res* **106**, 87–102.

118 Medina, G., Zhang, Y., Tang, Y., Gottwein, E., Vana, M.L., Bouamr, F., Leis, J. and Carter, C.A. (2005) The functionally exchangeable L domains in RSV and HIV-1 Gag direct particle release through pathways linked by Tsg101. *Traffic* **6**, 880–894.

119 Blot, V., Perugi, F., Gay, B., Prevost, M.C., Briant, L., Tangy, F., Abriel, H., Staub, O., Dokhelar, M.C. and Pique, C. (2004) Nedd4.1-mediated ubiquitination and subsequent recruitment of Tsg101 ensure HTLV-1 Gag trafficking towards the multivesicular body pathway prior to virus budding. *J Cell Sci* **117**, 2357–2367.

120 Perry, W.L., Hustad, C.M., Swing, D.A., O'Sullivan, T.N., Jenkins, N.A. and Copeland, N.G. (1998) The itchy locus encodes a novel ubiquitin protein ligase that is disrupted in a18H mice. *Nat Genet* **18**, 143–146.

121 D'andrea A.D. and Serhan C.N. (1998) Relieving the Itch. *Nat Genet* **18**, 97–99.

122 Paul, W.E. and Seder, R.A. (1994) Lymphocyte responses and cytokines. *Cell* **76**, 241–251.

123 Mosmann, T.R. and Coffman, R.L. (1989) TH1 and TH2 cells: different patterns of lymphokine secretion lead to different functional properties. *Annu Rev Immunol* **7**, 145–173.

124 Fang, D., Elly, C., Gao, B., Fang, N., Altman, Y., Joazeiro, C., Hunter, T., Copeland, N., Jenkins, N. and Liu, Y.C. (2002) Dysregulation of T lymphocyte function in itchy mice: a role for Itch in TH2 differentiation. *Nat Immunol* **3**, 281–287.

125 Walker, L.S. and Abbas, A.K. (2002) The enemy within: keeping self-reactive T cells at bay in the periphery. *Nat Rev Immunol* **2**, 11–19.

126 Kamradt, T. and Mitchison, N.A. (2001) Tolerance and autoimmunity. *N Engl J Med* **344**, 655–664.

127 Mueller, D.L. (2004) E3 ubiquitin ligases as T cell anergy factors. *Nat Immunol* **5**, 883–890.

128 Macian, F., Garcia-Cozar, F., Im, S.H., Horton, H.F., Byrne, M.C. and Rao, A. (2002) Transcriptional mechanisms underlying lymphocyte tolerance. *Cell* **109**, 719–731.

129 Heissmeyer, V., Macian, F., Im, S.H., Varma, R., Feske, S., Venuprasad, K., Gu, H., Liu, Y.C., Dustin, M.L. and Rao, A. (2004) Calcineurin imposes T cell unresponsiveness through targeted proteolysis of signaling proteins. *Nat Immunol* **5**, 255–265.

130 Yang, C., Zhou, W., Jeon, M.S., Demydenko, D., Harada, Y., Zhou, H. and Liu, Y.C. (2006) Negative regulation of the E3 ubiquitin ligase Itch via Fyn-mediated tyrosine phosphorylation. *Mol Cell* **21**, 135–141.

131 Gao, M., Labuda, T., Xia, Y., Gallagher, E., Fang, D., Liu, Y.C. and Karin, M. (2004) Jun turnover is controlled through JNK-dependent phosphorylation of the E3 ligase Itch. *Science* **306**, 271–275.

132 Venuprasad, K., Elly, C., Gao, M., Salek-Ardakani, S., Harada, Y., Luo, J.L., Yang, C., Croft, M., Inoue, K., Karin, M., et al. (2006) Convergence of Itch-induced ubiquitination with MEKK1-JNK signaling in Th2 tolerance and airway inflammation. *J Clin Invest* **116**, 1117–1126.

133 Zhu, H., Kavsak, P., Abdollah, S., Wrana, J.L. and Thomsen, G.H. (1999) A SMAD ubiquitin ligase targets the BMP pathway and affects embryonic pattern formation. *Nature* **400**, 687–693.

134 Zhang, Y., Chang, C., Gehling, D.J., Hemmati-Brivanlou, A. and Derynck, R. (2001) Regulation of Smad degradation and activity by Smurf2, an E3 ubiquitin ligase. *Proc Natl Acad Sci USA* **98**, 974–979.

135 Lin, X., Liang, M. and Feng, X.H. (2000) Smurf2 is a ubiquitin E3 ligase mediating proteasome-dependent degradation of Smad2 in transforming growth factor-beta signaling. *J Biol Chem* **275**, 36818–36822.

136 Bonni, S., Wang, H.R., Causing, C.G., Kavsak, P., Stroschein, S.L., Luo, K. and Wrana, J.L. (2001) TGF-beta induces assembly of a Smad2-Smurf2 ubiquitin ligase complex that targets SnoN for degradation. *Nat Cell Biol* **3**, 587–595.

137 Lo, R.S. and Massague, J. (1999) Ubiquitin-dependent degradation of TGF-beta-activated smad2. *Nat Cell Biol* **1**, 472–478.

138 Fukuchi, M., Imamura, T., Chiba, T., Ebisawa, T., Kawabata, M., Tanaka, K. and Miyazono, K. (2001) Ligand-dependent degradation of Smad3 by a ubiquitin ligase complex of ROC1 and associated proteins. *Mol Biol Cell* **12**, 1431–1443.

139 Ogunjimi, A.A., Briant, D.J., Pece-Barbara, N., Le Roy, C., Di Guglielmo, G.M., Kavsak, P., Rasmussen, R.K., Seet, B.T., Sicheri, F. and Wrana, J.L. (2005) Regulation of Smurf2 ubiquitin ligase activity by anchoring the E2 to the HECT domain. *Mol Cell* **19**, 297–308.

140 Kavsak, P., Rasmussen, R.K., Causing, C.G., Bonni, S., Zhu, H., Thomsen, G. H. and Wrana, J.L. (2000) Smad7 binds to Smurf2 to form an E3 ubiquitin ligase that targets the TGF beta receptor for degradation. *Mol Cell* **6**, 1365–1375.

141 Di Guglielmo, G.M., Le Roy, C., Goodfellow, A.F. and Wrana, J.L. (2003) Distinct endocytic pathways regulate TGF-beta receptor signalling and turnover. *Nat Cell Biol* **5**, 410–421.

142 Ebisawa, T., Fukuchi, M., Murakami, G., Chiba, T., Tanaka, K., Imamura, T. and Miyazono, K. (2001) Smurf1 interacts with transforming growth factor-beta type I receptor through Smad7 and induces receptor degradation. *J Biol Chem* **276**, 12477–12480.

143 Tajima, Y., Goto, K., Yoshida, M., Shinomiya, K., Sekimoto, T., Yoneda, Y., Miyazono, K. and Imamura, T. (2003) Chromosomal region maintenance 1 (CRM1)-dependent nuclear export of Smad ubiquitin regulatory factor 1 (Smurf1) is essential for negative regulation of transforming growth factor-beta signaling by Smad7. *J Biol Chem* **278**, 10716–10721.

144 Suzuki, C., Murakami, G., Fukuchi, M., Shimanuki, T., Shikauchi, Y., Imamura, T. and Miyazono, K. (2002) Smurf1 regulates the inhibitory activity of Smad7 by targeting Smad7 to the plasma membrane. *J Biol Chem* **277**, 39919–39925.

145 Murakami, G., Watabe, T., Takaoka, K., Miyazono, K. and Imamura, T. (2003) Cooperative inhibition of bone morphogenetic protein signaling by Smurf1 and inhibitory Smads. *Mol Biol Cell* **14**, 2809–2817.

146 Izzi, L. and Attisano, L. (2004) Regulation of the TGFbeta signalling pathway by ubiquitin-mediated degradation. *Oncogene* **23**, 2071–2078.

147 De Caestecker, M.P., Piek, E. and Roberts, A.B. (2000) Role of transforming growth factor-beta signaling in cancer. *J Natl Cancer Inst* **92**, 1388–1402.

148 Derynck, R., Akhurst, R.J. and Balmain, A. (2001) TGF-beta signaling in tumor suppression and cancer progression. *Nat Genet* **29**, 117–129.

149 Moren, A., Itoh, S., Moustakas, A., Dijke, P. and Heldin, C.H. (2000) Functional consequences of tumorigenic missense mutations in the amino-terminal domain of Smad4. *Oncogene* **19**, 4396–4404.

150 Moren, A., Hellman, U., Inada, Y., Imamura, T., Heldin, C.H. and Moustakas, A. (2003) Differential ubiquitination defines the functional status of the tumor suppressor Smad4. *J Biol Chem* **278**, 33571–33582.

151 Xu, J. and Attisano, L. (2000) Mutations in the tumor suppressors Smad2 and Smad4 inactivate transforming growth factor beta signaling by targeting Smads to the ubiquitin-proteasome pathway. *Proc Natl Acad Sci USA* **97**, 4820–4825.

152 Maurice, D., Pierreux, C.E., Howell, M., Wilentz, R.E., Owen, M.J. and Hill, C.S. (2001) Loss of Smad4 function in pancreatic tumors: C-terminal truncation leads to decreased stability. *J Biol Chem* **276**, 43175–43181.

153 Eppert, K., Scherer, S.W., Ozcelik, H., Pirone, R., Hoodless, P., Kim, H., Tsui, L.C., Bapat, B., Gallinger, S., Andrulis, I.L., et al. (1996) MADR2 maps to 18q21 and encodes a TGFbeta-regulated MAD-related protein that is functionally mutated in colorectal carcinoma. *Cell* **86**, 543–552.

154 Dumont, E., Lallemand, F., Prunier, C., FerrAnd, N., Guillouzo, A., Clement, B., Atfi, A. and Theret, N. (2003) Evidence for a role of Smad3 and Smad2 in stabilization of the tumor-derived mutant Smad2.Q407R. *J Biol Chem* **278**, 24881–24887.

155 Fukuchi, M., Fukai, Y., Masuda, N., Miyazaki, T., Nakajima, M., Sohda, M., Manda, R., Tsukada, K., Kato, H. and Kuwano, H. (2002) High-level expression of the Smad ubiquitin ligase Smurf2 correlates with poor prognosis in patients with esophageal squamous cell carcinoma. *Cancer Res* **62**, 7162–7165.

156 Akhurst, R.J. and Derynck, R. (2001) TGF-beta signaling in cancer – a double-edged sword. *Trends Cell Biol* **11**, S44–S51.

157 Subramaniam, V., Li, H., Wong, M., Kitching, R., Attisano, L., Wrana, J., Zubovits, J., Burger, A.M. and Seth, A. (2003) The RING-H2 protein RNF11 is overexpressed in breast cancer and is a target of Smurf2 E3 ligase. *Br J Cancer* **89**, 1538–1544.

158 Karsenty, G. and Wagner, E.F. (2002) Reaching a genetic and molecular understanding of skeletal development. *Dev Cell* **2**, 389–406.

159 Ying, S.X., Hussain, Z.J. and Zhang, Y.E. (2003) Smurf1 facilitates myogenic differentiation and antagonizes the bone morphogenetic protein-2-induced osteoblast conversion by targeting Smad5 for degradation. *J Biol Chem* **278**, 39029–39036.

160 Zhao, M., Qiao, M., Harris, S.E., Oyajobi, B.O., Mundy, G.R. and Chen, D. (2004) Smurf1 inhibits osteoblast differentiation and bone formation *in vitro* and in vivo. *J Biol Chem* **279**, 12854–12859.

161 Zhao, M., Qiao, M., Oyajobi, B.O., Mundy, G.R. and Chen, D. (2003) E3 ubiquitin ligase Smurf1 mediates core-binding factor alpha1/Runx2 degradation and plays a specific role in osteoblast differentiation. *J Biol Chem* **278**, 27939–27944.

162 Shen, R., Chen, M., Wang, Y.J., Kaneki, H., Xing, L., O'Keefe R.J. and Chen, D. (2006) Smad6 interacts with Runx2 and mediates Smad ubiquitin regulatory factor 1-induced Runx2 degradation. *J Biol Chem* **281**, 3569–3576.

163 Yamashita, M., Ying, S.X., Zhang, G.M., Li, C., Cheng, S.Y., Deng, C.X. and Zhang, Y.E. (2005) Ubiquitin ligase Smurf1 controls osteoblast activity and bone homeostasis by targeting MEKK2 for degradation. *Cell* **121**, 101–113.

5
Ubiquitin-independent Mechanisms of Substrate Recognition and Degradation by the Proteasome

Martin A. Hoyt and Philip Coffino

5.1
Introduction

As its name implies, the marking of proteasome substrates by ubiquitin conjugation is a central feature of the ubiquitin–proteasome system (UPS), and a topic which has been the subject of much investigation. There are, however, examples of proteins whose degradation by the proteasome has been divorced (or at least partially separated) from the process of ubiquitination. Given the predominance of ubiquitin-mediated targeting in proteasome function, and the central role many of these targets play in both regulation and maintenance of cellular homeostasis, it is fair to question the significance of a few exceptional cases. While one may be tempted to ascribe special import to such cases, and such perceptions may hold true, here we advocate a more modest view: that by examining the mechanisms of ubiquitin-dependent and -independent targeting systems, and identifying their point of convergence on the proteasome, we may illuminate something of general interest about proteasome biology.

What constitutes evidence for ubiquitin-independent degradation via the proteasome? A preliminary and obvious criterion is the lack of detectable ubiquitin conjugates of a particular proteasome substrate. However, since rapidly degraded proteins often exist at extremely low steady-state levels in the cell, these conjugates can be difficult to detect, even for proteasome substrates whose degradation has been demonstrated to require ubiquitination. Ubiquitin conjugates commonly represent only a small fraction of the steady-state population, and proteasome inhibition is often required for such conjugates to accumulate to detectable levels. Furthermore, existing polyubiquitin chains can be rapidly removed from substrate proteins by the action of cellular deubiquitinating enzymes. Given the difficulties of detection, the lack of detectable ubiquitinated species, is by itself insufficient to establish the ubiquitin-independence of protein degradation.

In addition to the problem of detection of ubiquitin conjugates, it is worth noting that ubiquitination is not a sufficient criterion to establish that a particular protein

Protein Degradation, Vol. 4: The Ubiquitin-Proteasome System and Disease.
Edited by R. J. Mayer, A. Ciechanover, M. Rechsteiner

ISBN: 978-3-527-31436-2

is a substrate for proteasomal degradation. As an example, ubiquitination is required for the function of the Met4 transcription factor of the yeast *Saccharomyces cerevisiae*, but this modification does not necessarily lead to its degradation [1–4]. This is true in spite of the observation that the Met4 ubiquitin chain is extended through lysine 48 of ubiquitin [1], a canonical linkage involved in the formation of the ubiquitin proteasomal recognition signal [5]. Additionally, as we will discuss below, some proteasome substrates are degraded by both ubiquitin-dependent and -independent pathways. The coincidence of ubiquitination and proteasome-mediated degradation is therefore not necessarily a causal relationship, as others have noted [6].

A second experimental approach used to establish that degradation of a particular proteasome substrate is independent of its ubiquitination, involves the perturbation of the ubiquitination machinery itself. It is anticipated that lesions which prevent or greatly impair ubiquitination should have little effect on the turnover of substrates not dependent on this modification. Implicit in this assumption is that ubiquitination is directly involved in the degradation of the substrate protein, and that secondary effects of the global inhibition of ubiquitination will be minor. This approach is limited by the lack of drugs that target the ubiquitination process specifically. Instead genetic approaches are most often used. Temperature-sensitive mutants of the ubiquitin-activating enzyme (E1) are available in animal cells [7, 8], and in *S. cerevisiae* a hypomorphic allele of the E1 gene (*UBA1*) was generated by a mutagenic insertion of a minitransposon in its promoter region [9]. Mutants of any particular ubiquitin-conjugating enzyme (E2) or ubiquitin ligase (E3) are also available in yeast.

As an alternative to global perturbations of the ubiquitination machinery, the ubiquitination of a particular substrate can be prevented by the removal of the target lysine residues within that substrate via mutation. When utilizing this methodology, consideration must be given to the effects such potentially extensive mutation could have on the structure and function of the protein in question. In the case of p21 (see Section 5.2.2), Sheaff and co-workers [10] demonstrated that when all six lysines residues were mutated to arginines, the mutant protein could bind, inhibit, and be phosphorylated by cyclin–Cdk complexes, and thus its normal function and structure were not severely affected. In addition to the structural consequences of lysine mutation, attention must also be given to sites of potential ubiquitin modification other than lysine residues. A lysine-less mutant of the muscle-specific transcription factor MyoD is ubiquitinated and targeted for proteasomal degradation through its N-terminus when no other modification site is available [11]. Human $p14^{Arf}$, which lacks any lysine residues, and its mouse homolog ($p19^{Arf}$) that contains a single lysine, were also shown to be ubiquitinated at their N-termini [12]. Other naturally occurring lysine-less proteins also appear to be ubiquitinated at their N-termini [13]. Ubiquitination can occur even when internal lysines or an accessible N-terminus are unavailable. Cadwell and Coscoy [14] recently demonstrated that MIR1, an E3 ubquitin ligase encoded by the Kaposi's sarcoma-associated herpesvirus, can mediate the ubiquitination of major histocompatibility complex class I (MHC I) molecules at an available cysteine residue via a thiol-ester bond.

Lastly, the ubiquitin-independence of protein degradation can be demonstrated by reconstitution using purified components in an *in-vitro* assay system. While this approach is more rigorous in that the presence of ubiquitin can be completely excluded by using recombinant proteins produced in prokaryotes, it also requires that all the necessary components have been identified beforehand, insuring that each can be purified in the active form and that they can be combined effectively. *In-vitro* assays are also useful in distinguishing whether the 20S or 26S form of the proteasome is the relevant protease, since this distinction is difficult to ascertain *in vivo*.

5.2 Ubiquitin-independent Proteasome Substrates

In discussing ubiquitin-independent substrates, we will first describe various substrates and the evidence that their turnover occurs independently of ubiquitin modification, followed by a discussion of these potential alternate mechanisms of proteasome recognition, and what can be learned through using such substrates.

5.2.1 Ornithine Decarboxylase

Ornithine decarboxylase (ODC) is a homodimeric enzyme that catalyzes an initial and key regulatory step in the biosynthesis of polyamines. ODC is a remarkably labile enzyme. Its activity undergoes marked and rapid changes in response to various biological perturbations, most prominently brought about by signals that promote cell growth [15]. Although subject to regulation by transcriptional and translational means, most of the dynamic changes in ODC protein levels depend on changes in stability [16]. A clue to the source of ODC lability came from comparing the protein sequences of vertebrate ODCs to those of an African trypanosome [17, 18]. Although vertebrate and parasite ODCs exhibit similar specific activities and a sequence homology of approximately 70%, the vertebrate proteins uniformly include a C-terminal extension of 35–40 amino acids, here termed cODC, that is absent in the corresponding parasite enzyme. Transfection of mammalian cells with DNAs encoding mouse and trypanosomal ODCs revealed that the former enzyme turned over with a short half-live, but that the latter is stable. Revealingly, truncated mouse ODC lacking cODC proved to be stable [17]. Conversely, appending cODC to the trypanosomal enzyme made it labile [18]. When tested within a common cellular context, cultured mammalian cells, the presence of cODC proved to be the necessary and sufficient determinant of rapid turnover. Degradation was found to be mediated by the proteasome, as shown by both biochemical studies using purified components and by genetics [19, 20]. The conclusion that cODC mediates degradation by the proteasome proved true in broader structural and biological contexts. Appending cODC to a variety of stable proteins is sufficient to destabilize them [21, 22]. The cODC degradation signal is recognized in eukaryotic organisms ranging from man to fungi and plants [23, 24].

The accumulation of excess cellular polyamines leads to a rapid reduction in ODC activity mediated by its regulatory protein antizyme (AZ) [25]. AZ inhibits ODC activity by dissociating the ODC homodimer to form catalytically inactive ODC-AZ heterodimers, and by facilitating the degradation of ODC by the 26S proteasome through the increased exposure of its C-terminal degradation signal. Although AZ augments the degradation of ODC, it is worth emphasizing that the cODC signal is both autonomous and portable as it destabilizes proteins with no capacity to associate with AZ and in biochemical and cellular contexts devoid of AZ.

An alternate pathway for the ubiquitin-independent degradation of ODC by the 20S proteasome, which appears to be elicited under conditions of oxidative stress, has recently been described [26]. This process is inhibited by NAD(P)H quinone oxidoreductase 1 (NQO1), an enzyme involved in the reduction of various quinones that can use NADH or NADPH as electron donors, which was also shown to inhibit the ubiquitin-independent degradation of p53 (see Section 5.2.4). This NQO1-mediated inhibition is alleviated by dicoumarol, a specific inhibitor of NQO1. The dicoumarol-induced degradation of ODC is accelerated by AZ, but apparently does not require cODC. The relevant effect of AZ in this case appears to be its ability to dissociate ODC homodimers: the degradation of a double mutant form of ODC, unable to bind AZ or to form homodimers, was accelerated by dicoumarol treatment, while degradation of a homodimeric ODC that could not interact with AZ was unaffected. The degradation of the monomeric ODC mutant, *in vitro*, was carried out by 20S (but not 26S) proteasomes, and inhibited by the interaction of NQO1 with ODC.

5.2.2
p21$^{Waf1/Cip1}$

p21 is a member of the Cip/Kip family of cyclin-dependent kinase (Cdk) inhibitors, which block progression through the G_1/S transition of the cell cycle [27]. Accordingly, the amount of p21 is often elevated in cells that are undergoing exit from the cell cycle, such as during terminal differentiation. Cellular levels of p21 are regulated both by its rate of synthesis and degradation. Treatment of cultured mammalian cells with proteasome inhibitors leads to the accumulation of ubiquitinated forms of p21 and a concurrent stabilization of the protein, without changes in its rate of synthesis. However, mutation of all six lysine residues, which could serve as sites of ubiquitin conjugation, prevented the formation of ubiquitinated p21 species, but had no effect on the half-life of the mutant protein [10]. The N-terminus was ruled out as a possible alternate site of ubiquitin attachment due to acetylation of the terminal residue [10, 28]. Furthermore, turnover of p21 was unaffected in ts20TG^R and tsBN75 cell lines harboring temperature-sensitive alleles of the ubiquitin-activating enzyme E1, at both the permissive and restrictive temperatures [28].

In contrast proteins whose turnover was known to be dependent on ubiquitination, such as the cyclins D1 and E, were stabilized in the mutant cell line.

The ubiquitin-independent degradation of p21 appears to occur through a direct interaction of its C-terminus with the C8 ($\alpha7$) subunit of the 20S proteasome [29]. Deletion mutants of p21 lacking this C-terminal domain stabilize p21 *in vivo*, and prevent its degradation by purified 20S proteasomes *in vitro*. The E3 ubiquitin ligase MDM2 targets both itself and p53 for ubiquitin-dependent degradation via the UPS [30]. Recently, MDM2 was found to regulate p21 activity in cells by mediating its proteasomal turnover independently of p21 ubiquitination [31]. Ectopic expression of MDM2 in cultured mammalian cells resulted in the accelerated degradation of p21 even when MDM2 mutants lacking ubiquitin ligase activity, or the lysine-less version of p21 were used [31]. It has not been determined whether the MDM2-mediated degradation of p21 depends on the interaction of p21 with the C8 subunit of the proteasome, however.

5.2.3
Retinoblastoma Protein

The human retinoblastoma protein (Rb) is a nuclear protein that inhibits the expression of genes dependent on the E2F family of transcription factors [32]. Rb can be targeted for proteasomal degradation by the human papillomavirus E7 oncoprotein [33], and by the pp71 protein of human cytomegalovirus [34]. Several observations imply that the latter process is ubiquitin independent: the absence of high molecular weight ubiquitinated forms of Rb, even following treatment with proteasome inhibitors; the lack of an effect of the dominant negative ubiquitin K48R mutant on Rb turnover; and the capacity of pp71 to direct the degradation of Rb in a cell line bearing a temperature-sensitive E1 (ts20) [35].

Recently MDM2 overexpression was shown to lead to a reduction in the half-life of the Rb protein, by a mechanism strikingly similar to that of MDM2-mediated turnover of p21 [36]. The MDM2-mediated turnover of Rb was shown to be ubiquitin-independent *in vivo*, as judged by criteria similar to those used by Kalejta and Shenk [35]. The degradation of Rb *in vitro* was accomplished by the 20S proteasome, and Rb co-fractionated with the 20S proteasome in cell lysates during gel filtration chromatography. Both Rb and MDM2 were shown to interact with the C8 subunit of the 20S proteasome, and MDM2 was shown to facilitate the Rb–C8 interaction [36]. The results of this study and those regarding p21 suggest a model whereby MDM2 functions to tether substrates to the 20S proteasome to facilitate their degradation. This model is supported by the observation that MDM2 mutants that bind C8, but not Rb, fail to degrade Rb [36].

5.2.4
p53 and p73

The tumor suppressor p53 is a labile protein that is targeted for destruction by the action of the MDM2 ubiquitin ligase. Like ODC, the stability of p53, and the related p73 protein, can be negatively regulated by NQO1 and dicoumarol treatment reversed this effect [37]. NQO1 does not inhibit MDM2 ubiquitination-mediated degradation of p53 [38], rather it blocks a ubiquitin-independent degradation pathway for p53 [37]. The ubiquitin-independence was demonstrated using a ubiquitin-depleted reticulocyte lysate-derived degradation system following dicoumarol treatment, and in a cell line bearing a temperature-sensitive E1 mutant.

As with ODC, the 20S proteasome was shown to be responsible for the ubiquitin-independent degradation of p53 and p73 [39]. The p73 gene is translated into p73α and p73β isoforms by alternate splicing of its transcript. While the degradation of p53 and the p73α isoform was inhibited by NQO1, the degradation of p73β, which lacks the C-terminal SAM domain, present in p73α, was not inhibited by NQO1. Both p53 and p73α were able to bind NQO1, while p73β was not. Because NQO1 was shown to co-fractionate with 20S proteasomes in this study, the authors proposed a model whereby NQO1 prevents unrestrained proteolysis by the 20S proteasome [39].

5.2.5 Human Thymidylate Synthase

Thymidylate synthase (TS) is an enzyme responsible for the formation of dTMP by the reductive methylation of dUMP, and is an essential source of nucleotides used in DNA synthesis. Ligand binding to TS, following treatment with inhibitors, results in both a change from an "open" to a "closed" conformation, and stabilization of the enzyme [40]. When compared to prokaryote TS, the human enzyme contains a 29-amino acid N-terminal extension that is structurally disordered and is dispensable for normal catalytic activity. Truncations of this N-terminal extension were found to have varying effects on the stability of the enzyme depending on the extent of the deletion [41]. Deletion of amino acids 2–7 from the N-terminus resulted in almost complete stabilization of the thymidylate synthase protein, and further analysis identified Pro2 as a critical residue [42]. Degradation of TS was mediated by the 26S proteasome, but ubiquitin-conjugated forms of the enzyme were undetectable in cells [41]. Experiments using either a temperature-sensitive E1 allele, or a "lysine-less" mutant of the enzyme, supported the conclusion that ubiquitination was not required for TS degradation.

5.2.6 Rpn4

Rpn4 is a transcriptional activator of genes encoding the proteasome subunits of the yeast *Saccharomyces cerevisiae* [43, 44]. It is itself a target for rapid degradation by the proteasome [45], and thus forms a feedback loop in which decreased proteasome activity leads to Rpn4 accumulation and increased expression of proteasomal genes. Degradation of Rpn4 by the proteasome occurs by both ubiquitin-dependent and -independent mechanisms, and the degradation signals for both pathways map to the N-terminal region of Rpn4 [46]. Ubiquitination occurs on lysines in the N-terminal 229 amino acids of Rpn4, and is mediated by the Ubr2 ubiquitin ligase [47]. However, mutating all 11 lysines in an N-terminal fragment of Rpn4 only partially stabilized the truncated protein [46]. Similarly, the turnover of the full-length protein was largely unaffected in a yeast *uba1-2* mutant defective in the activity of the ubiquitin-activating enzyme, E1 [45], pointing to the existence of a ubiquitin-independent degradation pathway. The ubiquitin-independent degradation signal resides at or proximal to the Rpn4 N-terminus, since either extension by the addition of an epitope tag, or deletion of the first 10 amino acids of the native N-terminus leads to partial stabilization, and these modifications combined with lysine mutations in the truncated protein lead to complete stabilization [46].

5.2.7 NF-κB and IκBα

The vertebrate transcription factor NF-κB functions in a number of signaling pathways including those involved in immune and inflammatory responses. Ubiquitination plays a central role in the regulation of the NF-κB pathway in mammalian cells (reviewed in [48]). However, ubiquitin-independent proteasome-mediated proteolysis also plays a role in this pathway. The IκBα inhibitory protein binds to and sequesters the NF-κB dimer in the cytoplasm by masking its nuclear localization signal. In response to appropriate stimuli, IκBα is phosphorylated and subsequently targeted for proteasomal degradation by ubiquitination, allowing NF-κB to transit to the nucleus. Apart from the ubiquitin-dependent degradation of IκBα in stimulated cells, basal turnover of this protein occurs when it is in its monomeric form, not associated with p65. The rapid turnover of monomeric IκBα by the proteasome in unstimulated cells was shown to be ubiquitin-independent by the following criteria: (1) No high molecular weight ubiquitin conjugates of IκBα were detected in unstimulated cells. (2) Mutants lacking a C-terminal domain required for signal-dependent ubiquitination and degradation of IκBα retained their basal instability. (3) A mutant form of the IκBα protein in which all the lysine residues had been removed or mutated to arginines, also did not affect the basal instability of the protein. (4) The constitutive turnover of IκBα, like that following signal-dependent stimulation, was sensitive to proteasome inhibition [49].

NF-κB is synthesized as a p105 precursor protein which is processed to the mature p50 form. Limited proteasome-mediated proteolysis of the C-terminal domain of p105 appears to be the mechanism by which p105 is processed [50, 51]. Remarkably, a recent study [52] provides evidence that p105 processing can be carried out by the 20S proteasome in the absence of ubiquitination. In this study processing of NF-κB to the p50 form was accomplished in a purified *in-vitro* system using only the 20S core particle of the proteasome. The processing *in vivo* also appears to occur independently of prior ubiquitination. Processing was unaffected in a temperature-sensitive E1 mutant cell line, and following mutation of "critical" lysine residues around the region required for processing. It should be noted that in this study, not all lysine residues were mutated, thus the possible role of the unmutated lysines as sites of ubiquitination cannot be dismissed. This model of p105 degradation is supported by the results of an earlier study which showed that the viral Tax protein facilitates interactions between NF-κB and an α and β subunit of the 20S proteasome [53].

The conclusion that the processing of the p105 form of NF-κB is ubiquitin independent must be reconciled with earlier findings which strongly implicated ubiquitination in this process [50]. Using cell-free systems to study p105 processing, it was shown that the process was ATP dependent; that p105 could be polyubiquitinated in the *in-vitro* system; that methylated ubiquitin (lacking free amine groups) inhibited both p105 polyubiquitination and processing; and that the activity of a specific ubiquitin-conjugating enzyme (E2) and a ubiquitin ligase (E3) was required.

5.2.8
Steroid Receptor Co-activator-3

The steroid co-activator receptor-3 (SRC-3 or AIB1) is a transcriptional co-activator that is encoded by an oncogene which is frequently amplified in breast cancer cell lines [54]. SRC-3 interacts with steroid hormone receptors and plays a role in proliferation, specifically in response to estrogen [55]. SRC-3 is degraded, in a ubiquitin-independent manner, by the REGγ (PA28γ) regulatory complex of the 20S proteasome [56]. REGγ is a member of a family of related 11S proteasome activators. In contrast to the related REGα and REGβ proteins that form a heteroheptameric regulatory complex, REGγ forms a homoheptameric complex and confers different peptide substrate specificities than the REGαβ complex [57]. Li and coworkers [56] found that SRC-3 could be immunoprecipitated in complexes containing REGγ and that SRC-3 interacted with REGγ through its histone acetyltransferase (HAT) domain (residues 1081–1417). Reducing REGγ expression in cells by means of RNA interference reduced SRC-3 turnover, and REGγ overexpression enhanced SRC-3 degradation. Degradation required the interaction of REGγ with SRC-3, since deletion of the SRC-3 HAT domain prevented SRC-3 degradation. Furthermore, a REGγ mutant incapable of activating the 20S proteasome also prevented SRC-3 degradation. The degradation of SRC-3, but not the related SRC-1 protein, by REGγ and the 20S proteasome could be reconstituted *in vitro* using purified components in the absence of either ATP or prior ubiquitination.

5.2.9
c-Jun

c-Jun is a member of the AP-1 family of transcription factors that must homodimerize or heterodimerize with other factors in order to recognize specific DNA-binding sites. c-Jun is a short-lived protein that is ubiquitinated *in vivo* [58]. Jariel-Encontre and co-workers [59] demonstrated that c-Jun could be degraded *in vitro* using fractionated rat liver lysates depleted of ubiquitin or the ubiquitin-activating enzyme E1. These extracts were shown to be incapable of modifying their endogenous proteins with radiolabeled ubiquitin, or of directing the degradation of recombinant p53, which depends on ubiquitination. This degradation of c-Jun in the absence of ubiquitin was mediated by the proteasome, since it was both ATP dependent and inhibited by the immunodepletion of proteasomes. Furthermore, c-Jun degradation could be accomplished using purified 26S proteasomes alone, but not by the 20S form, indicating that no other factors in the lysate were required for ubiquitin-independent degradation.

5.3 Mechanisms of Ubiquitin-independent Degradation

One question is central to the discussion of both ubiquitin-dependent and -independent pathways of proteasome-mediated proteolysis: what are the minimal structural elements a protein must have to be recognized and processed by the proteasome? For the majority of substrates, the simple conventional answer is that conjugation to a polyubiquitin chain containing Lys48 linkages is sufficient. However, as we described above, a wide variety of proteasome substrates can potentially be degraded in the absence of such ubiquitin modification. This leads to the fundamental question regarding ubiquitin-independent proteolysis: are ubiquitin-independent substrates inherently "ubiquitin-like" (that is recognized by same receptor), or is their mechanism of recognition by the proteasome completely divorced from the ubiquitin pathway? Given the evidence at hand, it seems that either alternative is possible, depending on the substrate in question.

The ubiquitin-independent substrates we have described can be roughly divided into two categories: those recognized and degraded by the 20S core particle alone, and those that require the intact 26S holoenzyme. Some fall into both categories. The requirements for degradation of a substrate by the 20S proteasome are likely to differ greatly from those processed by the 26S proteasome.

For those substrates that depend on the proteasome holoenzyme (or more specifically the 19S regulatory particle), one simple explanation of their lability is that they contain ubiquitin-like domains that mediate their interaction with the proteasome. Recent biochemical evidence suggests the cODC domain, in spite of lacking any obvious homology to ubiquitin, is such a domain [60]. The capacity of cODC to direct degradation depends on its mimicry of a polyubiquitin chain. Using purified rat 26S proteasomes, protein substrates carrying cODC and those with a ubiquitin oligomer attached were found to act as mutual competitive inhibitors, these experiments demonstrated that cODC and the ubiquitin conjugate co-occupy a common proteasome element required for substrate recognition. Alternately, the ubiquitin and cODC receptors might occur at distinct, but proximal, sites within the proteasome such that occupancy at one site prevents access to the other. Whether polyubiquitin chains compete with other ubiquitin-independent substrates, such as c-Jun, for the 26S proteasome has not yet been determined.

What sequences in ubiquitin-independent proteasome substrates act as signals for degradation by the 20S proteasome? In the case of ODC, the cODC signal required for degradation by the 26S proteasome is wholly dispensable for degradation by the 20S form under conditions of oxidative stress [26]. The observation that monomerization of ODC is required for degradation under these conditions suggests that the relevant alternative signal resides at, or proximal to, the dimer interface.

Both p21 and Rb interact with the C8/α7 subunit of the 20S proteasome and these interactions appear to be facilitated by the MDM2 ubiquitin ligase [29, 36]. The 31 C-terminal residues (134–164) of p21 are required for its interaction with C8, and this sequence was sufficient to confer this binding ability on p27, which normally does not interact with C8 when attached to its C-terminus [29]. The residues required for C8 interaction mapped to the C-terminal RbC pocket of Rb (residues 772–928) and this same region is required for binding MDM2 [61]. MDM2 itself appears to interact with C8 through its RING-finger domain, since a point mutation in this domain (C464A) greatly impairs MDM2–C8 interaction [36]. Whether, the NQO1-inhibited degradation of ODC and p53 by the 20S proteasome also involves such interactions with the C8 subunit remains to be determined.

How do substrates of the 20S proteasome gain access to catalytic sites sequestered within the core particle? The axial channel of the 20S core particle is normally occluded by interactions of the N-termini of its α subunits when the 20S core is not associated with regulatory complexes [62]. This gating impairs the degradation of both peptide and protein substrates by the 20S proteasome. An associated substrate must therefore be able to disrupt these interactions in order to gain access to the axial channel and the internal active sites. In the proteasome holoenzyme, association of the 19S or 11S/PA28 regulatory complexes reorganizes the N-terminal residues of the α subunits and leads to an "open-gate" conformation [62, 63]. It was found that purified 20S proteasomes progressively lose the auto-inhibition of peptidase activity by the N-termini of the α subunits unless they are maintained in the presence of potassium ions [64]. It therefore seems possible that the degradation of some substrates by the 20S proteasome might be non-specific and due to the spontaneous activation of the α subunits if these proteasome are not purified and stored in buffers containing potassium ions.

In addition to activating the 20S core particle to an open-gate conformation, substrate degradation also requires the unfolding of substrates to accommodate their insertion through the narrow (~13 Å) axial channel. For substrates degraded by the 26S proteasome, this unfolding is thought to be accomplished by the ATPase subunits of the 19S regulatory particle. For 20S substrates two possibilities seem obvious: first, a substrate may be unstructured, thus obviating the need for unfolding prior to internalization. A protein may either be inherently unstructured, such as p21, which lacks an ordered conformation unless bound by Cdk2 [65], or alternatively, a substrate may be rendered unstructured by modification prior to proteolysis [66, 67]. Second, unfolding a structured 20S substrate could be accomplished by associated proteins not intrinsic to the 19S regulatory particle. For example, the degradation of oxidized calmodulin requiring the presence of the chaperone Hsp90, is stimulated by ATP, and is sensitive to Hsp90 inhibitors [68]. The proteolysis of peptide substrates was unaffected by Hsp90 in these experiments, indicating that the chaperone is not acting to alter the gating conformation of the 20S proteasome, but more likely assisting in proteasome association and unfolding prior to entry. In contrast, the processing of the NF-κB p105 precursor by the 20S proteasome occurs in the absence of associated chaperones [52]. In this case, the absence of other proteasome-associated factors was established rigorously by mass spectrometric analysis. Apparently p105 is sufficient to enable processing by highly purified 20S complexes. This is likely a property of the C-terminal half of the precursor, since this region in isolation is degraded by purified 20S proteasomes following endoproteolytic cleavage from NF-κB p50 [52].

Is proteasome association itself sufficient to render a protein a substrate for proteolysis? There is an entire class of proteasome-associated "adaptor" proteins, such as the Rad23 and Dsk2 proteins of *S. cerevisiae*, which interact with the proteasome through ubiquitin-like domains but are not themselves subjected to degradation [69, 70]. This suggests that proteasome interaction by itself is insufficient to initiate destruction (see below). In the case of ubiquitinated substrates, native substrates cannot be used to resolve this question without further experimental manipulation. When subjected to such tests, a Lys48-linked polyubiquitin chain *per se* has proven insufficient (for examples, see [71, 72]). These data imply that the position of substrate conjugation matters, as may local sequence contiguous to the site of conjugation. Is that because of an effect of localization on the capacity of the ubiquitin chain to act as an association element with the proteasome? More specifically, is proteasome association but one attribute conferred by polyubiquitin, or is it sufficient on its own?

In principle, the question can be answered by providing an alternative method for delivering a protein to the proteasome, one that bypasses a need for ubiquitin conjugates. If any stable protein can be delivered to proteasomes without using ubiquitin and thereby undergo degradation, proteasome association must be the only requirement for degradation. This strategy was tested by conditionally tethering the non-substrate protein His3 to a non-essential intrinsic proteasome protein, Rpn10 [73]. Localizing His3 to the proteasome in this way converted His3 into a substrate. This result implies that localization promotes degradation, but it may nonetheless be insufficient: other components of the delivery system used in these experiments (or His3 itself) may provide further signals or structures required for degradation. Tethering provides a very high local concentration, which, in the example cited, was sufficient, but generality of this conclusion requires additional experimental testing. On the face of it, this cannot be a full description of substrate specification, or else the proteasome would be continually digesting bits of itself. And, as mentioned above, there exists numerous examples of proteasome-associated proteins that are not subjected to proteolysis [69, 70]. This conclusion is further supported by the recent work of Matouschek and colleagues that suggests that proteasome association cannot be the whole story [74]. They provided evidence instead for a two-element model, whereby an association tag, usually polyubiquitin, collaborates with an unstructured region that is required as a proteasome entry site. The question of whether localization is sufficient remains unresolved.

The degradation tag of ODC, cODC, provides a favorable test bed to answer these questions. A cysteine thiol of cODC, at position 441 of native mouse ODC, proved critical in determining the molecular basis for cODC action[75]. Removing that thiol or replacing it with a hydroxyl abolished both degradation and the ability of cODC to act as a competitive inhibitor of native ODC degradation. C441 is therefore essential for localization to the proteasome of proteins containing cODC. To assess whether association is sufficient, we performed experiments using constructs containing GFP, a compact single-domain globular protein lacking prominent unstructured protrusions, to dissect cODC function [76]. A fusion of GFP to cODC (GFP-cODC) is degraded by proteasomes, but the identical molecule with C441 thiol deleted (GFP-cODC^{C441A}) is not degraded, nor is it recognized by proteasomes. Fusing the non-essential proteasome protein Rpn10 to GFP results in the association of this fusion protein (Rpn10–GFP) with the proteasome. Despite its association with proteasomes, Rpn10–GFP is stable and not rapidly degraded. Adding cODC to Rpn10–GFP, to make Rpn10–GFP–cODC, produces a protein that is degraded, as expected; this protein has two potential interaction elements for the proteasome, Rpn10 and cODC. If the thiol of C441 within cODC is removed by mutation to make Rpn10–GFP–cODC^{C441A}, degradation persists. Because cODC^{C441A} cannot mediate docking, it must provide some other essential function, a function that accounts for the different properties of Rpn10–GFP (stable) versus Rpn10–GFP–cODC^{C441A} (unstable). The cODC^{C441A} element provides no specific structural information necessary for degradation in the context of Rpn10–GFP–cODC^{C441A}; a variety of alternate carboxyl termini inserted in place of cODC^{C441A} can support degradation. We infer that interaction alone does not make a substrate; association must be supplemented by the presence of an unstructured region. This finding is consistent with the two-element model of Matouschek [74].

5.4 Conclusion

An increasing interest in the role of proteolysis in the regulation of gene expression has led not only to a better understanding of the ubiquitin–proteasome system, but also to an increasing awareness of those substrates that are (in whole or in part) not completely dependent on the canonical degradation pathway. Our understanding of substrate–proteasome interactions has been broadened and deepened by extending investigations beyond ubiquitin conjugation and the 26S proteasome. A consideration of the variety of mechanisms utilized by the proteasome not only provides an appreciation of the complexities of the proteolytic machine, but potentially provides novel tools with which to dissect those mechanisms.

References

1 Flick, K., Ouni, I., Wohlschlegel, J.A., Capati, C., McDonald, W.H., Yates, J.R. and Kaiser, P. (2004) *Nat Cell Biol.* **6**, 634–641.

2 Kaiser, P., Flick, K., Wittenberg, C. and Reed, S.I. (2000) *Cell* **102**, 303–314.

3 Menant, A., Baudouin-Cornu, P., Peyraud, C., Tyers, M. and Thomas, D. (2006) *J Biol Chem.* **281**, 11744–11754.

4 Rouillon, A., Barbey, R., Patton, E.E., Tyers, M. and Thomas, D. (2000) *EMBO J.* **19**, 282–294.

5 Johnson, E.S., Ma, P.C., Ota, I.M. and Varshavsky, A. (1995) *J Biol Chem.* **270**, 17442–17456.

6 Verma, R. and Deshaies, R.J. (2000) *Cell* **101**, 341–344.

7 Finley, D., Ciechanover, A. and Varshavsky, A. (1984) *Cell* **37**, 43–55.

8 Kulka, R.G., Raboy, B., Schuster, R., Parag, H.A., Diamond, G., Ciechanover, A. and Marcus, M. (1988) *J Biol Chem.* **263**, 15726–15731.

9 Swanson, R. and Hochstrasser, M. (2000) *FEBS Lett.* **477**, 193–198.

10 Sheaff, R.J., Singer, J.D., Swanger, J., Smitherman, M., Roberts, J.M. and Clurman, B.E. (2000) *Mol Cell.* **5**, 403–410.

11 Breitschopf, K., Bengal, E., Ziv, T., Admon, A. and Ciechanover, A. (1998) *EMBO J.* **17**, 5964–5973.

12 Kuo, M.L., den Besten, W., Bertwistle, D., Roussel, M.F. and Sherr, C.J. (2004) *Genes Dev.* **18**, 1862–1874.

13 Ben-Saadon, R., Fajerman, I., Ziv, T., Hellman, U., Schwartz, A.L. and Ciechanover, A. (2004) *J Biol Chem.* **279**, 41414–41421.

14 Cadwell, K. and Coscoy, L. (2005) *Science* **309**, 127–130.

15 Pegg, A.E. (2006) *J Biol Chem.* **281**, 14529–14532.

16 van Daalen Wetters, T., Macrae, M., Brabant, M., Sittler, A. and Coffino, P. (1989) *Mol Cell Biol.* **9**, 5484–5490.

17 Ghoda, L., van Daalen Wetters, T., Macrae, M., Ascherman, D. and Coffino, P. (1989) *Science* **243**, 1493–1495.

18 Ghoda, L., Phillips, M.A., Bass, K.E., Wang, C.C. and Coffino, P. (1990) *J Biol Chem.* **265**, 11823–11826.

19 Murakami, Y., Matsufuji, S., Kameji, T., Hayashi, S., Igarashi, K., Tamura, T., Tanaka, K. and Ichihara, A. (1992) *Nature* **360**, 597–599.

20 Murakami, Y., Matsufuji, S., Tanaka, K., Ichihara, A. and Hayashi, S. (1993) *Biochem J.* **295**, 305–308.

21 Li, X., Zhao, X., Fang, Y., Jiang, X., Duong, T., Fan, C., Huang, C.C. and Kain, S.R. (1998) *J Biol Chem.* **273**, 34970–34975.

22 Loetscher, P., Pratt, G. and Rechsteiner, M. (1991) *J Biol Chem.* **266**, 11213–11220.

23 DeScenzo, R.A. and Minocha, S.C. (1993) *Plant Mol Biol.* **22**, 113–127.

24 Hoyt, M.A., Zhang, M. and Coffino, P. (2003) *J Biol Chem.* **278**, 12135–12143.

25 Coffino, P. (2001) *Nat Rev Mol Cell Biol.* **2**, 188–194.

26 Asher, G., Bercovich, Z., Tsvetkov, P., Shaul, Y. and Kahana, C. (2005) *Mol Cell.* **17**, 645–655.

27 Sherr, C.J. and Roberts, J.M. (1995) *Genes Dev.* **9**, 1149–1163.

28 Chen, X., Chi, Y., Bloecher, A., Aebersold, R., Clurman, B.E. and Roberts, J.M. (2004) *Mol Cell.* **16**, 839–847.

29 Touitou, R., Richardson, J., Bose, S., Nakanishi, M., Rivett, J. and Allday, M.J. (2001) *EMBO J.* **20**, 2367–2375.

30 Haupt, Y., Maya, R., Kazaz, A. and Oren, M. (1997) *Nature* **387**, 296–299.

31 Jin, Y., Lee, H., Zeng, S.X., Dai, M.S. and Lu, H. (2003) *EMBO J.* **22**, 6365–6377.

32 Stevaux, O. and Dyson, N.J. (2002) *Curr Opin Cell Biol.* **14**, 684–691.

33 Boyer, S.N., Wazer, D.E. and Band, V. (1996) *Cancer Res.* **56**, 4620–4624.

34 Kalejta, R.F., Bechtel, J.T. and Shenk, T. (2003) *Mol Cell Biol.* **23**, 1885–1895.

35 Kalejta, R.F. and Shenk, T. (2003) *Proc Natl Acad Sci USA* **100**, 3263–3268.

36 Sdek, P., Ying, H., Chang, D.L., Qiu, W., Zheng, H., Touitou, R., Allday, M.J. and Xiao, Z.X. (2005) *Mol Cell.* **20**, 699–708.

37 Asher, G., Lotem, J., Kama, R., Sachs, L. and Shaul, Y. (2002) *Proc Natl Acad Sci USA* **99**, 3099–3104.

38 Asher, G., Lotem, J., Sachs, L., Kahana, C. and Shaul, Y. (2002) *Proc Natl Acad Sci USA* **99**, 13125–13130.

39 Asher, G., Tsvetkov, P., Kahana, C. and Shaul, Y. (2005) *Genes Dev.* **19**, 316–321.

40 Kitchens, M.E., Forsthoefel, A.M., Rafique, Z., Spencer, H.T. and Berger, F.G. (1999) *J Biol Chem.* **274**, 12544–12547.

41 Forsthoefel, A.M., Pena, M.M., Xing, Y.Y., Rafique, Z. and Berger, F.G. (2004) *Biochemistry* **43**, 1972–1979.

42 Pena, M.M., Xing, Y.Y., Koli, S. and Berger, F.G. (2006) *Biochem J.* **394**, 355–363.

43 Leggett, D.S., Hanna, J., Borodovsky, A., Crosas, B., Schmidt, M., Baker, R.T., Walz, T., Ploegh, H. and Finley, D. (2002) *Mol Cell.* **10**, 495–507.

44 Mannhaupt, G., Schnall, R., Karpov, V., Vetter, I. and Feldmann, H. (1999) *FEBS Lett.* **450**, 27–34.

45 Xie, Y. and Varshavsky, A. (2001) *Proc Natl Acad Sci USA* **98**, 3056–3061.

46 Ju, D. and Xie, Y. (2004) *J Biol Chem.* **279**, 23851–23854.

47 Wang, L., Mao, X., Ju, D. and Xie, Y. (2004) *J Biol Chem.* **279**, 55218–55223.

48 Krappmann, D. and Scheidereit, C. (2005) *EMBO Rep.* **6**, 321–326.

49 Krappmann, D., Wulczyn, F.G. and Scheidereit, C. (1996) *EMBO J.* **15**, 6716–6726.

50 Orian, A., Whiteside, S., Israel, A., Stancovski, I., Schwartz, A.L. and Ciechanover, A. (1995) *J Biol Chem.* **270**, 21707–21714.

51 Palombella, V.J., Rando, O.J., Goldberg, A.L. and Maniatis, T. (1994) *Cell* **78**, 773–785.

52 Moorthy, A.K., Savinova, O.V., Ho, J.Q., Wang, V.Y., Vu, D. and Ghosh, G. (2006) *EMBO J.* **25**, 1945–1956.

53 Rousset, R., Desbois, C., Bantignies, F. and Jalinot, P. (1996) *Nature* **381**, 328–331.

54 Anzick, S.L., Kononen, J., Walker, R.L., Azorsa, D.O., Tanner, M.M., Guan, X.Y., Sauter, G., Kallioniemi, O.P., Trent, J.M. and Meltzer, P.S. (1997) *Science* **277**, 965–968.

55 McKenna, N.J. and O'Malley, B.W. (2002) *Cell* **108**, 465–474.

56 Li, X., Lonard, D.M., Jung, S.Y., Malovannaya, A., Feng, Q., Qin, J., Tsai, S.Y., Tsai, M.J. and O'Malley, B.W. (2006) *Cell* **124**, 381–392.

57 Realini, C., Jensen, C.C., Zhang, Z., Johnston, S.C., Knowlton, J.R., Hill, C.P. and Rechsteiner, M. (1997) *J Biol Chem.* **272**, 25483–25492.

58 Treier, M., Staszewski, L.M. and Bohmann, D. (1994) *Cell* **78**, 787–798.

59 Jariel-Encontre, I., Pariat, M., Martin, F., Carillo, S., Salvat, C. and Piechaczyk, M. (1995) *J Biol Chem.* **270**, 11623–11627.

60 Zhang, M., Pickart, C.M. and Coffino, P. (2003) *EMBO J.* **22**, 1488–1496.

61 Sdek, P., Ying, H., Zheng, H., Margulis, A., Tang, X., Tian, K. and Xiao, Z.X. (2004) *J Biol Chem.* **279**, 53317–53322.

62 Groll, M., Bajorek, M., Kohler, A., Moroder, L., Rubin, D.M., Huber, R., Glickman, M.H. and Finley, D. (2000) *Nat Struct Biol.* **7**, 1062–1067.

63 Forster, A., Masters, E.I., Whitby, F.G., Robinson, H. and Hill, C.P. (2005) *Mol Cell.* **18**, 589–599.

64 Kohler, A., Cascio, P., Leggett, D.S., Woo, K.M., Goldberg, A.L. and Finley, D. (2001) *Mol Cell.* **7**, 1143–1152.

65 Kriwacki, R.W., Hengst, L., Tennant, L., Reed, S.I. and Wright, P.E. (1996) *Proc Natl Acad Sci USA* **93**, 11504–11509.

66 Katznelson, R. and Kulka, R.G. (1985) *Eur J Biochem.* **146**, 437–442.

67 Michalek, M.T., Grant, E.P. and Rock, K.L. (1996) *J Immunol.* **157**, 617–624.

68 Whittier, J.E., Xiong, Y., Rechsteiner, M.C. and Squier, T.C. (2004) *J Biol Chem.* **279**, 46135–46142.

69 Elsasser, S. and Finley, D. (2005) *Nat Cell Biol.* **7**, 742–749.

70 Schmidt, M., Hanna, J., Elsasser, S. and Finley, D. (2005) *Biol Chem.* **386**, 725–737.

71 Lee, J.N., Gong, Y., Zhang, X. and Ye, J. (2006) *Proc Natl Acad Sci USA* **103**, 4958–4963.

72 Petroski, M.D. and Deshaies, R.J. (2003) *Mol Cell.* **11**, 1435–1444.

73 Janse, D.M., Crosas, B., Finley, D. and Church, G.M. (2004) *J Biol Chem.* **279**, 21415–21420.

74 Prakash, S., Tian, L., Ratliff, K.S., Lehotzky, R.E. and Matouschek, A. (2004) *Nat Struct Mol Biol.* **11**, 830–837.

75 Miyazaki, Y., Matsufuji, S., Murakami, Y. and Hayashi, S. (1993) *Eur J Biochem* **214**, 837–844.

76 Takeuchi, J., Chen, H. and Coffino, P. (2007) *EMBO J.* **26**, 123–131.

6
Endoplasmic Reticulum Protein Quality Control and Degradation

Antje Schäfer, Zlatka Kostova and Dieter H. Wolf

6.1
Introduction

Life is full of risks. This holds true even for the most important molecules of a cell, i.e. its proteins. Many steps that can go wrong lie ahead of a protein, starting from its birth or synthesis, all through its development into a mature, biologically active entity, [1, 2]. Problems include premature inhibition of synthesis, improper folding, incorrect maturation and, in the case of oligomeric proteins, lack of interacting partners resulting in orphan proteins. The cell has to make sure that these abnormal or orphan proteins are rapidly eliminated. In mammalian cells, failure to do so may lead to severe protein folding diseases such as Parkinson's disease, Alzheimer's disease, Huntington's disease, Creutzfeldt Jacob disease, bovine spongiform encephalophathy (BSE, cattle) and many others [3–8]. In eukaryotic cells, aside from a small portion of protein synthesis within the mitochondria, the majority of proteins are synthesized on cytoplasmic ribosomes. Secretory proteins, proteins of the cell membrane, the lysosome (vacuole), the Golgi apparatus and the endoplasmic reticulum (ER) have to be imported from the cytoplasm into the ER in an unfolded state, folded within this organelle and subsequently delivered to predetermined cellular sites in their biologically-active forms [9, 10]. About one-quarter of the proteome traverses the secretory pathway [11]. The ER provides the cell with an optimized environment that deals with the heavy load of folding work with its high concentration of chaperones. Nevertheless, folding of secretory proteins often fails. The ER fights this problem with two distinct, but interconnected mechanisms. The first is an ER-dedicated stress response that remodels the ER in such a way that its folding capacity increases and this is known as the unfolded protein response (UPR) [12, 13]. The second is a strict ER protein quality control system (ERQC) followed by the elimination of improperly folded proteins and protein complexes unable to assemble into higher order structures (ERAD) [14–22]. Both events are components of what has been called the ER quality control and degradation (ERQD) process [23]. UPR and ERQD are tightly interconnected: UPR induction increases the capacity of ERQD and failure of ERQD leads to UPR induction. ERQD requires an initial

Protein Degradation, Vol. 4: The Ubiquitin-Proteasome System and Disease.
Edited by R. J. Mayer, A. Ciechanover, M. Rechsteiner

ISBN: 978-3-527-31436-2

recognition step, which retains and/or retrieves improperly folded proteins in the ER and decides whether "misfolding" has occurred. In a second step, the misfolded protein is handed over to an elimination system. Due to the prevailing dogma of the time that secretory proteins, once imported into the ER, were trapped in the secretory pathway, unable to return into the cytoplasm [24], the idea of an ER-localized proteolytic system for degradation of the misfolded proteins was proposed, but was difficult to conceive [25]. Shortly thereafter it was shown that misfolded proteins of the ER membrane can become targets of the cytosolic ubiquitin–proteasome system [26–29]. The delivery mechanism of misfolded proteins of the ER to the ubiquitin–proteasome system, however, remained an enigma. In a study on yeast, using a misfolded mutant of the vacuolar enzyme Carboxypeptidase yscY (CPY*) Hiller et al. showed that after translocation into the ER, the mutant protein was fully glycosylated, then retro-translocated out of the ER into the cytoplasm where it was ubiquitinated and degraded by the proteasome [30, 31]. At nearly the same time, a virus-induced "dislocation" and proteasomal degradation of the major histocompatibility complex (MHC) class I molecule from the ER was reported [32]. In addition, a report concerning the retro-translocation of a mutated yeast pheromone peptide, α-factor, from the ER to the cytoplasm and its degradation by the proteasome, was published [33]. These studies overturned the existing dogma and completely reshaped our thinking with regard to the mechanism of elimination of misfolded proteins from the ER. The model eukaryote *Saccharomyces cerevisiae,* easily amenable to genetic, molecular biological and biochemical experimentation, has been a pacemaker in the study of ER protein quality control and degradation (ERQD). Since ERQD is a "housekeeping" mechanism common to all eukaryotic cells, research in yeast is likely to uncover the basic principles conserved in all mammalian cells and will continue to pave the way for the elucidation of the ERQD mechanism and for our understanding of this central cellular process.

6.2 ER-import, Folding and the Unfolded Protein Response

Proteins destined for secretion or for residence within the compartments of the secretory pathway enter the ER via a translocation channel known as the Sec61 translocon. The translocon is composed of three integral membrane proteins, Sec61, Sbh1 and Sss1 in yeast and Sec61α, Sec61β and Sec61γ in mammalian cells [9, 34]. Nascent polypeptides enter the ER in an unfolded state. Their folding in the ER lumen, accompanied by chemical modifications such as disulfide bond formation and addition of carbohydrates, requires a multitude of enzymes and chaperones. Among those are peptidyl-prolyl isomerases, disulfide bond modifying enzymes (Ero1p, PDI and others), classical chaperones (BiP, Kar2 in yeast) and co-chaperones, the oligosaccharyltransferase complex, N-glycan-modifying enzymes such as glucosidases I and II, and α-1,2-mannosidase and, in mammalian cells, UDP-Glc: glycoprotein glucosyltransferase as well as lectin-like chaperones (calnexin, calreticulin) [8, 35]. Properly modified and folded proteins are packed into vesicles and transported to the Golgi apparatus from where they proceed towards their final destination [8, 11, 14]. A highly sophisticated protein quality control system residing in the ER scans proteins for correct folding. If anything goes wrong with folding or membrane insertion, the ER reacts with the two interconnected mechanisms: the unfolded protein response (UPR) and, if all measures fail, the ER-associated protein quality control and degradation (ERQD) pathway. These mechanisms work together: the unfolded protein response (UPR) provides the ER with an increased capacity to fold proteins and control their folding [12, 13, 36] and upregulates the components of the machinery capable of degrading misfolded proteins [37–39]. Loss of ERAD leads to constitutive induction of UPR. Loss of UPR and ERAD results in dramatically decreased cell viability [36, 38, 39].

In yeast, a highly conserved transmembrane kinase of the ER, Ire1, monitors the folding capacity of the ER. Ire1 is composed of an ER-lumenal domain that senses the presence of misfolded proteins and a cytoplasmic tail consisting of a kinase and an endoribonuclease domain. ER stress, resulting from the accumulation of misfolded proteins, leads to oligomerization and autophosphorylation of Ire1, which activates the kinase function. Ire1 is targeted to the inner nuclear membrane via a nuclear targeting sequence [40]. Here, the endoribonuclease domain of Ire1 participates in the nonconventional splicing of the b-ZiP transcription factor HAC1 (XBP-1 in metazoans) mRNA. Splicing of the HAC1 mRNA leads to the synthesis of an active transcription factor that triggers the synthesis of UPRE-controlled genes, including genes encoding components of the ERQD machinery. It was initially proposed that Ire1 is activated when the major Hsp70 of the ER, BiP (Kar2 in yeast), usually bound to the ER-lumenal domain of Ire1, is titrated out by increasing amounts of misfolded proteins [36]. The finding that the conserved core region of the lumenal domain of dimerized Ire1 forms a deep hydrophobic groove has led to the proposal that Ire1 binds misfolded proteins directly [41]. The discussion continues as recent data indicate that this groove is too narrow for peptide binding and that the dimerization of Ire1 does not require the direct binding of unfolded proteins [42]. However, regardless of whether Ire1 senses the presence of misfolded proteins directly or indirectly, it is certain that UPR responds to subtle changes in the misfolded protein content of the ER. Mammalian cells express two other sensors in addition to IRE1: (i) the ER transmembrane kinase PERK containing a cytoplasmic eIF2α kinase domain and (ii) the ER transmembrane transcription factor ATF6. Binding of misfolded proteins to PERK activates the kinase domain. This leads to inhibition of translation and upregulation of the transcription factor ATF4 that initiates the transcription of some UPR target genes. Transmembrane ATF6, on the other hand, reaches the Golgi apparatus when there is an accumulation of misfolded proteins, where a proteolytic cleavage releases the cytoplasmic transcription factor domain. The transcription factor enters the nucleus to initiate gene transcription. Equipped with a set of three distinct sensors for misfolded proteins, IRE1, PERK and ATF6, mammalian cells respond to the accumulation of misfolded proteins in the ER in a more nuanced fashion than yeast. These responses range from increased ERQD, in everyday life, to induction of apoptosis under conditions of permanent stress [13].

6.3 General Principles and Components of ERQD (Endoplasmic Reticulum Quality Control and Protein Degradation)

The initial components of ER-associated protein degradation were, to a large extent, discovered via yeast genetics studies. The first indication that the cytoplasmic ubiquitin–proteasome system participated in the degradation of a misfolded ER protein came from studies on a yeast *sec61-2* mutant harboring a defective ER-translocation channel that, at restrictive temperatures, was degraded following ubiquitination: a mutation in the ubiquitin-conjugating enzyme Ubc6 restored the growth of *sec61-2* cells at the restrictive temperature leading to the conclusion that the Sec61-2 protein was not degraded in this mutant under these conditions [29]. The role of the cytoplasmic proteasome in the degradation of an ER membrane protein was discovered when the fate of a mutated ΔF508 cystic fibrosis transmembrane conductance regulator (CFTR) was studied in human cells. The ΔF508 mutation renders the CFTR protein unable to leave the ER to reach the plasma membrane. Trapped in the ER, mutant CFTR undergoes rapid degradation by the proteasome which results in the manifestation of the disease state, cystic fibrosis [27, 28]. Up till then, the ubiquitin–proteasome system was known to be an essential and selective protein degradation machinery required for signal-induced protein elimination as well as for removal of misfolded cytoplasmic proteins [43–46]. Proteins destined for degradation by the proteasome were known to be polyubiquitinated via a cascade of three enzymes. First, an ubiquitin-activating enzyme utilizing ATP to form an anhydride bond between the carboxyl of the C-terminal glycine of ubiquitin and the phosphate group of AMP, transfers this activated ubiquitin onto the active site cysteine of the enzyme to form a thioester bond. Second, the thioester-bonded ubiquitin is transferred to a ubiquitin-conjugation enzyme (Ubc, E2) retaining the active thioester bond. Finally a ubiquitin protein ligase (E3) links the ubiquitin in an isopeptide bond to the -amino group of a lysine side chain of the protein to be degraded, either directly or via the E2. When internal lysines are not available, ubiquitin can be linked with a peptide bond to the N-terminus of a protein [47]. For degradation by the proteasome, a polyubiquitin chain has to be formed on the lysine 48 residue of each preceding ubiquitin. Ubiquitin chains of four and more units are recognized by the proteasome, initiating protein degradation [44, 48]. The finding that mutated ER membrane proteins are degraded by the ubiquitin–proteasome system [26–29] did not give any information about the elimination mechanism of these proteins. A mutational analysis of yeast expressing a mutant and, therefore, misfolded vacuolar Carboxypeptidase yscY (CPY*) disclosed the mechanistic steps of ERAD. It was discovered that mutant CPY* is imported into the ER completely, it is fully N-glycosylated, recognized as unable to fold properly, retrograde transported back into the cytoplasm in its glycosylated form, ubiquitinated and, finally, degraded by the proteasome [19–21, 30, 31, 49]. The ubiquitin-conjugating enzymes Ubc6 and Ubc7 [31] and, later, Ubc1 [39] were found to catalyze the ubiquitination of CPY*. Genetic and biochemical analyses of yeast strains exhibiting defective CPY* degradation resulted in the discovery of the following additional ERQD components, listed in chronological order: Der1 (degradation of the ER), a protein spanning the ER membrane four times with its N- and C-termini located in the cytosol [37, 50]; soluble ER α-mannosidase I [51]; Cue1, a type I protein of the ER [52]; Kar2p, the major Hsp70 chaperone of the ER lumen [53] and its J-domain co-chaperones Jem1 and Scj1 [54]; Der3, a RING-finger ubiquitin-protein ligase (E3) spanning the ER membrane six times with the N- and C-termini facing the cytosol [55–57]; Der5/Pmr1, an ER/Golgi-located Ca^{2+} pump [58]; Png1, a cytoplasmic peptide N-glycanase [59]; Htm1/Mnl1, an ER lectin [60, 61]; the trimeric AAA-ATPase complex of the cytoplasm consisting of Cdc48, Ufd1 and Npl4 [62–64]; Dsk2 and Rad23, two UBA-UBL domain proteins [65]; Der7/α-glucosidase I involved in N-glycan trimming [66]; Yos9, a lectin-like protein of the ER lumen [67–70]; Ubx2 a protein containing an UBX domain and spanning the ER membrane twice, the N- and C-termini reaching into the cytosol [71, 72] and Usa1, another double-pass ER membrane protein [73, 74]. Studies on the regulated degradation of the HMG-CoA-reductase isozyme 2 in yeast [75] revealed that its downregulation is dependent on several *HRD* gene products (HMG-CoA reductase degradation) and the proteasome. *HRD1* was found to be identical to *DER3,* encoding the ER ubiquitin protein ligase involved in CPY* degradation [57, 75]. *HRD3* encodes a single-pass transmembrane protein with a long N-terminal portion in the ER lumen and a short C-terminal tail in the cytoplasm. Hrd3 is also required for the degradation of misfolded CPY* [76]. Degradation of HMG-CoA reductase 2 was also dependent on the Golgi P-type ATPase, Cod1/Spf1 [77]. Like Pmr1, Cod1, is also required for CPY* degradation [78]. The overlap of the "*DER*" genes with the "*HRD*" genes and the subsequent finding that most of the components required for ERAD are also required for the downregulation of HMG-CoA-reductase 2 shows that the cell uses the degradation machinery of the ER not only for the elimination of misfolded secretory proteins but also for the regulation of central metabolic enzymes located in the ER. A screen for mutants defective in the degradation of fusion proteins carrying the Deg1 degradation signal of the MATα2 repressor yielded Doa10, a novel ubiquitin-protein-ligase of the ER [79]. Like Hrd1/Der3, Doa10 is an ER membrane E3 containing a RING-finger domain. The protein spans the membrane 14 times with both termini facing the cytoplasm [80].

6.4
Mechanism of ERQD

The involvement of carbohydrate chains in the recognition of misfolded proteins was revealed by the finding that a mutation in α-mannosidase I resulted in considerable retardation of CPY* degradation [51]. Upon entry into the ER, pre-assembled oligosaccharide chains of the structure $Glc_3Man_9GlcNAc_2$ are transferred onto asparagine residues within the consensus Asn-X-Ser(Thr) sequence of the nascent polypeptide chain. While the polypeptide chain folds, the Glc_3-Man_9-$GlcNAc_2$ structure undergoes trimming. Sequential action of glucosidases I and II leads to the removal of the two terminal glucose residues, resulting in a Glc_1Man_9-$GlcNAc_2$ structure. In mammalian cells, the chaperones calnexin and calreticulin specifically recognize and bind this mono-glucosylated oligosaccharide. These lectin chaperones and foldases such as peptidyl-proline isomerases and disulfide isomerases bind and facilitate folding of the protein into its correct native structure. During this process the last and innermost glucose residue is also removed by glucosidase II. If the protein has not folded properly by the end of the time allocated for glucose-trimming, the de-glucosylated oligosaccharide is re-glucosylated by the folding sensor UDP: glucose glycoprotein glucosyltransferase (UGGT) to restore the $Glc_1Man_9GlcNAc_2$ structure and allow time for a new round of folding. If folding remains unsuccessful, α-mannosidase I removes the outermost mannose residue from the de-glucosylated oligosaccharide, thus preventing secretion of the protein from the ER and committing it to ERAD [14, 81]. In yeast, the re-glucosylation mechanism by UGGT does not exist. Instead, following removal of the three glucose residues by glucosidases I and II [66, 82], the slow acting α-mannosidase I [51] seems to be the only timer for the decision process of secretion or degradation. Terminally misfolded proteins containing the $Man_8GlcNAc_2$ structure are retained in the ER to undergo ERAD.

The first protein assembly required for the degradation of soluble, ER lumenal misfolded CPY* and ER membrane HMG-CoA reductase was described in yeast as a complex between the E3 ubiquitin protein ligase Hrd1/Der3 and its partner Hrd3 [55, 83, 84]. The ubiquitin-conjugating enzyme Ubc7, recruited to the ER membrane by Cue1p [52] was found to interact with the RING-finger domain of Hrd1/Der3 [55], thus expanding the membrane complex to a Ubc7–Cue1–Hrd1/Der3-Hrd3 structure. Later, the membrane protein Usa1 and the first identified ERAD component Der1 [51] were shown to be part of the ERAD complex [73, 74, 85, 86]. More recently, transmembrane Ubx2 was identified as the link between the Ubc7–Cue1–Hrd1/Der3–Hrd3–Usa1–Der1 complex of the ER membrane and the trimeric Cdc48–Ufd1–Npl4 complex of the cytosol [71, 72] (Figure 6.1). Similarly, in the lumen of the ER, interactions of the lectin-like protein Yos9 and the major Hsp70 of the ER, Kar2 (BiP in mammalian cells) with Hrd3 were uncovered [73, 74, 85]. A detailed picture of the degradation pathway of N-glycosylated misfolded ER-lumenal and lumenal lesion-containing ER membrane proteins is slowly emerging. After synthesis in the cytoplasm secretory proteins are translocated into the ER where, during/after N-glycosylation, they undergo folding. An important component of the ER folding machinery is the Hsp70 chaperone Kar2, which on one hand assists folding, and on the other, keeps the protein that has not reached its native structure, in a soluble form [54, 87]. During the folding process, glucosidases I and II cleave the three terminal glucose residues of the $Glc_3Man_9GlcNAc_2$ glycan. Then, misfolded proteins become substrates of α-mannosidase I which trims the glycans further to create a $Man_8GlcNAc_2$ structure;

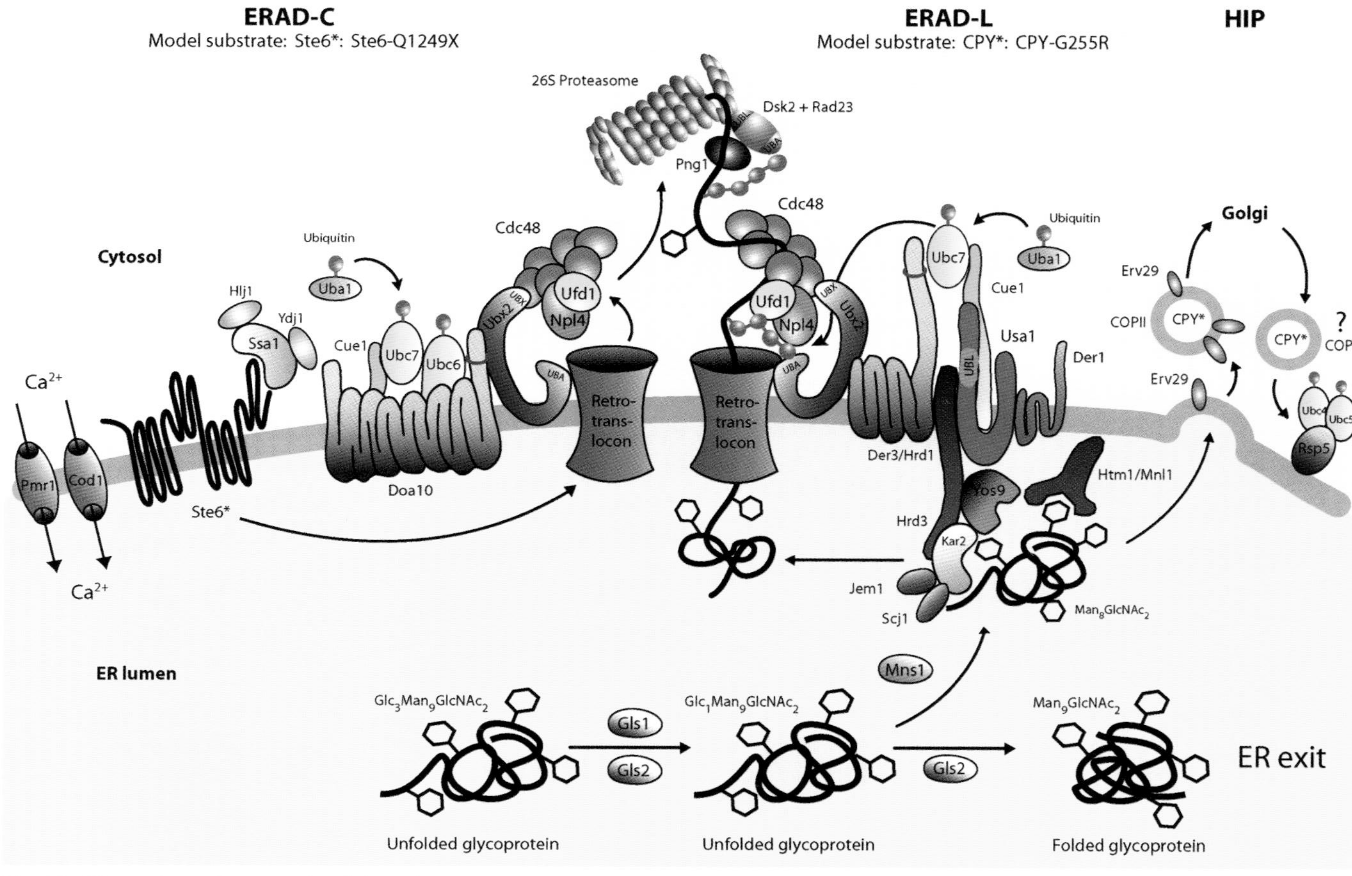

Fig. 6.1. ER protein quality control and degradation (ERQD). ER-associated degradation pathways: ERAD-C, ERAD-L and HiP.

the misfolded proteins are then recognized by and interact with the lectin-like proteins Yos9 and Htm1. Kar2 and Yos9 interact with the N-terminal lumenal tail of Hrd3 [73, 85] (Figure 6.1). One may speculate that Kar2 and Yos9 bind hydrophobic patches of proteins undergoing folding and deliver these to the lumenal tail of Hrd3, which acts as a receptor for misfolded proteins exposing hydrophobic amino acid stretches. So far, Yos9 is known to affect only the degradation of glycosylated proteins [67–70] even though the initial step of Yos9 binding to misfolded proteins is independent of their glycosylation status and the interaction of Yos9 with Hrd3 [73]. It is conceivable that the lectin domain of Yos9 scans the bound substrate, which is also interacting with Kar2 and Hrd3, for the presence of the $Man_8GlcNAc_2$ structure indicative of improper folding, for delivery to the ubiquitin-protein ligase Hrd1/Der3. Such a bipartite control mechanism, inspecting both the hydrophobicity and N-glycan structure, would ensure that only misfolded proteins are polyubiquitinated and degraded. It is interesting to note that the location of carbohydrates on a misfolded protein seems to be important for its degradation [88, 89]. The relationship between the N-glycan position and degradation is presently not understood. Similarly, the function of the second lectin-like protein involved in ERAD, Htm1/Mnl1p, [60, 61] is unclear. One may speculate that it also has a N-glycan screening function. While the role of most components of the Hrd/Der surveillance complex is emerging, the function of Der1 is still not understood. In yeast ERQD, Der1 is required only for the degradation of soluble substrates like CPY* [37, 50, 87]. Der1 orthologs (known as Derlins) have been found in mammalian cells and also shown to be involved in ERAD [90–92]. Derlin-1 is required for the degradation of MHC class I heavy chains induced by the human cytomegalovirus (HCMV) protein US11. Interestingly, MHC class I heavy chain degradation induced by the HCMV protein US2 is independent of Derlin-1 [92, 93]. The formation of homo-oligomers of Derlin-1 that span the ER membrane several times has given grounds for the speculation that Derlin-1 might be part of the retro-translocation channel [91, 92]. The link between the ER-membrane Hrd/Der ubiquitination assembly and the proteasome is a trimeric chaperone complex consisting of the AAA-ATPase Cdc48 (p97 in mammalian cells), Ufd1 and Npl4. It is believed that this complex pulls the polyubiquitinated proteins out of the ER or away from the ER membrane [62–64]. Delivery of ER-released polyubiquitinated substrates to the proteasome requires the UBA-UBL domain proteins Dsk2 and Rad23 to function, presumably, as adaptors preventing the dissociation and accumulation of free, unfolded protein chains in the cytosol [65, 94] (Figure 6.1). The ERQD mechanism, as described, leaves out Doa10, the second ERAD ubiquitin protein ligase. Doa10 has 14 transmembrane domains and resides both in the ER and the inner nuclear membrane [79, 80, 95]. The enzyme is not involved in the degradation of ER luminal-soluble CPY* [79]. While CPY* and membrane proteins carrying a misfolded domain in the ER lumen are degraded via the Hrd1/Der3 ubiquitin ligase pathway known as the ERAD-L (luminal) pathway [87, 96, 97], degradation of proteins carrying lesions on the cytoplasmic side of the ER membrane (such as Ste6*, a truncated version of the yeast a-factor transporter Ste6) were shown to be substrates of the Doa10 ubiquitin ligase [96, 98]. Accordingly, this degradation pathway was called the ERAD-C (cytosolic) pathway [96]. As for the Hrd1/-dependent ERAD-L pathway, the main ubiquitin-conjugating enzymes required for polyubiquitination of ERAD-C substrates are Ubc6 and Ubc7 bound to Cue1 [96, 98]. The two pathways, ERAD-L and ERAD-C, converge at the Cdc48–Ufd1–Npl4 complex and the proteasome [74, 96, 98] (Figure 6.1). The degradation of Ste6*, with the truncated protein domain in the cytosol, requires the action of the cytoplasmic Hsp70 chaperone Ssa1 as well as the Hsp40 co-chaperones Ydj1 and Hdj1 (Figure 6.1). This requirement is similar to that of some misfolded cytoplasmic proteins, as discovered recently [99, 100]. The cytoplasmic Hsp70 chaperone machinery is, most likely, involved in the discovery of the misfolded, cytoplasmic protein domain. It is unclear if this machinery is also active in the unfolding of misfolded membrane-embedded proteins. Also recently, the existence of a third "pathway", ERAD-M, responsible for the degradation of membrane proteins with destabilized transmembrane domains has been postulated [74]. For such substrates degradation depends on the Hrd1/Der3 ubiquitin ligase and Hrd3 of the ERAD-L pathway, but it does not always require Der1. ERAD-M seems to converge with ERAD-L and ERAD-C at the Cdc48–Ufd1–Npl4 complex. Der1 is also not required for the degradation of CTG*, a chimeric ERAD substrate containing misfolded CPY* tethered to the ER membrane by the last transmembrane domain of the multidrug resistance transporter Pdr5 and fused to GFP on the cytoplasmic side of the ER [87]. It is unclear how or whether ER-lumenal CPY* distorts the transmembrane domain to make CTG* into an ERAD-M substrate. On the other hand, a process requiring Der1 is the degradation of Hrd1/Der3, which occurs when Hrd3 is absent [76]. Given the heterogeneity of both substrates and components, a simple classification of misfolded protein degradation within the ER into distinct pathways, such as ERAD-L, ERAD-C and ERAD-M, may not provide all the answers.

6.5 "Overflow" Degradation Pathways: ER-to-Golgi Transport and Autophagocytosis

A portion of the ER-luminal ERAD substrate CPY* [49], was shown to escape into the Golgi apparatus from where it is retrieved back to the ER. The secretory competence of the ER is essential for efficient ER-associated degradation [101]. Deletion of Der1 induces the unfolded protein response [37] and significantly increases the escape of CPY* into the Golgi apparatus. A second, Hrd1/Der3-independent degradation pathway (HIP: HRD independent proteolysis) has also been described, which becomes operative when CPY* is considerably overexpressed. Overexpression of CPY* likely saturates the HRD/DER pathway and leads to the induction of the HIP pathway which, like ERAD, is also regulated by the UPR. HIP requires ER to Golgi transport, the HECT domain ubiquitin-protein ligase Rsp5 and the ubiquitin-conjugating enzymes Ubc4 and Ubc5 instead of Ubc7 and Ubc1. Whether misfolded proteins enter the cytosol from the Golgi apparatus or the ER is not clear [102]. Transport of overexpressed misfolded proteins to the vacuole has been reported as an overflow pathway [103]. Recent experiments suggest that only proteins carrying an intact ER exit signal are prone to transport to the Golgi apparatus when misfolded. If, despite the misfolding of parts of the protein, the ER exit signal remains correctly folded, it can be recognized by the Erv29 cargo receptor and channeled into the secretory pathway. From here, the misfolded protein is directed to the HIP pathway or possibly to the vacuole for degradation. This model proposes competition between the ERQD machinery and ER exit facilitators for binding to misfolded proteins. The relative strength of the export signal on an aberrant protein against the affinity for the ERQD machinery would then determine what fraction, if any, of a misfolded protein exits the ER [104]. To completely block the degradation of CPY* both the HIP and the HRD/DER pathways have to be eliminated, emphasizing that both pathways contribute to the disposal of CPY*. Recently, autophagocytosis has been described as an additional overflow pathway for misfolded proteins that functions in coordination with ER-associated degradation. In yeast, expression of the Z-variant of the human α-1 proteinase inhibitor (A1PiZ) responsible for human liver disease, leads to saturation of ERQD and transport of the excess protein to the vacuole. A portion of A1PiZ reaches the vacuole via the secretory pathway. However, another portion, which forms aggregates, reaches the vacuole via autophagocytosis [105]. The mechanistic diversity in a cell which ensures the disposal of mutated and misfolded proteins when the ERQD machinery is overloaded reflects the importance given to the avoidance of the accumulation of disease-causing protein "garbage".

6.6
The Retrotranslocation Channel

One of the missing links in the current knowledge of ERQD is the nature of the retro translocation channel that delivers the misfolded proteins of the ER lumen or membrane back into the cytosol. The existence of a protein import channel in the ER membrane containing Sec61 as the channel-forming component drew attention to Sec61 as a possible component of the export channel. Indeed, genetic [53, 76, 106] and biochemical [107] experiments in yeast and mammalian cells [32, 108, 109] indicated the involvement of Sec61 in the retrotranslocation of the substrates studied. However, none of the studies undertaken for the isolation of ERAD-specific protein complexes containing Hrd1/Der3 or Doa10 have identified Sec61 as a component of the system [74, 86]. It is, therefore, unlikely that protein import and retrograde export use one and the same Sec61 channel to enable bi-directional transport. Rather, Sec61 import channels may associate either with the Hrd/Der or the Doa10 complexes to gain directionality for retrograde transport. Alternatively, Sec61 may form a hybrid channel with the Hrd/Der or Doa10 complexes which have the capacity to form channel-like entities due to their polytopic nature. One may also envisage that only certain misfolded proteins, i.e. the hydrophilic soluble proteins of the ER lumen, require the help of Sec61 for retrograde transport. The failure to find a biochemical connection between one of the ERAD membrane complexes and Sec61 may be due to the high mobility of Sec61, shuttling between protein import and retrograde transport channels. Finally, Sec61 may not be involved in retro translocation at all and the export channels may be solely composed of the Hrd/Der and/or Doa10 complexes. Since genetic data involve Sec61 in retrotranslocation [53, 76] this would indicate an indirect effect of Sec61 mutations on the retrotranslocation process. This may be due to the fact that the Sec61-2 mutant used in the genetic studies is itself degraded via ERAD [26, 29]. Experiments carried out under conditions that would arrest the channel in the process of substrate retrotranslocation may ultimately provide us with the correct picture. The discussion still continues as a recent study has shown that the Sec61 protein import channel binds 26S proteasomes and suggested that it acts as a proteasome receptor on the ER membrane [110].

Some studies indicate that not all ERAD substrates exploit the canonical ERAD machinery for the retrotranslocation step. The yeast pro-α-factor carrying mutated glycosylation sites is retrograde transported out of the ER and degraded by the proteasome without ubiquitination [33] or involvement of the AAA-ATPase Cdc48 [111]. *In vitro* studies have shown that the 19S cap of the proteasome was able to retrotranslocate mutant α-factor. Addition of 20S proteasome core particles to this *in vitro* system led to degradation of the substrate, indicating that the two processes are uncoupled [111]. Interestingly, a recent *in vitro* study in mammalian cells provides evidence that the Cdc48 (p97) machinery is not essential but rather facilitates the degradation of one of the most studied mammalian ERAD substrates, the cystic fibrosis transmembrane conductance regulator (CFTR) [112]. This observation suggests that the AAA-ATPases of Cdc48 and the proteasome 19S cap cooperate in the unfolding and extraction of the polytopic membrane protein out of the ER membrane.

6.7 Metazoan ERQD

ERQD and UPR are “housekeeping” processes essential for cell survival that have evolved early in the history of eukaryotic life. It is, therefore, expected that mechanisms similar to those identified in yeast operate in higher eukaryotic cells. Indeed, many of the yeast proteins involved in ERQD and UPR have mammalian counterparts with similar functions. Table 6.1 lists the mammalian counterparts of the known yeast ERAD components. As might be expected, due to the specialized needs of mammalian cells, several components of ERQD have multiplied and diverged in substrate specificity. For instance, while only one Der1 protein seems to operate in yeast ERAD (a homolog, Dfm1, has not been found to operate in ERQD) [21, 37, 50], three Der1 orthologs, Derlin-1, Derlin-2 and Derlin-3, have been discovered in mammalian cells. While Derlin-1 is required for human cytomegalovirus-US11-triggered elimination of MHC class I heavy chains [92, 93], Derlin-2 and Derlin-3, but not Derlin-1, are involved in the degradation of the null Hong Kong (NHK) mutant of α1-proteinase inhibitor [90]. Two orthologs of the

Table 6.1. ERAD components in yeast and their mammalian counterparts.

Yeast	Mammalian	References
Kar2	BiP/GRP78	117–119
Yos9	OS-9 (?)	120–122
Htm1/Mnl1	EDEM1, EDEM2, EDEM3	123
Der1	Derlin-1, Derlin-2, Derlin-3	90, 124, 125
Usa1	HERP	126–128
Hrd3	SEL1L	129
Hrd1/Der3	HRD1 (Synoviolin), Gp78	115, 130–133, 113, 134–136
Doa10	TEB4 (MARCH-VI)	80, 137
Ubc6	Ube2g1	138–141
Ubc7	Ube2g2	113, 142, 143
Ubx2/Sel1	KIAA0887 (?)	74
Cdc48	p97/VCP	124, 144, 145
Ufd1	UFD1	146–148
Npl4	NPL4	146–149
Dsk2	PLIC-1 and 2 (Ubiquilin1,2)	150, 151
Rad23	hHR23A and B	152, 153

yeast ubiquitin ligase Hrd1/Der3 were identified in mammalian cells, gp78 [113, 114] and Hrd1 [115]. The two enzymes show different substrate specificity: gp78, but not Hrd1, is involved in the regulation of mammalian HMG-CoA reductase degradation [116]. There is strong evidence that mammalian ERQD pathways merge at the Cdc48–(p97)–Ufd1–Npl4 machinery [91] prior to degradation of the selected misfolded proteins via the proteasome. Further studies will gradually complete the present mosaic of ERQD and finally provide us with the whole picture of this life-saving process.

Acknowledgements

This work was supported by the Deutsche Forschungsgemeinschaft, Bonn; the German–Israeli Project Cooperation (DIP) of the German Federal Ministry of Education and Research, Bonn; the German–Israeli Foundation for Scientific Research and Development, Jerusalem; and the Fonds der Chemischen Industrie, Frankfurt.

References

1 Goldberg, A.L. (2003) Protein degradation and protection against misfolded or damaged proteins. *Nature* **426**, 895–899.

2 Dobson, C.M. (2003) Protein folding and misfolding. *Nature* **426**, 884–890.

3 Aridor, M. and Hannan, L.A. (2000) Traffic jam: a compendium of human

diseases that affect intracellular transport processes. *Traffic* **1**, 836–851.
4 Rutishauser, J. and Spiess, M. (2002) Endoplasmic reticulum storage diseases. *Swiss Med Wkly* **132**, 211–222.
5 Ciechanover, A. and Schwartz, A.L. (2004) The ubiquitin system: pathogenesis of human diseases and drug targeting. *Biochim Biophys Acta* **1695**, 3–17.
6 Tanaka, K., Suzuki, T., Hattori, N. and Mizuno, Y. (2004) Ubiquitin, proteasome and parkin. *Biochim Biophys Acta* **1695**, 235–247.
7 Helenius, A. (2001) Quality control in the secretory assembly line. *Philos Trans R Soc Lond B Biol Sci* **356**, 147–150.
8 Kostova, Z. and Wolf, D.H. (2002) Protein quality control in the export pathway: the endoplasmic reticulum and its cytoplasmic proteasome connection, in *Protein Targeting, Transport and Translocation* (eds R.E. Dalbey and G. von Heijne), Academic Press, London–New York, pp. 180–213.
9 Haigh, N.G. and Johnson, A.E. (2002) Protein sorting at the membrane of the endoplasmic reticulum, in *ProteinTargeting, Transport and Translocation* (eds R.E. Dalbey and G. von Heijne), Academic Press, London–New York, pp. 74–106.
10 Johnson, A.E. and van Waes, M.A. (1999) The translocon: a dynamic gateway at the ER membrane. *Annu Rev Cell Dev Biol* **15**, 799–842.
11 van Anken, E. and Braakman, I. (2005) Versatility of the endoplasmic reticulum protein folding factory. *Crit Rev Biochem Mol Biol* **40**, 191–228.
12 Patil, C. and Walter, P. (2001) Intracellular signaling from the endoplasmic reticulum to the nucleus: the unfolded protein response in yeast and mammals. *Curr Opin Cell Biol* **13**, 349–355.
13 Schroder, M. and Kaufman, R.J. (2005) The mammalian unfolded protein response. *Annu Rev Biochem* **74**, 739–789.
14 Ellgaard, L. and Helenius, A. (2003) Quality control in the endoplasmic reticulum. *Nat Rev Mol Cell Biol* **4**, 181–191.
15 Sitia, R. and Braakman, I. (2003) Quality control in the endoplasmic reticulum protein factory. *Nature* **426**, 891–894.
16 Ellgaard, L., Molinari, M. and Helenius, A. (1999) Setting the standards: quality control in the secretory pathway. *Science* **286**, 1882–1888.
17 Trombetta, E.S. and Parodi, A.J. (2003) Quality control and protein folding in the secretory pathway. *Annu Rev Cell Dev Biol* **19**, 649–676.
18 Brodsky, J.L. and McCracken, A.A. (1999) ER protein quality control and proteasome-mediated protein degradation. *Semin Cell Dev Biol* **10**, 507–513.
19 Sommer, T. and Wolf, D.H. (1997) Endoplasmic reticulum degradation: reverse protein flow of no return. *FASEB J* **11**, 1227–1233.
20 Plemper, R.K. and Wolf, D.H. (1999) Retrograde protein translocation: ERADication of secretory proteins in health and disease. *Trends Biochem Sci* **24**, 266–270.
21 Kostova, Z. and Wolf, D.H. (2003) For whom the bell tolls: protein quality control of the endoplasmic reticulum and the ubiquitin-proteasome connection. *EMBO J* **22**, 2309–2317.
22 Hirsch, C., Jarosch, E., Sommer, T. and Wolf, D.H. (2004) Endoplasmic reticulum-associated protein degradation-one model fits all? *Biochim Biophys Acta* **1695**, 215–223.
23 Schäfer, A. and Wolf, D.H. (2005) Yeast genomics in the elucidation of endoplasmic reticulum (ER) quality control and associated protein degradation (ERQD). *Methods Enzymol* **399**, 459–468.
24 Blobel, G. (1995) Unidirectional and bidirectional protein traffic across membranes. *Cold Spring Harb Symp Quant Biol* **60**, 1–10.
25 Bonifacino, J.S. and Klausner, R.D. (1994) Degradation of proteins retained in the endoplasmic reticulum, in *Modern Cell Biology, Cellular Proteolytic Systems*, Vol. 15 (eds A.J. Ciechanover and A.L. Schwartz), John Wiley and Sons, New York, pp. 137–160.
26 Biederer, T., Volkwein, C. and Sommer, T. (1996) Degradation of subunits of the

Sec61p complex, an integral component of the ER membrane, by the ubiquitin-proteasome pathway. *EMBO J* **15**, 2069–2076.

27 Jensen, T.J., Loo, M.A., Pind, S., Williams, D.B., Goldberg, A.L. and Riordan, J.R. (1995) Multiple proteolytic systems, including the proteasome, contribute to CFTR processing. *Cell* **83**, 129–135.

28 Ward, C.L., Omura, S. and Kopito, R.R. (1995) Degradation of CFTR by the ubiquitin-proteasome pathway. *Cell* **83**, 121–127.

29 Sommer, T. and Jentsch, S. (1993) A protein translocation defect linked to ubiquitin conjugation at the endoplasmic reticulum. *Nature* **365**, 176–179.

30 Hochstrasser, M. (2006) Ubiquitination and protein turnover, in *Landmark Papers in Yeast Biology* (eds P. Lindner, D. Shore and M.N. Hall), Cold Spring Harbor Laboratory Press, Cold Spring Harbor, New York, pp. 273–274.

31 Hiller, M.M., Finger, A., Schweiger, M. and Wolf, D.H. (1996) ER degradation of a misfolded luminal protein by the cytosolic ubiquitin-proteasome pathway. *Science* **273**, 1725–1728.

32 Wiertz, E.J., Tortorella, D., Bogyo, M., Yu, J., Mothes, W., Jones, T.R., Rapoport, T.A. and Ploegh, H.L. (1996) Sec61-mediated transfer of a membrane protein from the endoplasmic reticulum to the proteasome for destruction. *Nature* **384**, 432–438.

33 Werner, E.D., Brodsky, J.L. and McCracken, A.A. (1996) Proteasome-dependent endoplasmic reticulum-associated protein degradation: an unconventional route to a familiar fate. *Proc Natl Acad Sci USA* **93**, 13797–13801.

34 Corsi, A.K. and Schekman, R. (1996) Mechanism of polypeptide translocation into the endoplasmic reticulum. *J Biol Chem* **271**, 30299–30302.

35 Zapun, A., Jakob, C.A., Thomas, D.Y. and Bergeron, J.J. (1999) Protein folding in a specialized compartment: the endoplasmic reticulum. *Structure* **7**, R173–R182.

36 Sidrauski, C., Brickner, J.H. and Walter, P. (2002) The unfolded protein response, in *Protein Targeting, Transport & Translocation* (eds R.E. Dalbey and G.V. Heijne), Academic Press, London-New York, pp. 151–179.

37 Knop, M., Finger, A., Braun, T., Hellmuth, K. and Wolf, D.H. (1996) Der1, a novel protein specifically required for endoplasmic reticulum degradation in yeast. *EMBO J* **15**, 753–763.

38 Travers, K.J., Patil, C.K., Wodicka, L., Lockhart, D.J., Weissman, J.S. and Walter, P. (2000) Functional and genomic analyses reveal an essential coordination between the unfolded protein response and ER-associated degradation. *Cell* **101**, 249–258.

39 Friedländer, R., Jarosch, E., Urban, J., Volkwein, C. and Sommer, T. (2000) A regulatory link between ER-associated protein degradation and the unfolded-protein response. *Nat Cell Biol* **2**, 379–384.

40 Goffin, L., Vodala, S., Fraser, C., Ryan, J., Timms, M., Meusburger, S., Catimel, B., Nice, E.C., Silver, P.A., Xiao, C.Y., Jans D.A. and Gething M.J. (2006) The unfolded protein response transducer Ire1p contains a nuclear localization sequence recognized by multiple Beta importins. *Mol Biol Cell* **17**, 5309–5323.

41 Credle, J.J., Finer-Moore, J.S., Papa, F.R., Stroud, R.M. and Walter, P. (2005) On the mechanism of sensing unfolded protein in the endoplasmic reticulum. *Proc Natl Acad Sci USA* **102**, 18773–18784.

42 Zhou, J., Liu, C.Y., Back, S.H., Clark, R.L., Peisach, D., Xu, Z. and Kaufman, R.J. (2006) The crystal structure of human IRE1 luminal domain reveals a conserved dimerization interface required for activation of the unfolded protein response. *Proc Natl Acad Sci USA* **103**, 14343–14348.

43 Varshavsky, A. (2005) Regulated protein degradation. *Trends Biochem Sci* **30**, 283–286.

44 Pickart, C.M. and Eddins, M.J. (2004) Ubiquitin: structures, functions, mechanisms. *Biochim Biophys Acta* **1695**, 55–72.

45 Wolf, D.H. and Hilt, W. (2004) The proteasome: a proteolytic nanomachine of cell regulation and waste disposal. *Biochim Biophys Acta* **1695**, 19–31.

46 Hershko, A. and Ciechanover, A. (1998) The ubiquitin system. *Annu Rev Biochem* **67**, 425–479.

47 Ciechanover, A. and Ben-Saadon, R. (2004) N-terminal ubiquitination: more protein substrates join in. *Trends Cell Biol* **14**, 103–106.

48 Fang, S. and Weissman, A.M. (2004) A field guide to ubiquitylation. *Cell Mol Life Sci* **61**, 1546–1561.

49 Wolf, D.H. and Schäfer, A. (2005) CPY* and the power of yeast genetics in the elucidation of quality control and associated protein degradation of the endoplasmic reticulum, in *Current Topics in Microbiology and Immunology*, Vol. 300, (eds E. Wiertz and M. Kikkert), Springer, Heidelberg, pp. 41–56.

50 Hitt, R. and Wolf, D.H. (2004) Der1p, a protein required for degradation of malfolded soluble proteins of the endoplasmic reticulum: topology and Der1-like proteins. *FEMS Yeast Res* **4**, 721–729.

51 Knop, M., Hauser, N. and Wolf, D.H. (1996) N-Glycosylation affects endoplasmic reticulum degradation of a mutated derivative of carboxypeptidase yscY in yeast. *Yeast* **12**, 1229–1238.

52 Biederer, T., Volkwein, C. and Sommer, T. (1997) Role of Cue1p in ubiquitination and degradation at the ER surface. *Science* **278**, 1806–1809.

53 Plemper, R.K., Böhmler, S., Bordallo, J., Sommer, T. and Wolf, D.H. (1997) Mutant analysis links the translocon and BiP to retrograde protein transport for ER degradation. *Nature* **388**, 891–895.

54 Nishikawa, S., Fewell, S.W., Kato, Y., Brodsky, J.L. and Endo, T. (2001) Molecular chaperones in the yeast endoplasmic reticulum maintain the solubility of proteins for retrotranslocation and degradation. *J Cell Biol* **153**, 1061–1070.

55 Deak, P.M. and Wolf, D.H. (2001) Membrane topology and function of Der3/Hrd1p as a ubiquitin-protein ligase (E3) involved in endoplasmic reticulum degradation. *J Biol Chem* **276**, 10663–10669.

56 Bordallo, J. and Wolf, D.H. (1999) A RING-H2 finger motif is essential for the function of Der3/Hrd1 in endoplasmic reticulum associated protein degradation in the yeast Saccharomyces cerevisiae. *FEBS Lett* **448**, 244–248.

57 Bordallo, J., Plemper, R.K., Finger, A. and Wolf, D.H. (1998) Der3p/Hrd1p is required for endoplasmic reticulum-associated degradation of misfolded lumenal and integral membrane proteins. *Mol Biol Cell* **9**, 209–222.

58 Dürr, G., Strayle, J., Plemper, R., Elbs, S., Klee, S.K., Catty, P., Wolf, D.H. and Rudolph, H.K. (1998) The medial-Golgi ion pump Pmr1 supplies the yeast secretory pathway with Ca^{2+} and Mn2+ required for glycosylation, sorting and endoplasmic reticulum-associated protein degradation. *Mol Biol Cell* **9**, 1149–1162.

59 Suzuki, T., Park, H., Hollingsworth, N.M., Sternglanz, R. and Lennarz, W.J. (2000) PNG1, a yeast gene encoding a highly conserved peptide:N-glycanase. *J Cell Biol* **149**, 1039–1052.

60 Nakatsukasa, K., Nishikawa, S., Hosokawa, N., Nagata, K. and Endo, T. (2001) Mnl1p, an alpha -mannosidase-like protein in yeast *Saccharomyces cerevisiae*, is required for endoplasmic reticulum-associated degradation of glycoproteins. *J Biol Chem* **276**, 8635–8638.

61 Jakob, C.A., Bodmer, D., Spirig, U., Battig, P., Marcil, A., Dignard, D., Bergeron, J.J., Thomas, D.Y. and Aebi, M. (2001) Htm1p, a mannosidase-like protein, is involved in glycoprotein degradation in yeast. *EMBO Rep* **2**, 423–430.

62 Rabinovich, E., Kerem, A., Frohlich, K.U., Diamant, N. and Bar-Nun, S. (2002) AAA-ATPase p97/Cdc48p, a cytosolic chaperone required for endoplasmic reticulum-associated protein degradation. *Mol Cell Biol* **22**, 626–634.

63 Ye, Y., Meyer, H.H. and Rapoport, T.A. (2001) The AAA ATPase Cdc48/p97 and its partners transport proteins from the ER into the cytosol. *Nature* **414**, 652–656.

64 Jarosch, E., Taxis, C., Volkwein, C., Bordallo, J., Finley, D., Wolf, D.H. and Sommer, T. (2002) Protein dislocation from the ER requires polyubiquitination and the AAA- ATPase Cdc48. *Nat Cell Biol* **4**, 134–139.

65 Medicherla, B., Kostova, Z., Schaefer, A. and Wolf, D.H. (2004) A genomic screen identifies Dsk2p and Rad23p as essential components of ER-associated degradation. *EMBO Rep* **5**, 692–697.

66 Hitt, R. and Wolf, D.H. (2004) DER7, encoding alpha-glucosidase I is essential for degradation of malfolded glycoproteins of the endoplasmic reticulum. *FEMS Yeast Res* **4**, 815–820.

67 Bhamidipati, A., Denic, V., Quan, E.M. and Weissman, J.S. (2005) Exploration of the topological requirements of ERAD identifies Yos9p as a lectin sensor of misfolded glycoproteins in the ER lumen. *Mol Cell* **19**, 741–751.

68 Szathmary, R., Bielmann, R., Nita-Lazar, M., Burda, P. and Jakob, C.A. (2005) Yos9 protein is essential for degradation of misfolded glycoproteins and may function as lectin in ERAD. *Mol Cell* **19**, 765–775.

69 Kim, W., Spear, E.D. and Ng, D.T. (2005) Yos9p detects and targets misfolded glycoproteins for ER-associated degradation. *Mol Cell* **19**, 753–764.

70 Buschhorn, B., Kostova, Z., Medicherla, B. and Wolf, D.H. (2004) A genome wide screen identifies Yos9p as a new lectin essential for ER-associated degradation (ERAD) of glycoproteins. *FEBS Lett* **577**, 422–426.

71 Neuber, O., Jarosch, E., Volkwein, C., Walter, J. and Sommer, T. (2005) Ubx2 links the Cdc48 complex to ER-associated protein degradation. *Nat Cell Biol* **7**, 993–998.

72 Schuberth, C. and Buchberger, A. (2005) Membrane-bound Ubx2 recruits Cdc48 to ubiquitin ligases and their substrates to ensure efficient ER-associated protein degradation. *Nat Cell Biol* **7**, 999–1006.

73 Denic, V., Quan, E.M. and Weissman, J.S. (2006) A luminal surveillance complex that selects misfolded glycoproteins for ER-associated degradation. *Cell* **126**, 349–359.

74 Carvalho, P., Goder, V. and Rapoport, T.A. (2006) Distinct ubiquitin-ligase complexes define convergent pathways for the degradation of ER proteins. *Cell* **126**, 361–373.

75 Hampton, R.Y., Gardner, R.G. and Rine, J. (1996) Role of 26S proteasome and HRD genes in the degradation of 3-hydroxy-3-methylglutaryl-CoA reductase, an integral endoplasmic reticulum membrane protein. *Mol Biol Cell* **7**, 2029–2044.

76 Plemper, R.K., Bordallo, J., Deak, P.M., Taxis, C., Hitt, R. and Wolf, D.H. (1999) Genetic interactions of Hrd3p and Der3p/Hrd1p with Sec61p suggest a retro-translocation complex mediating protein transport for ER degradation. *J Cell Sci* **112**, 4123–4134.

77 Cronin, S.R., Khoury, A., Ferry, D.K. and Hampton, R.Y. (2000) Regulation of HMG-CoA reductase degradation requires the P-type ATPase Cod1p/Spf1p. *J Cell Biol* **148**, 915–924.

78 Vashist, S., Frank, C.G., Jakob, C.A. and Ng, D.T. (2002) Two distinctly localized p-type ATPases collaborate to maintain organelle homeostasis required for glycoprotein processing and quality control. *Mol Biol Cell* **13**, 3955–3966.

79 Swanson, R., Locher, M. and Hochstrasser, M. (2001) A conserved ubiquitin ligase of the nuclear envelope/endoplasmic reticulum that functions in both ER-associated and Matalpha2 repressor degradation. *Genes Dev* **15**, 2660–2674.

80 Kreft, S.G., Wang, L. and Hochstrasser, M. (2006) Membrane topology of the yeast endoplasmic reticulum-localized ubiquitin ligase Doa10 and comparison with its human ortholog TEB4 (MARCH-VI). *J Biol Chem* **281**, 4646–4653.

81 Schrag, J.D., Procopio, D.O., Cygler, M., Thomas, D.Y. and Bergeron, J.J. (2003) Lectin control of protein folding and sorting in the secretory pathway. *Trends Biochem Sci* **28**, 49–57.

82 Jakob, C.A., Burda, P., Roth, J. and Aebi, M. (1998) Degradation of misfolded endoplasmic reticulum glycoproteins in

Saccharomyces cerevisiae is determined by a specific oligosaccharide structure. *J Cell Biol* **142**, 1223–1233.

83 Bays, N.W., Gardner, R.G., Seelig, L.P., Joazeiro, C.A. and Hampton, R.Y. (2001) Hrd1p/Der3p is a membrane-anchored ubiquitin ligase required for ER- associated degradation. *Nat Cell Biol* **3**, 24–29.

84 Gardner, R.G., Swarbrick, G.M., Bays, N.W., Cronin, S.R., Wilhovsky, S., Seelig, L., Kim, C. and Hampton, R.Y. (2000) Endoplasmic reticulum degradation requires lumen to cytosol signaling. Transmembrane control of Hrd1p by Hrd3p. *J Cell Biol* **151**, 69–82.

85 Gauss, R., Jarosch, E., Sommer, T. and Hirsch, C. (2006) A complex of Yos9p and the HRD ligase integrates endoplasmic reticulum quality control into the degradation machinery. *Nat Cell Biol* **8**, 849–854.

86 Gauss, R., Sommer, T. and Jarosch, E. (2006) The Hrd1p ligase complex forms a linchpin between ER-lumenal substrate selection and Cdc48p recruitment. *EMBO J* **25**, 1827–1835.

87 Taxis, C., Hitt, R., Park, S.H., Deak, P.M., Kostova, Z. and Wolf, D.H. (2003) Use of modular substrates demonstrates mechanistic diversity and reveals differences in chaperone requirement of ERAD. *J Biol Chem* **278**, 35903–35913.

88 Spear, E.D. and Ng, D.T. (2005) Single, context-specific glycans can target misfolded glycoproteins for ER-associated degradation. *J Cell Biol* **169**, 73–82.

89 Kostova, Z. and Wolf, D.H. (2005) Importance of carbohydrate positioning in the recognition of mutated CPY for ER-associated degradation. *J Cell Sci* **118**, 1485–1492.

90 Oda, Y., Okada, T., Yoshida, H., Kaufman, R.J., Nagata, K. and Mori, K. (2006) Derlin-2 and Derlin-3 are regulated by the mammalian unfolded protein response and are required for ER-associated degradation. *J Cell Biol* **172**, 383–393.

91 Ye, Y., Shibata, Y., Kikkert, M., van Voorden, S., Wiertz, E. and Rapoport, T.A. (2005) Inaugural Article: Recruitment of the p97 ATPase and ubiquitin ligases to the site of retrotranslocation at the endoplasmic reticulum membrane. *Proc Natl Acad Sci USA* **102**, 14132–14138.

92 Ye, Y., Shibata, Y., Yun, C., Ron, D. and Rapoport, T.A. (2004) A membrane protein complex mediates retro-translocation from the ER lumen into the cytosol. *Nature* **429**, 841–847.

93 Lilley, B.N. and Ploegh, H.L. (2004) A membrane protein required for dislocation of misfolded proteins from the ER. *Nature* **429**, 834–840.

94 Richly, H., Rape, M., Braun, S., Rumpf, S., Hoege, C. and Jentsch, S. (2005) A series of ubiquitin binding factors connects CDC48/p97 to substrate multiubiquitylation and proteasomal targeting. *Cell* **120**, 73–84.

95 Deng, M. and Hochstrasser, M. (2006) Spatially regulated ubiquitin ligation by an ER/nuclear membrane ligase. *Nature* **443**, 827–831.

96 Vashist, S. and Ng, D.T. (2004) Misfolded proteins are sorted by a sequential checkpoint mechanism of ER quality control. *J Cell Biol* **165**, 41–52.

97 Plemper, R.K., Egner, R., Kuchler, K. and Wolf, D.H. (1998) Endoplasmic reticulum degradation of a mutated ATP-binding cassette transporter Pdr5 proceeds in a concerted action of Sec61 and the proteasome. *J Biol Chem* **273**, 32848–32856.

98 Huyer, G., Piluek, W.F., Fansler, Z., Kreft, S.G., Hochstrasser, M., Brodsky, J.L. and Michaelis, S. (2004) Distinct machinery is required in *Saccharomyces cerevisiae* for the endoplasmic reticulum-associated degradation of a multispanning membrane protein and a soluble luminal protein. *J Biol Chem* **279**, 38369–38378.

99 Park, S.H., Bolender, N., Eisele, F., Kostova, Z., Takeuchi, J., Coffino, P. and Wolf, D.H. (2006) The cytoplasmic Hsp70 chaperone machinery subjects misfolded and ER import incompetent proteins to degradation via the ubiquitin–proteasome system. *Mol Biol Cell* **18**, 153–165.

100 McClellan, A.J., Scott, M.D. and Frydman, J. (2005) Folding and quality control of the VHL tumor suppressor proceed through distinct chaperone pathways. *Cell* **121**, 739–748.

101 Taxis, C., Vogel, F. and Wolf, D.H. (2002) ER-Golgi traffic is a prerequisite for efficient ER degradation. *Mol Biol Cell* **13**, 1806–1818.

102 Haynes, C.M., Caldwell, S. and Cooper, A.A. (2002) An HRD/DER-independent ER quality control mechanism involves Rsp5p-dependent ubiquitination and ER-Golgi transport. *J Cell Biol* **158**, 91–102.

103 Spear, E.D. and Ng, D.T. (2003) Stress tolerance of misfolded carboxypeptidase Y requires maintenance of protein trafficking and degradative pathways. *Mol Biol Cell* **14**, 2756–2767.

104 Kincaid, M.M. and Cooper, A.A. (2006) Misfolded proteins traffic from the ER due to ER export signals. *Mol Biol Cell* **18**, 455–463.

105 Kruse, K.B., Brodsky, J.L. and McCracken, A.A. (2006) Characterization of an ERAD gene as VPS30/ATG6 reveals two alternative and functionally distinct protein quality control pathways: one for soluble Z variant of human alpha-1 proteinase inhibitor (A1PiZ) and another for aggregates of A1PiZ. *Mol Biol Cell* **17**, 203–212.

106 Zhou, M. and Schekman, R. (1999) The engagement of Sec61p in the ER dislocation process. *Mol Cell* **4**, 925–934.

107 Pilon, M., Schekman, R. and Romisch, K. (1997) Sec61p mediates export of a misfolded secretory protein from the endoplasmic reticulum to the cytosol for degradation. *EMBO J* **16**, 4540–4548.

108 Bebök, Z., Mazzochi, C., King, S.A., Hong, J.S. and Sorscher, E.J. (1998) The mechanism underlying cystic fibrosis transmembrane conductance regulator transport from the endoplasmic reticulum to the proteasome includes Sec61beta and a cytosolic, deglycosylated intermediary. *J Biol Chem* **273**, 29873–29878.

109 de Virgilio, M., Weninger, H. and Ivessa, N.E. (1998) Ubiquitination is required for the retro-translocation of a short-lived luminal endoplasmic reticulum glycoprotein to the cytosol for degradation by the proteasome. *J Biol Chem* **273**, 9734–9743.

110 Kalies, K.U., Allan, S., Sergeyenko, T., Kroger, H. and Romisch, K. (2005) The protein translocation channel binds proteasomes to the endoplasmic reticulum membrane. *EMBO J* **24**, 2284–2293.

111 Lee, R.J., Liu, C.W., Harty, C., McCracken, A.A., Latterich, M., Romisch, K., DeMartino, G.N., Thomas, P.J. and Brodsky, J.L. (2004) Uncoupling retro-translocation and degradation in the ER-associated degradation of a soluble protein. *EMBO J* **23**, 2206–2215.

112 Carlson, E.J., Pitonzo, D. and Skach, W.R. (2006) p97 functions as an auxiliary factor to facilitate TM domain extraction during CFTR ER-associated degradation. *EMBO J* **25**, 4557–4566.

113 Chen, B., Mariano, J., Tsai, Y.C., Chan, A.H., Cohen, M. and Weissman, A.M. (2006) The activity of a human endoplasmic reticulum-associated degradation E3, gp78, requires its Cue domain, RING finger and an E2-binding site. *Proc Natl Acad Sci USA* **103**, 341–346.

114 Fang, S., Ferrone, M., Yang, C., Jensen, J.P., Tiwari, S. and Weissman, A.M. (2001) The tumor autocrine motility factor receptor, gp78, is a ubiquitin protein ligase implicated in degradation from the endoplasmic reticulum. *Proc Natl Acad Sci USA* **98**, 14422–14427.

115 Kikkert, M., Doolman, R., Dai, M., Avner, R., Hassink, G., van Voorden, S., Thanedar, S., Roitelman, J., Chau, V. and, Wiertz, E. (2004) Human HRD1 is an E3 ubiquitin ligase involved in degradation of proteins from the endoplasmic reticulum. *J Biol Chem* **279**, 3525–3534.

116 Song, B.L., Sever, N. and DeBose-Boyd, R.A. (2005) Gp78, a membrane-anchored ubiquitin ligase, associates with Insig-1 and couples sterol-regulated ubiquitination to degradation of HMG CoA reductase. *Mol Cell* **19**, 829–840.

117 Hegde, N.R., Chevalier, M.S., Wisner, T.W., Denton, M.C., Shire, K., Frappier, L. and Johnson, D.C. (2006) The role of BiP in endoplasmic reticulum-associated degradation of major histocompatibility complex class I heavy chain induced by cytomegalovirus proteins. *J Biol Chem* **281**, 20910–20919.

118 Elkabetz, Y., Shapira, I., Rabinovich, E. and Bar-Nun, S. (2004) Distinct steps in dislocation of luminal endoplasmic reticulum-associated degradation substrates: roles of endoplasmic reticulum-bound p97/Cdc48p and proteasome. *J Biol Chem* **279**, 3980–3989.

119 Fewell, S.W., Travers, K.J., Weissman, J.S. and Brodsky, J.L. (2001) The action of molecular chaperones in the early secretory pathway. *Annu Rev Genet* **35**, 149–191.

120 Friedmann, E., Salzberg, Y., Weinberger, A., Shaltiel, S. and Gerst, J.E. (2002) YOS9, the putative yeast homolog of a gene amplified in osteosarcomas, is involved in the endoplasmic reticulum (ER)-Golgi transport of GPI-anchored proteins. *J Biol Chem* **277**, 35274–35281.

121 Litovchick, L., Friedmann, E. and Shaltiel, S. (2002) A selective interaction between OS-9 and the carboxyl-terminal tail of meprin beta. *J Biol Chem* **277**, 34413–34423.

122 Baek, J.H., Mahon, P.C., Oh, J., Kelly, B., Krishnamachary, B., Pearson, M., Chan, D.A., Giaccia, A.J. and Semenza, G.L. (2005) OS-9 interacts with hypoxia-inducible factor 1alpha and prolyl hydroxylases to promote oxygen-dependent degradation of HIF-1alpha. *Mol Cell* **17**, 503–512.

123 Ruddock, L.W. and Molinari, M. (2006) N-glycan processing in ER quality control. *J Cell Sci* **119**, 4373–4380.

124 Wang, X., Ye, Y., Lencer, W. and Hansen, T.H. (2006) The viral E3 ubiquitin ligase mK3 uses the Derlin/p97 endoplasmic reticulum-associated degradation pathway to mediate down-regulation of major histocompatibility complex class I proteins. *J Biol Chem* **281**, 8636–8644.

125 Sun, F., Zhang, R., Gong, X., Geng, X., Drain, P.F. and Frizzell, R.A. (2006) Derlin-1 promotes the efficient degradation of the cystic fibrosis transmembrane conductance regulator (CFTR) and CFTR folding mutants. *J Biol Chem* **281**, 36856–36863.

126 Liang, G., Audas, T.E., Li, Y., Cockram, G.P., Dean, J.D., Martyn, A.C., Kokame, K. and Lu, R. (2006) Luman/CREB3 induces transcription of the endoplasmic reticulum (ER) stress response protein Herp through an ER stress response element. *Mol Cell Biol* **26**, 7999–8010.

127 Nogalska, A., Engel, W.K., McFerrin, J., Kokame, K., Komano, H. and Askanas, V. (2006) Homocysteine-induced endoplasmic reticulum protein (Herp) is up-regulated in sporadic inclusion-body myositis and in endoplasmic reticulum stress-induced cultured human muscle fibers. *J Neurochem* **96**, 1491–1499.

128 Schulze, A., Standera, S., Buerger, E., Kikkert, M., van Voorden, S., Wiertz, E., Koning, F., Kloetzel, P.M. and Seeger, M. (2005) The ubiquitin-domain protein HERP forms a complex with components of the endoplasmic reticulum associated degradation pathway. *J Mol Biol* **354**, 1021–1027.

129 Mueller, B., Lilley, B.N. and Ploegh, H.L. (2006) SEL1L, the homologue of yeast Hrd3p, is involved in protein dislocation from the mammalian ER. *J Cell Biol* **175**, 261–270.

130 Omura, T., Kaneko, M., Okuma, Y., Orba, Y., Nagashima, K., Takahashi, R., Fujitani, N., Matsumura, S., Hata, A., Kubota, K., Murahashi, K., Uehara, T. and Nomura, Y. (2006) A ubiquitin ligase HRD1 promotes the degradation of Pael receptor, a substrate of Parkin. *J Neurochem* **99**, 1456–1469.

131 Kaneko, M., Ishiguro, M., Niinuma, Y., Uesugim, M. and Nomura, Y. (2002) Human HRD1 protects against ER stress-induced apoptosis through ER-associated degradation. *FEBS Lett* **532**, 147–152.

132 Nadav, E., Shmueli, A., Barr, H., Gonen, H., Ciechanover, A. and Reiss, Y. (2003) A novel mammalian endoplasmic reticulum ubiquitin ligase homologous to the yeast Hrd1. *Biochem Biophys Res Commun* **303**, 91–97.

133 Yang, H., Zhong, X., Ballar, P., Luo, S., Shen, Y., Rubinsztein, D.C., Monteiro, M.J. and Fang, S. (2006) Ubiquitin ligase Hrd1 enhances the degradation and suppresses the toxicity of polyglutamine-expanded huntingtin. *Exp Cell Res* **313**, 538–550.

134 Yamasaki, S., Yagishita, N., Sasaki, T., Nakazawa, M., Kato, Y., Yamadera, T., Bae, E., Toriyama, S., Ikeda, R., Zhang, L., Fujitani, K., Yoo, E., Tsuchimochi, K., Ohta, T., Araya, N., Fujita, H., Aratani, S., Eguchi, K., Komiya, S., Maruyama, I., Higashi, N., Sato, M., Senoo, H., Ochi, T., Yokoyama, S., Amano, T., Kim, J., Gay, S., Fukamizu, A., Nishioka, K., Tanaka, K. and Nakajima, T. (2007) Cytoplasmic destruction of p53 by the endoplasmic reticulum-resident ubiquitin ligase "Synoviolin". *EMBO J* **26**, 113–122.

135 Lee, J.N., Song, B., DeBose-Boyd, R.A. and Ye, J. (2006) Sterol-regulated degradation of Insig-1 mediated by the membrane-bound ubiquitin ligase gp78. *J Biol Chem* **281**, 39308–39315.

136 Shen, Y., Ballar, P. and Fang, S. (2006) Ubiquitin ligase gp78 increases solubility and facilitates degradation of the Z variant of alpha-1-antitrypsin. *Biochem Biophys Res Commun* **349**, 1285–1293.

137 Hassink, G., Kikkert, M., van Voorden, S., Lee, S.J., Spaapen, R., van Laar, T., Coleman, C.S., Bartee, E., Fruh, K., Chau, V. and Wiertz, E. (2005) TEB4 is a C4HC3 RING finger-containing ubiquitin ligase of the endoplasmic reticulum. *Biochem J* **388**, 647–655.

138 Arteaga, M.F., Wang, L., Ravid, T., Hochstrasser, M. and Canessa, C.M. (2006) An amphipathic helix targets serum and glucocorticoid-induced kinase 1 to the endoplasmic reticulum-associated ubiquitin-conjugation machinery. *Proc Natl Acad Sci USA* **103**, 11178–11183.

139 Oh, R.S., Bai, X. and Rommens, J.M. (2006) Human homologs of Ubc6p ubiquitin-conjugating enzyme and phosphorylation of HsUbc6e in response to endoplasmic reticulum stress. *J Biol Chem* **281**, 21480–21490.

140 Lenk, U., Yu, H., Walter, J., Gelman, M.S., Hartmann, E., Kopito, R.R. and Sommer, T. (2002) A role for mammalian Ubc6 homologues in ER-associated protein degradation. *J Cell Sci* **115**, 3007–3014.

141 Tiwari, S. and Weissman, A.M. (2001) Endoplasmic reticulum (ER)-associated degradation of T cell receptor subunits. Involvement of ER-associated ubiquitin-conjugating enzymes (E2s). *J Biol Chem* **276**, 16193–16200.

142 Arai, R., Yoshikawa, S., Murayama, K., Imai, Y., Takahashi, R., Shirouzu, M. and Yokoyama, S. (2006) Structure of human ubiquitin-conjugating enzyme E2 G2 (UBE2G2/UBC7). *Acta Crystallograph Sect F Struct Biol Cryst Commun* **62**, 330–334.

143 Kim, B.W., Zavacki, A.M., Curcio-Morelli, C., Dentice, M., Harney, J.W., Larsen, P.R. and Bianco, A.C. (2003) Endoplasmic reticulum-associated degradation of the human type 2 iodothyronine deiodinase (D2) is mediated via an association between mammalian UBC7 and the carboxyl region of D2. *Mol Endocrinol* **17**, 2603–2612.

144 Kothe, M., Ye, Y., Wagner, J.S., De Luca, H.E., Kern, E., Rapoport, T.A. and Lencer, W.I. (2005) Role of p97 AAA-ATPase in the retrotranslocation of the cholera toxin A1 chain, a non-ubiquitinated substrate. *J Biol Chem* **280**, 28127–28132.

145 Nowis, D., McConnell, E. and Wojcik, C. (2006) Destabilization of the VCP-Ufd1-Npl4 complex is associated with decreased levels of ERAD substrates. *Exp Cell Res* **312**, 2921–2932.

146 Ye, Y., Meyer, H.H. and Rapoport, T.A. (2003) Function of the p97–Ufd1–Npl4 complex in retrotranslocation from the ER to the cytosol: dual recognition of nonubiquitinated polypeptide segments and polyubiquitin chains. *J Cell Biol* **162**, 71–84.

147 Meyer, H.H., Wang, Y. and Warren, G. (2002) Direct binding of ubiquitin conjugates by the mammalian p97 adaptor complexes, p47 and Ufd1-Npl4. *EMBO J* **21**, 5645–5652.

148 Meyer, H.H., Shorter, J.G., Seemann, J., Pappin, D. and Warren, G. (2000) A complex of mammalian ufd1 and npl4 links the AAA-ATPase, p97, to ubiquitin and nuclear transport pathways. *EMBO J* **19**, 2181–2192.

149 Botta, A., Tandoi, C., Fini, G., Calabrese, G., Dallapiccola, B. and Novelli, G. (2001) Cloning and characterization of the gene encoding human NPL4, a protein interacting with the ubiquitin fusion-degradation protein (UFD1L). *Gene* **275**, 39–46.

150 Kleijnen, M.F., Alarcon, R.M. and Howley, P.M. (2003) The ubiquitin-associated domain of hPLIC-2 interacts with the proteasome. *Mol Biol Cell* **14**, 3868–3875.

151 Kleijnen, M.F., Shih, A.H., Zhou, P., Kumar, S., Soccio, R.E., Kedersha, N.L., Gill, G. and Howley, P.M. (2000) The hPLIC proteins may provide a link between the ubiquitination machinery and the proteasome. *Mol Cell* **6**, 409–419.

152 Kaur, M., Pop, M., Shi, D., Brignone, C. and Grossman, S.R. (2006) hHR23B is required for genotoxic-specific activation of p53 and apoptosis. *Oncogene* **26**, 1231–1237.

153 Chen, L. and Madura, K. (2006) Evidence for distinct functions for human DNA repair factors hHR23A and hHR23B. *FEBS Lett* **580**, 3401–3408.

7
Interactions between Viruses and the Ubiquitin–Proteasome System

Jessica M. Boname and Paul J. Lehner

7.1
Introduction

Ubiquitination plays an increasingly important role in the regulation of many essential cellular processes. Like phosphorylation, glycosylation, methylation and acetylation, ubiquitination is a post-translational modification that tags a protein allowing recognition by host cellular machinery that will in turn, direct its subcellular localization and fate. The cellular pathways regulated by ubiquitination are diverse and include proteolysis, membrane protein trafficking, transcription, cell cycle control and cell signaling [1–3]. In addition, ubiquitination is also involved in important viral processes such as entry and egress.

A productive viral infection requires the effective manipulation of host functions by different viral genes. As obligate intracellular parasites, viruses have co-evolved with their hosts and adapted many cellular pathways for their own requirements. Dissecting the role of viral proteins in disease pathogenesis has led to not only a greater understanding of the function of the viral proteins, but also allowed identification of cellular homologs and an improved understanding of normal cell physiology. The ubiquitin–proteasome system (UPS) is no exception to this. Indeed the UPS provides many fine examples of how different viruses manipulate the host ubiquitin machinery resulting in altered protein function. Since the best recognized function of the UPS is protein degradation, it is not surprising that viruses are particularly adept at exploiting this pathway, leading to the accelerated degradation of cellular proteins which may interfere with viral fitness – examples include the virally-induced degradation of p53, MHC class I and APOBEC3G. Viral proteins may also manipulate the UPS to affect membrane trafficking, control of cell cycle, DNA repair, alterations of the immune system and may disrupt virtually every pathway involved in the UPS. A detailed review of all these pathways is beyond the scope of this chapter; however a brief overview of the viral proteins known to interact with the UPS will be followed by a discussion of viral E3 ligases and viral proteins that recruit cellular E3 ligases.

Protein Degradation, Vol. 4: The Ubiquitin-Proteasome System and Disease.
Edited by R. J. Mayer, A. Ciechanover, M. Rechsteiner

ISBN: 978-3-527-31436-2

7.2
Overview of Viruses and the Ubiquitin–Proteasome System

7.2.1
Proteolysis

Proteolysis within the cell is closely linked to the UPS, and viruses can interfere with many stages of protein degradation including the proteasome, the ER-associated degradation pathway (ERAD) and lysosomal degradation pathways. The proteasome is the major non-lysosomal site of protein degradation within the cell. Lysine-48-linked polyubiquitinated proteins are normally targeted for degradation by the proteasome, but during certain viral infections, the normal proteasome-mediated degradation process is perturbed. To avoid presentation to cytotoxic T lymphocytes (CTL), the Epstein-Barr virus (EBV) encoded EBNA-1 gene product contains an internal glycine–alanine repeat motif that not only prevents degradation by the proteasome but also reduces its rate of translation, blocking the formation of "Defective Ribosomal Products" (DRiPs) and therefore prevents the subsequent release of peptides for binding MHC class I molecules [4–6]. Similarly, latency-associated nuclear antigen-1 (LANA-1) of Kaposi's sarcoma-associated virus (KSHV or HHV-8) contains a strongly acidic string of amino acids that will also block the presentation of *cis* encoded peptides to CTL [7]. The matrix phosphoprotein pp65 of human cytomegalovirus (HCMV) is a major virion component that also blocks proteasome-mediated peptide generation. As an abundant component of the virion, pp65 is already present in high concentrations when the virus enters the cell, and does not require *de novo* synthesis. pp65 decreases the presentation of peptides *in trans* derived from the major immediate early transactivator (IE1) of HCMV [8]. Hepatitis B virus protein X (HBX) binds the proteasome and inhibits its protease and chymotryptic peptidase functions, leading to enhanced virus replication [9–11]. Tat, the transcriptional activator encoded by human immunodeficiency virus (HIV), inhibits the peptidase activity of the 20S proteasome and interferes with formation of the 20S proteasome–11S regulator complex by interfering with levels of LMP2, LMP7 and MECL1 transcripts – the components of the immunoproteasome [12, 13].

7.2.2
Viruses and the ERAD Pathway

Endoplasmic Reticulum Associated Degradation (ERAD) is a quality control step that involves the dislocation or retrograde translocation of misfolded proteins from the lumen of the ER to the cytosol where they are destined for proteasome-mediated degradation [14]. Misfolded proteins must be selected by ER-resident chaperones and brought to sites of retro-translocation, where they are then transferred back across the ER membrane via a protein-conducting channel. This retro-translocation requires polyubiquitination and the cytosolic ATPase p97/Cdc48. Insight into this pathway was initially achieved through study of the US2 and US11 gene products of human cytomegalovirus (HCMV). US2 and US11 were originally identified as proteins that downregulate major histocompatibility (MHC) class I molecules from the surface of infected cells by catalyzing the rapid retro-translocation of MHC class I heavy chains from the lumen of the ER back to the cytosol [15–17]. By using these viral proteins as a model system, many details of ERAD have been elucidated. Polyubiquitination is required for US2- and US11-dependent degradation of class I heavy chains [18, 19]. However, while ubiquitination of the class I molecules themselves is required for US2-mediated dislocation, a "lysineless" class I molecule can be dislocated from the ER in the presence of US11. While there is still a requirement for ubiquitination in the dislocation process, the target of this ubiquitination has not been identified [20].

Derlin-1 was identified as a binding partner of US11, but not of US2, required for dislocation [21, 22]. It is a tetraspanning ER membrane protein, proposed to act as a channel for the dislocation of misfolded proteins from the ER [23]. Another protein associated with ERAD is p97, a cytosolic AAA-ATPase required to provide energy for driving proteins through the dislocation channel or for releasing dislocated proteins from the cytosolic face of the ER membrane [24]. Although Sec61 was initially identified as a component of the retro-translocation machinery involved in US2-dependent class I degradation [17], more stringent conditions of analysis have not confirmed this. Further analysis of US2-associated proteins required for ER dislocation identified a role for the signal peptide peptidase (SPP), an intramembrane-cleaving aspartic protease of the presenilin family [25, 26]. While depletion of SPP by RNA interference blocks US2-mediated class I heavy chain degradation, it remains unclear whether it is the protease activity of SPP, or an additional function that is required for class I dislocation.

7.2.3
Membrane Protein Trafficking and Endosomal Sorting

Ubiquitination plays a role in the regulation of both the endocytic and exocytic pathways. The direct ubiquitination of a cell surface receptor, as well as its adaptor protein, is used as a mechanism for receptor internalization [27]. The direct effect of surface receptor ubiquitination is cargo recognition, internalization and sorting by the cellular trafficking machinery. Cargo sorting based on ubiquitination may result in either recycling to the plasma membrane or lysosomal degradation. Fine regulation occurs by a combination of ubiquitination and deubiquitination of components of the endocytic pathway and the cargo itself, although the details remain to be defined [28, 29]. For some receptors monoubiquitination is a sufficient stimulus for internalization [30, 31]. However, an increasingly important role for lysine-63-linked polyubiquitination is being currently recognized [32–34].

The highly conserved Endosomal Sorting Complex Required for Transport (ESCRT) machinery is recruited from the cytosol for the sorting of predominantly ubiquitinated proteins to multivesicular bodies (MVB) formed from membranes of the late endosomal compartments [35]. Tumor Susceptibility gene 101 (TSG101), a component of ESCRT-I, is essential for the sorting of ubiquitinated proteins to the MVB [36]. Depletion of TSG101 rescues the downregulation of MHC class I molecules caused by the K3 gene product of Kaposi's sarcoma-associated herpesvirus (KSHV), by allowing recycling rather than lysosomal degradation of the polyubiquitinated class I molecules [37]. Ubiquitin-mediated receptor downregulation by related viral E3 ligases will be discussed in more detail below. Murine cytomegalovirus (MCMV) encodes several proteins known to downregulate NKG2D ligands on the surface of infected cells in order to avoid activating natural killer (NK) cells – the first line of defence against MCMV infection. Glycoprotein 40 (gp40) is expressed from the m152 gene of MCMV and is known to target RAE-1 [38], while the related gene, m155, encodes a protein shown to downregulate H60 in a UPS-dependent manner [39]. The adenovirus gene products 10.4 K and 14.5 K form a heterotrimeric complex known as receptor internalization and degradation (RID). RID expression results in the endocytosis and degradation of several important death receptors including tumor necrosis factor receptor 1 (TNFR1), TNF-related apoptosis-inducing ligand (TRAIL) receptors 1 and 2 (TR1, TR2) and FAS, thereby protecting adenovirus-infected cells from apoptosis. Although not much is known about the mechanism of action of RID, it is known that targeting of TNFR1 involves the mu2 subunit of adaptor protein 2 and that clathrin-dependent internalization proceeds at a normal rate. However, increased degradation probably results from RID-promoted sorting into the endosomal/lysosomal compartment [40, 41].

7.2.4
Viral Entry and Egress

The UPS is involved in many stages of viral pathogenesis. Two key aspects of viral biology are entry and egress from the cells in which they replicate. Proteasome inhibition blocks the transfer of mouse hepatitis virus particles from endosomes into the cytosol suggesting a role for the UPS in the entry of this virus into cells [42]. Nuclear penetration of certain parvoviruses, including the minute virus of mice and canine parvovirus was also shown to involve the UPS. Inhibition of the proteasome by MG132 resulted in an accumulation of virus particles in the perinuclear region of the cell [43]. Sumoylation of the capsid protein of murine leukemia virus (MLV) by Ubc9 and PIASy was required for an early stage of virus replication; in the absence of capsid sumoylation, reverse-transcribed viral genomes were unable to circularize and enter the nucleus for integration into the host genome [44].

Many viruses recruit the ubiquitin-dependent ESCRT machinery to their site of release at the plasma membrane to promote viral budding and their egress from the cell. Smaller enveloped viruses including retroviruses (HIV-1, HTLV-1, MLV, Rous sarcoma virus, equine infectious anemia virus), filoviruses (Ebola), rhabdoviruses (vesicular stomatitis virus, rabies virus), arenaviruses (Lassa fever virus, lymphocytic choriomeningitis virus) and paramyxoviruses (simian virus 5, human parainfluenza virus) all require virally encoded sequences called late assembly or L-domains for separation of virus from host cells. Various L-domain motifs have been documented including PTAP, PPxY, FPIV, LxxLF and YPxL [45]. For example the PTAP motif within the p6 domain of HIV-1 Gag binds to TSG101 [46] and indeed the complete ESCRT-1 is required for PTAP-mediated virus budding [47]. Tal, the TSG101-associated ligase, is necessary for the multiple monoubiquitination of TSG101 and regulates retrovirus budding [48]. Thus virus budding is akin to MVB formation with membrane invagination and pinching off of vesicles into a non-cytoplasmic compartment. A second L-domain motif, PPxY recruits the cellular E3 ligase Nedd4 through interactions with this HECT E3 ligase's WW domain [49]. Both the PTAP and PPxY motifs are involved in Ebola virus and human T lymphotropic virus type 1 (HTLV-1) budding [50, 49]. Two HBV proteins, core and the envelope protein L, interact with gamma2-adaptin, while the core protein also interacts with the E3 Nedd4 through its L-domain-like PPAY motif. Mature virus production was also decreased when a catalytically inactive form of Nedd4 was expressed [51]. YPxL L-domains recruit the E3 ligase AIP1 to facilitate budding of viruses including equine infectious anemia virus [52], Sendai virus [53] and HIV-1 [54]. Ubiquitination of foamy virus glycoprotein LP leads to an increase in subviral particle release [55].

7.2.5 Transcriptional Regulation

The UPS is required for transcriptional regulation, and many viruses disrupt cellular transcription to favor virus replication and counter host anti-viral strategies. The herpesvirus transactivators are a case in point. ICP0 of herpes simplex virus type-1 (HSV-1) is a virally encoded E3 ligase that transactivates both viral and cellular genes and is discussed in more detail below. Ubiquitination of viral transactivators is also important as exemplified by a requirement for ubiquitination of the HSV-1 virion transactivator VP16 by the E3 ligase Met30 [56]. Ubiquitination of the HIV-1 transactivator Tat has been shown to increase its activity [57].

7.2.6 Cell Cycle Control

The cell cycle is controlled by the activity of cyclin-dependent kinases (cdks) which either drive or inhibit crucial events in cell division. Another key regulator of cell cycle progression is the anaphase-promoting complex (APC). Through its E3 ligase activity the APC targets cdks and other regulators of the cell cycle for degradation by the UPS [58]. Viruses often disrupt the cell cycle in order to upregulate cellular proteins required for virus replication. This may lead to dysregulation of the cell cycle and transformation. Indeed the tumor suppressor properties of two important cell regulatory proteins, p53 and retinoblastoma sensitivity protein (Rb), were recognized in part by their ability to suppress tumor formation caused by the expression of "oncogenic" proteins of certain viruses [59, 60].

The specific targeting of p53 for proteasomal degradation is a common theme for viruses. E6 of human papilloma virus (HPV) has long been known to target p53 for ubiquitin-dependent degradation [61]. E1B55K and E4orf6 of adenovirus recruit a cellular E3 to ubiquitinate p53 while the E3 ligase ICP0 of HSV-1 can ubiquitinate p53 [62, 63]. Similarly, E7 of HPV and EBV nuclear antigen 3C (EBNA3C) induce ubiquitin-dependent degradation of Rb [64, 65]. The viral trans-activator Tax of HTLV-1 prematurely activates the APC and disrupts mitosis by promoting the polyubiquitination of cyclin B1 [66]. ICP0 of HSV-1 also impacts on cell cycle progression, although reports differ as to whether or not ICP0 induces the degradation of the ubiquitin-conjugating enzyme (E2) cdc34 thereby stabilizing cyclins D1 and D3 [67, 68].

Programmed cell death or apoptosis is often a consequence of viral infection. The term "inhibitors of apoptosis" (IAP) was first used to describe a baculovirus protein that blocked host insect cell apoptosis during viral infection. This protein contained a zinc finger motif that was also found in other proteins known to regulate apoptosis [69, 70]. Now several virus families have been shown to encode IAP proteins with E3 activity as noted later in this chapter.

Gamma herpesviruses establish latency in cells of lymphoid origin, and must insure that their episomal genome is replicated during cell division. EBNA-1, a viral protein important for viral DNA replication and EBV genome segregation, binds a cellular deubiquitinating enzyme (DUB) – the ubiquitin-specific protease 7 (USP7) also known as the herpesvirus-associated USP (HAUSP) – although the functional significance of this is still under investigation [71].

7.2.7
Cell Signaling

During the course of infection viruses downregulate surface receptors in a variety of ways that often mimic normal cellular processes. Receptor down-modulation in response to ligand binding shows some similarities with viral down-modulation of class I MHC [72]. The process of viral interference with the class I presentation pathway has been extensively studied [73, 74] and there are many examples of ubiquitin-mediated regulation induced by viruses to circumvent cellular regulation. The RTA protein of KSHV induces ubiquitination and proteasome-mediated degradation of interferon regulatory factor 7 (IRF7) [75].

The Janus kinase/signal transducers and activators of transcription (Jak/STAT) pathway are important in regulating cytokine signal transduction, cell growth and cell survival. In HTLV-1-transformed T cells, induction of the deubiquitinating enzyme DUB-2 results in prolonged activation of the Jak/STAT pathway. This in turn suppresses apoptosis of the virally transformed cell [76]. As will be described below, many paramyxoviruses encode V proteins that control degradation of STAT proteins via the UPS system [77–79].

The Toll Like Receptors (TLRs) as extracellular pattern recognition receptors, together with the intracellular recognition receptors such as nucleotide-binding oligomerization domain (NOD) and protein kinase R (PKR), detect pathogen-associated molecular patterns (PAMPS) and initiate a cascade of cellular signals, culminating in the activation of NFκB and the production of proinflammatory cytokines, such as TNF-α, IFN-γ, and the release of interleukins. Activation of the NFκB signaling pathway therefore represents an important line of defence against virus infection. Regulation of this signal cascade by the UPS occurs at three points – degradation of the inhibitor of NFκB (IκB), processing of NFκB precursors and the activation of the IκB kinase (IKK) [80]. The HBV E5 protein disrupts the interaction of the E3 ligase cCbl with the epidermal growth factor receptor (EGFR). This leads to a decrease in EGFR ubiquitination and degradation, and thus to an increase in EGFR signaling [81]. There is also a report that ICP0, the E3 ligase encoded by HSV-1, complexes with cCbl and its adaptor protein CIN85 to increase the degradation of the EGFR in cells transfected with ICP0 alone, and in cells infected with HSV-1 [82]. ICP0 has also been reported to activate NFκB by catalyzing the polyubiquitination of IκB [83].

The Wnt/β-catenin signaling pathway is important during development and differentiation, and is often dysregulated in cancer. The polyubiquitination and degradation of β-catenin by the E3 ligase "seven in absentia homolog 1" (Siah-1) is important for turning off signal-transduction through growth factor receptors. Expression of latent membrane protein 1 (LMP1) of EBV decreases the levels of Siah-1. Thus β-catenin levels are stabilized to activate signaling pathways [84]. Another protein expressed in cells latently infected with EBV is LMP2A. LMP2A is important for maintaining latency in B cells through its interaction with Nedd4 family E3 ligases [85]. In epithelial cells, LMP2A inhibits differentiation and activates β-catenin signaling, perhaps contributing to carcinogenesis in these cells [86]. β-Catenin also accumulated when an additional E3 ligase complex, beta-transducin repeat-containing protein (βTrCP), was sequestered in the cytoplasm by Vpu of HIV-1 [87].

7.3 Viruses and E3 Ubiquitin-Protein Ligases

The ubiquitin reaction involves (i) ubiquitin activation via an E1 enzyme, (ii) transfer of the ubiquitin via a cysteine residue to an E2 ubiquitin-conjugating enzyme and (iii) targeting of the charged ubiquitin from the E2 to the lysine residue of the target protein. This latter reaction is catalyzed by the ubiquitin E3 ligase that associates with the substrate and thereby confers specificity to the ubiquitination reaction. Ubiquitin E3 ligases come in three different forms, containing a HECT (homologous to E6AP carboxyl terminus), a RING (really interesting new gene) or a U-box (UFD2 homology) domain [88]. Although the E3 ligase activity of APC and SCF complexes was well known, it was not until an analysis of the subunits of APC revealed homology with a subunit of SCF that the idea that the RING-finger motif might represent a superfamily of E3 ubiquitin ligases was appreciated [89, 90] whereas the concept of a HECT family of E3 ligases was recognized much earlier [91]. The E6 protein encoded by HPV associates with a cellular protein E6AP and, in oncogenic strains of the virus, the E6/E6AP complex binds and targets the p53 tumor-suppressor protein for ubiquitin-mediated proteolysis [61, 92]. It was found that the E6AP ubiquitin E3 ligase directly accepts ubiquitin from an E2 ubiquitin-conjugating enzyme in the form of a thioester and transfers the ubiquitin to targeted substrates. Subsequently the HECT family of related proteins with a C-terminus sequence motif similar to that of E6AP was identified and shown to have E3 ligase activity [91]. The HECT family has 28 family members.

In contrast to the HECT family, other E3 ligases do not directly bind ubiquitin, but recruit a ubiquitin-charged E2 conjugating enzyme and promote transfer of ubiquitin to the target protein after substrate binding. These are the RING family of E3 ligases [93]. RING domains contain a variable number of interspersed cysteine and histidine residues which bind the two zinc ions required for E3 ligase activity and fold to form a cross-braced structure [88]. The arrangement of cysteines and histidines can be C_3HC_4 (RING-HC) or $C_3H_2C_3$ (RING-H2) for the canonical RING-finger-containing E3s or C_4HC_3 (RING-CH) for the more recently described RING-variant structure [94, 95]. The SCF E3 ligases containing multiple subunits including a Skp-Cullin-F-box are a subset of the RING domain-containing E3s. The U-box-containing family is based on an E2 binding domain first described in the yeast Ufd2 protein [96]. U-box-containing proteins adopt a structure that is very similar to a RING domain without the zinc coordination, but rather than ubiquitinating substrates directly, they polyubiquitinate substrates of other E3s and are thus also referred to as E4s [88]. While viruses encode a number of RING-containing E3 ligases, as yet no virally encoded HECT or U-box domain-containing proteins have been described. Viruses do, however, recruit both RING- and HECT-containing cellular E3 protein-ubiquitin ligases.

7.3.1
ICP0 – A Viral RING E3 Ligase in HSV Activation

The most widely studied viral RING-containing protein is ICP0 of HSV-1. ICP0 is an immediate early protein required for the activation of most viral and many cellular genes and is critical for the reactivation of HSV-1 from latency. ICP0 has a RING domain near its N-terminus, encoded in exon 2, which confers E3 ligase activity and induces the proteasome-dependent degradation of substrates including components of centromeres and PML bodies [97–99], although its exact protein substrates have yet to be defined. The RING domain is required for the accumulation of conjugated ubiquitin species, as is the RING finger of ICP0-related proteins from other alphaherpesviruses [100, 101]. Controversy exists about the ability of ICP0 to affect cellular levels of cyclins D1 and D3 through the degradation of cdc34 (also known as UbcH3), the cellular E2 component of the SCF complexes that target these cyclins for degradation. It remains controversial whether additional ICP0 E3 ligase activity, associated with exon 3 of ICP0 rather than the RING domain encoded within exon 2, promotes the autoubiquitination of cdc34 [102, 103]. When conditions were adjusted to allow a similar rate of progression of virus infection in the presence and absence of ICP0, no evidence was found for ICP0-mediated protection of cyclins D1 and D3, nor the degradation of cdc34 [67]. *In vitro* E3 ligase activity of the RING domain encoded within exon 2 has been demonstrated using the E2 enzymes UbcH5a and Ubc6 [104, 105]. Subsequently, the degradation of PML and Sp100 *in vivo*, activities associated with the dispersal of ND10 and reactivation and lytic replication of HSV-1, was shown to depend on the RING-finger domain of ICP0 and the E2 ubiquitin-conjugating enzyme UbcH5a [106, 107]. The question of whether PML is a direct target of ICP0 and the relationship with SUMO-1-modified PML and the SUMO-specific protease SENP1 is complex and remains unresolved [108, 106].

Other targets of ICP0 include p53 [62], the translational repressor 4E-BP1 [109] and the cellular ubiquitin-specific protease USP7 [110]. Analysis of the interaction between ICP0 and USP7 is an elegant example of a viral gene product exploiting the cellular machinery to full use. USP7 binds ICP0 and protects ICP0 from autoubiquitination and subsequent degradation. This is important as autoubiquitination is a key feature of RING-containing E3 ligases and leads to the short half-life of these proteins. The interaction between ICP0 and USP7 is finely balanced. By binding USP7, ICP0 is stabilized and escapes autoubiquitination, but inevitably the interaction of these two protein leads to ubiquitination and degradation of USP7 itself. The stabilization of ICP0 by USP7 appears to be dominant over the ubiquitination and degradation of USP7 [111, 110], allowing the virus to synthesize only small amounts of ICP0 protein.

7.3.2
Preventing the Release of Interferon

All interferons (IFNs) induce an antiviral state in target cells causing impaired virus replication. Due to the effectiveness of the IFN response, many viruses encode gene products that utilize different mechanisms to counteract the release of IFN and development of the IFN-mediated antiviral state. IRF7, a key mediator of type I IFN induction, is targeted for degradation by binding RTA, the immediate-early nuclear transcription factor of KSHV. RTA has E3 ligase activity and blocks IRF7-mediated IFN-α and IFN-β mRNA production by promoting ubiquitination and degradation of the IRF7 protein in a proteasome-dependent fashion *in vivo* and *in vitro*. Like ICP0, RTA also autoregulates its own polyubiquitination and stability, and both activities require the N-terminal cysteine/histidine-rich domain of RTA [75]. Therefore, RTA manipulates the stability and function of IRF7 and provides a regulatory strategy for circumventing the innate immune defence system. Interestingly, RTA from the related gammaherpesvirus EBV is sumoylated by PIAS1 and Ubc9, and this enhances the transactivation function of RTA, although there have been no reports of E3 ligase activity for this protein [112].

Many other RING-containing proteins are encoded by viruses, including some that have been shown to have E3 ligase activity, such as those encoded by the baculoviruses. The Bombyx mori nucleopolyhedrovirus (BmNPV) has six predicted RING-finger proteins, four of these, IAP2, IE2, PE38 and CG30 can induce polyubiquitin chain formation *in vitro*, three in conjunction with the Ubc4/5 E2 conjugating enzymes [113]. The inhibitor of apoptosis Op-AIP3 from the related baculovirus Orgyia pseudotsugata M nucleopolyhedrovirus, was shown to have RING-dependent autoubiquitination activity and was able to ubiquitinate the host pro-apoptotic protein HID [114]. The white spot syndrome virus (WSSV) of shrimp encodes four proteins with RING-finger motifs one of which, WSSV249, has been shown to interact with and sequester the shrimp E2 PvUbc. It shows little specificity in that it will also ubiquitinate in the presence of UbcH1, UbcH2, UbcH5a, UbcH5b, UbcH5c, UbcH6 and UbcH10 [115]. A second WWSV RING-finger protein, WSSV222, functions to block apoptosis by causing the degradation of the shrimp turnover suppressor-like protein TSL [116].

Poxviruses encode RING-finger proteins with both the classical C_3HC_4 and the RING-CH sequence motif. Variola (the causative agent of smallpox) and ectromelia virus (mousepox) both encode a RING protein p28 that *in vitro* functions with Ubc4 and UbcH5c, and forms lysine-63 linked polyubiquitin chains in the presence of Ubc13 [34]. Although host cellular targets have yet to be identified, p28 is known to be an important virulence determinant for ectromelia virus [34, 117].

7.3.3
Viral E3 Ligases Ubiquitinate and Dispose of Critical Immune Receptors

The first E3 protein-ubiquitin ligase activity described for a RING-CH protein was that of the mK3 protein of the murine gammaherpesvirus 68 (MHV-68). MK3 is unusual in that as well as its RING-CH domain it is a type III transmembrane protein including two transmembrane domains with both the N and C termini in the cytosol. Encoded within the N-terminal domain of the protein is a C_4HC_3 zinc finger similar to that of a plant homeodomain (PHD) and thus different from the canonical zinc fingers of E3 ubiquitin-protein ligases – RING-HC (C_3HC_4) or RING-H2 ($C_3H_2C_3$). However, the NMR solution structure of the highly related zinc-finger from the K3 protein of KSHV was determined to be a RING-CH or RING-variant similar to those of E3 ligases rather than a PHD zinc finger [94]. This RING-variant domain in membrane-associated proteins has since been identified in other viruses and indeed in the human genome where a family of proteins (dubbed MARCH for membrane associated RING-CH) has been shown to have E3 ligase activity [118].

7.3.4
Degradation of MHC Class I Molecules by the mK3 Protein of MHV-68 Virus

CTL recognize and kill a virally-infected cell through identification of viral peptides bound to MHC class I molecules at the surface of the infected cell. The importance of this viral defence pathway is emphasized by the number of different viruses that encode many different gene products to prevent MHC class I-mediated antigen presentation [73]. Indeed, many viruses encode several unrelated gene products whose main function appears to be subversion of the antigen presentation pathway. As described earlier, involvement of the UPS is best characterized by the US2 and US11 gene products from HCMV which subvert the ER quality control pathway to dislocate MHC class I heavy chains from the ER to the cytosol where proteasome-mediated degradation ensues [24].

MK3 uses a related mechanism in that it directly ubiquitinates MHC class I molecules and other components of the antigen presentation pathway. MK3 was initially identified by screening a plasmid library of the MHV-68 genome for activities that downregulate cell surface MHC class I molecules. Overexpression of mK3 in cell culture resulted in a rapid proteasome-dependent degradation of MHC class I molecules and an inhibition of class I antigen presentation [119]. This proteasome-dependent degradation of conformational class I molecules was shown to require the N-terminal RING domain and immunoprecipitation of class I associated with mK3 in the presence of proteasome inhibitors yielded class I and a higher molecular weight "ladder" of ubiquitinated class I species [120]. Further experiments showed that mK3 associates with the ER-resident MHC class I peptide loading complex, and binding to the peptide loading complex was required for mK3 stabilization, thus preventing its rapid degradation [121, 122]. In addition to ubiquitination of MHC class I heavy chains, mK3 activity also results in degradation of the TAP peptide transporter and tapasin, both important components of the peptide loading complex [123, 124]. By targeting additional components of the peptide loading complex, as well as class I, mK3 is also able to degrade other immunoreceptors which are dependent on peptide loading, but lack cytoplasmic lysine residues and might otherwise escape ubiquitination. For example, the GPI-linked MHC class I like molecule Qa2 is effectively downregulated by mK3 but contains no relevant lysine residues.

The increased stability offered to mK3 by binding the peptide loading complex also allows mK3 to "buffer" cytokines such as IFN-γ which increase class I surface expression through upregulation of class I and components of the peptide loading complex – TAP and tapasin. The high levels of class I normally induced by virus infection through IFN-γ activity are balanced by an increase in mK3 stability on the peptide loading complex and therefore activity in degrading these components.

How does mK3-mediated ubiquitination of class I heavy chains from the ER lead to degradation by the proteasome? Like US11, the link between mK3-mediated MHC class I ubiquitination and subsequent proteasome-mediated degradation appears to be the ERAD complex containing Derlin1 and p97. In addition to the peptide loading complex, mK3 associates with Derlin1 and requires the AAA-ATPase activity of p97 for degradation of ubiquitinated class I molecules [125].

7.3.5
Degradation of Immunoreceptors by Kaposi's Sarcoma-associated Herpesvirus

Given the high degree of homology between mK3 and its homologs in KSHV, K3 and K5 [126, 127, 119], it was not surprising that these proteins were also found to function as E3 ubiquitin-protein ligases and to target specific immunoreceptors for degradation. However, a fundamental difference between these related viral ligases is that unlike mK3, K3 and K5 do not predominantly function in the ER but ubiquitinate their targets at the cell surface where they are internalized and in some cases degraded [128, 37]. More is known about how K3 downregulates its target, MHC class I, than K5 and its multiple targets. K3-dependent ubiquitination of class I molecules occurs in the late secretory pathway and is followed by internalization and sorting via the endosomal machinery leading to lysosomal degradation [37]. This endolysosomal sorting is clathrin dependent with a requirement for epsin – a clathrin adaptor [32]. Endolysosomal sorting is dependent on components of the ESCRT-I machinery [37], but independent of ESCRT-II [72]. For members of the tyrosine kinase growth factor receptor family, monoubiquitination provides a sufficient signal for internalization and degradation [129, 31]. While K3 recruitment of UbcH5 is necessary for monoubiquitination of surface MHC class I molecules this is not sufficient to signal class I degradation. Indeed K3 recruits the Ubc13 E2 conjugating enzyme which catalyzes the lysine-63-linked polyubiquitination of class I, leading to class I internalization and degradation [32]. K3 (but not K5) is also the first ubiquitin E3 ligase shown to ubiquitinate mutant class I molecules via a single cysteine residue in a thiol-ester linkage, as opposed to via a lysine residue [130].

K5 ubiquitinates and downregulates MHC class I molecules as well as other immunoreceptors including B7.2, ICAM-1, CD31 and CD1d [126, 127, 131, 132, 119]. The specificity of K5 activity and how it effectively targets so many different immunoreceptors is unclear. The functional consequences are a decreased activation of T cells [132], reduced endothelial cell migration [131] and inhibition of YTS (NK cell line) killing in an MHC class I unrestricted fashion [133]. Immunoprecipitation of MHC class I and B7.2 followed by western blotting with ubiquitin-specific antibody demonstrated K5-mediated ubiquitination *in vivo*, while *in vitro* ubiquitination was demonstrated with GST or GST-K5 plus ubiquitin, ATP, E1 and UbcH5a [128]. Endocytosis was implicated in the downregulation of ICAM-1 and B7.2 in B cells by the use of dominant negative dynamin [134]. While K5 expression results in the downregulation and lysosomal degradation of class I and B7.2, this is not the case for all targets; CD1d is downregulated but not degraded in cells expressing K5 [132], while CD31/PECAM is ubiquitinated and degraded by both the proteasome and lysosomes [131]. Further details about K5-mediated ubiquitination, including the E2 conjugating enzymes recruited, mono-versus-poly ubiquitination, and mechanism of target acquisition remain to be elucidated.

Other viral RING-CH E3 ligases have also been described. They all share a similar organization to the K3 family (a RING-CH domain followed by two transmembrane regions). The K3-related protein of myxoma virus (MV), M153R or MV-LAP, contains an N-terminal RING-CH domain and targets MHC class I, Fas and CD4 for internalization and lysosomal degradation [135–137]. Interestingly, optimal activity of M153R requires myxoma virus infection [138], suggesting the involvement of additional viral proteins. Some viral RING proteins function as part of larger SCF E3 protein-ubiquitin ligases as detailed below.

7.3.6
Viral SCF E3 Ligases

Some viral proteins function as part of the larger subunit type SCF ubiquitin ligases. The Cullin E3 ubiquitin ligases are a family of modular RING E3 ligases that consist of four main components: a Cullin (Cul1, 2, 3, 4a, 4b, 5, or 7), a RING-finger protein, an adaptor protein, and a substrate receptor. Cul1 serves as a scaffold for the assembly of the catalytic components consisting at its C-terminus of the RING-finger protein Rbx1, which binds and activates the E2 conjugating enzyme. At the N-terminus Cul1 binds Skp1 and an F-box family member that serves as a specificity factor for substrate binding [139]. This organization allows the F-box protein to bring the substrate protein into the proximity of the ubiquitination machinery. Like papillomaviruses, other DNA tumor viruses target p53 for inactivation or degradation. Adenovirus expresses two proteins, E1B-55K and E4orf6 that form a complex with Cul5, Rbx-1 and elongins B and C to target p53 for ubiquitination and degradation. Both viral proteins act independently to bind p53 and prevent gene expression. In combination they induce ubiquitin-dependent degradation [140, 141].

The V proteins of paramyxoviruses all interfere with the Jak/STAT signal transduction pathway. Type II human parainfluenza virus V protein forms a complex with the damaged DNA binding protein (DDB1) and Cul4a to polyubiquitinate STAT2 *in vivo*, while the V protein of simian virus 5 (SV5) functions as an adaptor to bind STAT2 and DBB1 in complex with Cul4a facilitating the polyubiquitination and degradation of STAT1. The loss of STAT1 in turn leads to a decrease in affinity of V for STAT2 and presumably a dissociation of the complex [77, 142]. The structure of the DDBA–Cul4a–V protein complex shows that the viral protein is inserted into the double propeller pocket of the DDB1 protein while the third propeller of DDB1 binds to Cul4a allowing the V protein to recruit alternative substrates to the E3 ligase complex [143].

7.3.7 HIV Vif and APOBEC Function

Cytidine deamination by host cell apolipoprotein B editing complex (APOBEC) proteins is a potent anti-retroviral strategy. DeoxyC to dU deamination of the noncoding (minus) strand of the genome results in G to A hypermutations in the coding strand that are lethal to the virus [144]. Elucidation of how the HIV-1 virion infectivity factor (Vif) overcomes APOBEC3G has revealed a link with the UPS. HIV-1 Vif co-opts the Cul5 E3 ubiquitin ligase, acting in effect as the substrate receptor for its target APOBEC3G. The new SCF E3 ligase complex containing Cul5, elongin B and C and Vif induce lysine-48 linked polyubiquitination and proteasome-mediated degradation of APOBEC3G [145–147]. Vif binds to elongin C through a novel "suppressor of cytokine signaling" (SOCS)-box motif. Two highly conserved cysteine residues in Vif, outside the SOCS-box, are required for Cul5 interaction [148, 149]. APOBEC3F molecules are also targeted by Vif for ubiquitination and degradation through a Cul5-dependent SCF E3 ligase complex [150]. If the activity of Vif fails to neutralize all APOBEC3F/G activity within the cell, an additional HIV-1 accessory protein Vpr, forms a complex with Cul1 and Cul4 to target cellular uracil DNA glycosylase (UNG) for ubiquitination and proteasomal degradation [151]. UNG removal of the uracil base leaves the viral reverse transcripts open to error-prone translesion repair. By inducing the degradation of UNG, the number of abasic sites in viral reverse transcripts may be decreased, increasing the viability of progeny virus. Indeed, Vpr^+ viruses in a Vif^- background replicated better than Vpr^- viruses highlighting the importance of cytidine deamination in the anti-viral arsenal of the host [151]. Given that the SCF type E3 ligases are a protein complex, it is sometimes difficult to distinguish a component of such a complex from a protein that recruits a complex. A discussion of viral proteins that recruit E3 ligases follows.

7.3.8
Viral Recruitment of E3 Ligases

Simian virus 40 large T antigen (SV40T) binds and inactivates the tumor suppressors p53 and Rb contributing to the transforming activity of this viral protein. SV40T associates with an SCF-like complex composed of Cul7, Rbx1, and the F box protein Fbw6 [152] suggesting that the UPS is involved in this activity.

EBV is an oncogenic virus that encodes several proteins which recruit cellular E3 ligases to disrupt the cell cycle and interfere with normal signaling pathways. EBV nuclear antigen EBNA3C associates with cyclinA/cdk2 complexes and recruits the SCF/Skp2 E3 ligase complex resulting in the ubiquitination and degradation of the kinase inhibitor p27 and dysregulation of the cell cycle [153]. EBV-encoded latent membrane protein 1 (LMP1) and latent membrane protein 2 (LMP2) are important oncogenic proteins that interfere with cell signaling in latently infected cells. LMP1 interaction with the SCF E3 ligase complex containing the "homolog of Slimb" (HOS) helps regulate NFκB activation. LMP1 interacts directly with HOS but is not a substrate of this E3 ligase complex. Mutations that abrogate HOS binding, increase LMP1-induced transformation by increasing IκB degradation and therefore NFκB transcriptional activity. Therefore, LMP1 sequestration of HOS may restrict NFκB activation in EBV immortalized cells to help minimize transformation [154]. LMP1 also enhances the stability of the E3 ubiquitin ligase Siah1 that leads to the proteasomal degradation of prolyl hydroxylases 1 and 3 that in turn protects hypoxia inducible factor 1 subunit alpha (HIF1α) from ubiquitination and degradation [155]. The net result is an increase in levels of HIF1α in EBV-infected cells.

EBV encoded proteins can also regulate B- and T-lymphocyte receptor signaling. LMP2A interferes with normal B cell receptor (BCR) signaling and provides a constitutively active survival signal in latently infected B cells [156]. The N-terminal cytoplasmic domain of LMP2A binds to the BCR-associated kinases Syk and Lyn and recruits the cellular E3 ligases AIP4 and KIAA0439, resulting in the ubiquitination and degradation of Syk and Lyn and the inhibition of BCR signaling [85, 157]. Similarly, LMP2A in association with AIP4 regulates T-cell receptor (TCR) levels providing a survival signal to T-cell tumors harbouring latent EBV [158].

HIV-1 and related retroviruses encode many proteins that interact with cellular E3 protein-ubiquitin ligases during the course of infection. HIV-1 integrase associates with the E3 ligase Rad18, involved in cellular DNA repair. Rad18 expression stabilizes the integrase and may play a role in integration of the HIV-1 genome [159]. HIV-tat stimulates the transcriptional elongation of the HIV-1 genome by recruiting the positive transcriptional elongation factor b (P-Tefb) to pause RNA polymerase II. Tat recruits the SCF E3 ligase Skp2 to ubiquitinate the catalytic subunit of P-Tefb, cdk9, which in turn allows optimal transactivation of the HIV-1 long terminal repeat [160]. Another HIV-1 accessory protein, Vpu binds to newly synthesized CD4 in the ER. Phosphorylated Vpu recruits the SCF βTrCP E3 ligase complex resulting in the ubiquitination, retro-translocation and proteasome-mediated degradation of CD4. In addition, Vpu sequesters the SCF βTrCP E3 ligase from its normal substrates which include β-catenin, IκBα and ATF4 [87, 161–163].

7.4 Conclusions

Viruses interact with and exploit the UPS at many points during their life cycle. Modification by ubiquitin plays a crucial role from initial entry of virus particles into the cell via ubiquitin-mediated endocytosis, through to assembly, egress and protection from immune surveillance. To replicate and avoid host anti-viral mechanisms viruses alternately mimic and interfere with host cellular processes, many of which are regulated by ubiquitination. Homology between viral and host gene products has and will continue to promote the use of viral proteins as tools and model systems to unlock the secrets of host protein function. This will both improve our understanding of viral pathogenesis and allow the development of superior anti-viral strategies.

References

1 Elsasser, S. and Finley, D. (2005) Delivery of ubiquitinated substrates to protein-unfolding machines. *Nat Cell Biol* **7**, 742–749.

2 Meusser, B., Hirsch, C., Jarosch, E. and Sommer, T. (2005) ERAD: the long road to destruction. *Nat Cell Biol* **7**, 766–772.

3 Roos-Mattjus, P. and Sistonen, L. (2004) The ubiquitin-proteasome pathway. *Ann Med* **36**, 285–295.

4 Levitskaya, J., Sharipo, A., Leonchiks, A., Ciechanover, A. and Masucci, M.G. (1997) Inhibition of ubiquitin/proteasome-dependent protein degradation by the Gly-Ala repeat domain of the Epstein-Barr virus nuclear antigen 1. *Proc Natl Acad Sci USA* **94**, 12616–12621.

5 Mukherjee, S., Trivedi, P., Dorfman, D.M., Klein, G. and Townsend, A. (1999) Murine cytotoxic T lymphocytes recognize an epitope in an EBNA-1 fragment, but fail to lyse EBNA-1-expressing mouse cells. *J Exp Med* **187**, 445–450.

6 Yin, Y., Manoury, B. and Fahraeus, R. (2003) Self-inhibition of synthesis and antigen presentation by Epstein-Barr virus-encoded EBNA1. *Science* **301**, 1371–1374.

7 Zaldumbide, A., Ossevoort, M., Wiertz, E.J. and Hoeben, R.C. (2007) In cis inhibition of antigen processing by the latency-associated nuclear antigen I of Kaposi sarcoma Herpes virus. *Mol Immunol* **44**, 1352–1360.

8 Gilbert, M.J., Riddell, S.R., Plachter, B. and Greenberg, P.D. (1996) Cytomegalovirus selectively blocks antigen processing and presentation of its

immediate-early gene product. *Nature* **383**, 720–722.

9 Hu, Z., Zhang, Z., Doo, E., Coux, O., Goldberg, A.L. and Liang, T.J. (1999) Hepatitis B virus X protein is both a substrate and a potential inhibitor of the proteasome complex. *J Virol* **73**, 7231–7240.

10 Stohwasser, R., Holzhutter, H.G., Lehmann, U., Henklein, P. and Kloetzel, P.M. (2003) Hepatitis B virus HBx peptide 116-138 and proteasome activator PA28 compete for binding to the proteasome alpha4/MC6 subunit. *Biol Chem* **384**, 39–49.

11 Zhang, Z., Protzer, U., Hu, Z., Jacob, J. and Liang, T.J. (2004) Inhibition of cellular proteasome activities enhances hepadnavirus replication in an HBX-dependent manner. *J Virol* **78**, 4566–4572.

12 Remoli, A.L., Marsili, G., Perrotti, E., Gallerani, E., Ilari, R., Nappi, F., Cafaro, A., Ensoli, B., Gavioli, R. and Battistini, A. (2006) Intracellular HIV-1 Tat protein represses constitutive LMP2 transcription increasing proteasome activity by interfering with the binding of IRF-1 to STAT1. *Biochem J* **396**, 371–380.

13 Seeger, M., Ferrell, K., Frank, R. and Dubiel, W. (1997) HIV-1 tat inhibits the 20 S proteasome and its 11 S regulator-mediated activation. *J Biol Chem* **272**, 8145–8148.

14 Romisch, K. (2005) Endoplasmic reticulum-associated degradation. *Annu Rev Cell Dev Biol* **21**, 435–456.

15 Jones, T.R., Hanson, L.K., Sun, L., Slater, J.S., Stenberg, R.M. and Campbell, A.E. (1995) Multiple independent loci within the human cytomegalovirus unique short region down-regulate expression of major histocompatibility complex class I heavy chains. *J Virol* **69**, 4830–4841.

16 Wiertz, E.J., Jones, T.R., Sun, L., Bogyo, M., Geuze, H.J. and Ploegh, H.L. (1996a) The human cytomegalovirus US11 gene product dislocates MHC class I heavy chains from the endoplasmic reticulum to the cytosol. *Cell* **84**, 769–779.

17 Wiertz, E.J., Tortorella, D., Bogyo, M., Yu, J., Mothes, W., Jones, T.R., Rapoport, T.A. and Ploegh, H.L. (1996b) Sec61-mediated transfer of a membrane protein from the endoplasmic reticulum to the proteasome for destruction. *Nature* **384**, 432–438.

18 Furman, M.H., Loureiro, J., Ploegh, H.L. and Tortorella, D. (2003) Ubiquitinylation of the cytosolic domain of a type I membrane protein is not required to initiate its dislocation from the endoplasmic reticulum. *J Biol Chem* **278**, 34804–34811.

19 Shamu, C.E., Flierman, D., Ploegh, H.L., Rapoport, T.A. and Chau, V. (2001) Polyubiquitination is required for US11-dependent movement of MHC class I heavy chain from endoplasmic reticulum into cytosol. *Mol Biol Cell* **12**, 2546–2555.

20 Hassink, G.C., Barel, M.T., Van Voorden, S.B., Kikkert, M. and Wiertz, E.J. (2006) Ubiquitination of MHC class I heavy chains is essential for dislocation by human cytomegalovirus-encoded US2 but not US11. *J Biol Chem* **281**, 30063–30071.

21 Lilley, B.N. and Ploegh, H.L. (2004) A membrane protein required for dislocation of misfolded proteins from the ER. *Nature* **429**, 834–840.

22 Ye, Y., Shibata, Y., Yun, C., Ron, D. and Rapoport, T.A. (2004) A membrane protein complex mediates retro-translocation from the ER lumen into the cytosol. *Nature* **429**, 841–847.

23 Lilley, B.N. and Ploegh, H.L. (2005) Multiprotein complexes that link dislocation, ubiquitination, and extraction of misfolded proteins from the endoplasmic reticulum membrane. *Proc Natl Acad Sci USA* **102**, 14296–14301.

24 Loureiro, J. and Ploegh, H.L. (2006) Antigen presentation and the ubiquitin-proteasome system in host–pathogen interactions. *Adv Immunol* **92**, 225–305.

25 Loureiro, J., Lilley, B.N., Spooner, E., Noriega, V., Tortorella, D. and Ploegh, H.L. (2006) Signal peptide peptidase is required for dislocation from the endoplasmic reticulum. *Nature* **441**, 894–897.

26 Wolfe, M.S. and Kopan, R. (2004) Intramembrane proteolysis: theme and variations. *Science* **305**, 1119–1123.

27 Hicke, L., Schubert, H.L. and Hill, C.P. (2005) Ubiquitin-binding domains. *Nat Rev Mol Cell Biol* **6**, 610–621.

28 Clague, M.J. and Urbe, S. (2006) Endocytosis: the DUB version. *Trends Cell Biol* **16**, 551–559.

29 Millard, S.M. and Wood, S.A. (2006) Riding the DUBway: regulation of protein trafficking by deubiquitylating enzymes. *J Cell Biol* **173**, 463–468.

30 Hoeller, D., Crosetto, N., Blagoev, B., Raiborg, C., Tikkanen, R., Wagner, S., Kowanetz, K., Breitling, R., Mann, M., Stenmark, H. and Dikic, I. (2006) Regulation of ubiquitin-binding proteins by monoubiquitination. *Nat Cell Biol* **8**, 163–169.

31 Mosesson, Y., Shtiegman, K., Katz, M., Zwang, Y., Vereb, G., Szollosi, J. and Yarden, Y. (2003) Endocytosis of receptor tyrosine kinases is driven by monoubiquitylation, not polyubiquitylation. *J Biol Chem* **278**, 21323–21326.

32 Duncan, L.M., Piper, S., Dodd, R.B., Saville, M.K., Sanderson, C.M., Luzio, J.P. and Lehner, P.J. (2006) Lysine-63-linked ubiquitination is required for endolysosomal degradation of class I molecules. *EMBO J*, **25**, 1635–1645.

33 Huang, F., Kirkpatrick, D., Jiang, X., Gygi, S. and Sorkin, A. (2006) Differential regulation of EGF receptor internalization and degradation by multiubiquitination within the kinase domain. *Mol Cell* **21**, 737–748.

34 Huang, J., Huang, Q., Zhou, X., Shen, M.M., Yen, A., Yu, S.X., Dong, G., Qu, K., Huang, P., anderson, E.M., Daniel-Issakani, S., Buller, R.M., Payan, D.G. and Lu, H.H. (2004) The poxvirus p28 virulence factor is an E3 ubiquitin ligase. *J Biol Chem* **279**, 54110–54116.

35 Katzmann, D.J., Odorizzi, G. and Emr, S.D. (2002) Receptor downregulation and multivesicular-body sorting. *Nat Rev Mol Cell Biol* **3**, 893–905.

36 Babst, M., Odorizzi, G., Estepa, E.J. and Emr, S.D. (2000) Mammalian tumor susceptibility gene 101 (TSG101) and the yeast homolog, Vps23p, both function in late endosomal trafficking. *Traffic* **1**, 248–258.

37 Hewitt, E.W., Duncan, L., Mufti, D., Baker, J., Stevenson, P.G. and Lehner, P.J. (2002) Ubiquitylation of MHC class I by the K3 viral protein signals internalization and TSG101-dependent degradation. *EMBO J*, **21**, 2418–2429.

38 Lodoen, M., Ogasawara, K., Hamerman, J.A., Arase, H., Houchins, J.P., Mocarski, E.S. and Lanier, L.L. (2003) NKG2D-mediated natural killer cell protection against cytomegalovirus is impaired by viral gp40 modulation of retinoic acid early inducible 1 gene molecules. *J Exp Med* **197**, 1245–1253.

39 Lodoen, M.B., Abenes, G., Umamoto, S., Houchins, J.P., Liu, F. and Lanier, L.L. (2004) The cytomegalovirus m155 gene product subverts natural killer cell antiviral protection by disruption of H60-NKG2D interactions. *J Exp Med* **200**, 1075–1081.

40 Chin, Y.R. and Horwitz, M.S. (2005) Mechanism for removal of tumor necrosis factor receptor 1 from the cell surface by the adenovirus RIDalpha/beta complex. *J Virol* **79**, 13606–13617.

41 Chin, Y.R. and Horwitz, M.S. (2006) Adenovirus RID complex enhances degradation of internalized tumor necrosis factor receptor 1 without affecting its rate of endocytosis. *J Gen Virol* **87**, 3161–3167.

42 Yu, G.Y. and Lai, M.M. (2005) The ubiquitin-proteasome system facilitates the transfer of murine coronavirus from endosome to cytoplasm during virus entry. *J Virol* **79**, 644–648.

43 Ros, C. and Kempf, C. (2004) The ubiquitin-proteasome machinery is essential for nuclear translocation of incoming minute virus of mice. *Virology* **324**, 350–360.

44 Yueh, A., Leung, J., Bhattacharyya, S., Perrone, L.A., de los Santos, K., Pu, S.Y. and Goff, S.P. (2006) Interaction of moloney murine leukemia virus capsid with Ubc9 and PIASy mediates SUMO-1 addition required early in infection. *J Virol* **80**, 342–352.

45 Bieniasz, P.D. (2006) Late budding domains and host proteins in enveloped virus release. *Virology* **344**, 55–63.

46 Garrus, J.E., von Schwedler, U.K., Pornillos, O.W., Morham, S.G., Zavitz, K.H., Wang, H.E., Wettstein, D.A., Stray, K.M., Cote, M., Rich, R.L., Myszka, D.G. and Sundquist, W.I. (2001) Tsg101 and the vacuolar protein sorting pathway are essential for HIV-1 budding. *Cell* **107**, 55–65.

47 Martin-Serrano, J., Zang, T. and Bieniasz, P.D. (2003) Role of ESCRT-I in retroviral budding. *J Virol* **77**, 4794–4804.

48 Amit, I., Yakir, L., Katz, M., Zwang, Y., Marmor, M.D., Citri, A., Shtiegman, K., Alroy, I., Tuvia, S., Reiss, Y., Roubini, E., Cohen, M., Wides, R., Bacharach, E., Schubert, U. and Yarden, Y. (2004) Tal, a Tsg101-specific E3 ubiquitin ligase, regulates receptor endocytosis and retrovirus budding. *Genes Dev* **18**, 1737–1752.

49 Yasuda, J., Nakao, M., Kawaoka, Y. and Shida, H. (2003) Nedd4 regulates egress of Ebola virus-like particles from host cells. *J Virol* **77**, 9987–9992.

50 Wang, H., Machesky, N.J. and Mansky, L.M. (2004a) Both the PPPY and PTAP motifs are involved in human T-cell leukemia virus type 1 particle release. *J Virol* **78**, 1503–1512.

51 Rost, M., Mann, S., Lambert, C., Doring, T., Thome, N. and Prange, R. (2006) Gamma-adaptin, a novel ubiquitin-interacting adaptor, and Nedd4 ubiquitin ligase control hepatitis B virus maturation. *J Biol Chem* **281**, 29297–29308.

52 Chen, C., Vincent, O., Jin, J., Weisz, O.A. and Montelaro, R.C. (2005) Functions of early (AP-2) and late (AIP1/ALIX) endocytic proteins in equine infectious anemia virus budding. *J Biol Chem* **280**, 40474–40480.

53 Sakaguchi, T., Kato, A., Sugahara, F., Shimazu, Y., Inoue, M., Kiyotani, K., Nagai, Y. and Yoshida, T. (2005) AIP1/Alix is a binding partner of Sendai virus C protein and facilitates virus budding. *J Virol* **79**, 8933–8941.

54 Strack, B., Calistri, A., Craig, S., Popova, E. and Gottlinger, H.G. (2003) AIP1/ALIX is a binding partner for HIV-1 p6 and EIAV p9 functioning in virus budding. *Cell* **114**, 689–699.

55 Stanke, N., Stange, A., Luftenegger, D., Zentgraf, H. and Lindemann, D. (2005) Ubiquitination of the prototype foamy virus envelope glycoprotein leader peptide regulates subviral particle release. *J Virol* **79**, 15074–15083.

56 Salghetti, S.E., Caudy, A.A., Chenoweth, J.G. and Tansey, W.P. (2001) Regulation of transcriptional activation domain function by ubiquitin. *Science* **293**, 1651–1653.

57 Bres, V., Kiernan, R.E., Linares, L.K., Chable-Bessia, C., Plechakova, O., Treand, C., Emiliani, S., Peloponese, J.M., Jeang, K.T., Coux, O., Scheffner, M. and Benkirane, M. (2003) A non-proteolytic role for ubiquitin in Tat-mediated transactivation of the HIV-1 promoter. *Nat Cell Biol* **5**, 754–761.

58 Thornton, B.R. and Toczyski, D.P. (2006) Precise destruction: an emerging picture of the APC. *Genes Dev* **20**, 3069–3078.

59 Levine, A.J. (1989) The p53 tumor suppressor gene and gene product. *Princess Takamatsu Symp* **20**, 221–230.

60 Stanbridge, E.J. (1989) The evidence for human tumor suppressor genes. *Princess Takamatsu Symp* **20**, 3–13.

61 Scheffner, M., Werness, B.A., Huibregtse, J.M., Levine, A.J. and Howley, P.M. (1990) The E6 oncoprotein encoded by human papillomavirus types 16 and 18 promotes the degradation of p53. *Cell* **63**, 1129–1136.

62 Boutell, C. and Everett, R.D. (2003) The herpes simplex virus type 1 (HSV-1) regulatory protein ICP0 interacts with and Ubiquitinates p53. *J Biol Chem* **278**, 36596–36602.

63 Querido, E., Blanchette, P., Yan, Q., Kamura, T., Morrison, M., Boivin, D., Kaelin, W.G., Conaway, R.C., Conaway, J.W. and Branton, P.E. (2001) Degradation of p53 by adenovirus E4orf6 and E1B55K proteins occurs via a novel mechanism involving a Cullin-containing complex. *Genes Dev* **15**, 3104–3117.

64 Boyer, S.N., Wazer, D.E. and Band, V. (1996) E7 protein of human papilloma virus-16 induces degradation of retinoblastoma protein through the ubiquitin-proteasome pathway. *Cancer Res* **56**, 4620–4624.

65 Knight, J.S., Sharma, N. and Robertson, E.S. (2005a) Epstein-Barr virus latent antigen 3C can mediate the degradation of the retinoblastoma protein through an SCF cellular ubiquitin ligase. *Proc Natl Acad Sci USA* **102**, 18562–18566.

66 Liu, B., Hong, S., Tang, Z., Yu, H. and Giam, C.Z. (2005a) HTLV-I Tax directly binds the Cdc20-associated anaphase-promoting complex and activates it ahead of schedule. *Proc Natl Acad Sci USA* **102**, 63–68.

67 Everett, R.D. (2004) Herpes simplex virus type 1 regulatory protein ICP0 does not protect cyclins D1 and D3 from degradation during infection. *J Virol* **78**, 9599–9604.

68 Hagglund, R. and Roizman, B. (2003) Herpes simplex virus 1 mutant in which the ICP0 HUL-1 E3 ubiquitin ligase site is disrupted stabilizes cdc34 but degrades D-type cyclins and exhibits diminished neurotoxicity. *J Virol* **77**, 13194–13202.

69 Clem, R.J., Hardwick, J.M. and Miller, L.K. (1996) Anti-apoptotic genes of baculoviruses. *Cell Death Differ* **3**, 9–16.

70 Crook, N.E., Clem, R.J. and Miller, L.K. (1993) An apoptosis-inhibiting baculovirus gene with a zinc finger-like motif. *J Virol* **67**, 2168–2174.

71 Holowaty, M.N., Zeghouf, M., Wu, H., Tellam, J., Athanasopoulos, V., Greenblatt, J. and Frappier, L. (2003) Protein profiling with Epstein-Barr nuclear antigen-1 reveals an interaction with the herpesvirus-associated ubiquitin-specific protease HAUSP/USP7. *J Biol Chem* **278**, 29987–29994.

72 Bowers, K., Piper, S.C., Edeling, M.A., Gray, S.R., Owen, D.J., Lehner, P.J. and Luzio, J.P. (2006) Degradation of endocytosed epidermal growth factor and virally ubiquitinated major histocompatibility complex class I is independent of mammalian ESCRTII. *J Biol Chem* **281**, 5094–5105.

73 Hewitt, E.W. (2003) The MHC class I antigen presentation pathway: strategies for viral immune evasion. *Immunology* **110**, 163–169.

74 Ploegh, H.L. (1999) Viral strategies of immune evasion. *Science* **280**, 248–253.

75 Yu, Y., Wang, S.E. and Hayward, G.S. (2005) The KSHV immediate-early transcription factor RTA encodes ubiquitin E3 ligase activity that targets IRF7 for proteosome-mediated degradation. *Immunity* **22**, 59–70.

76 Migone, T.S., Humbert, M., Rascle, A., Sanden, D., D'Andrea, A. and Johnston, J.A. (2001) The deubiquitinating enzyme DUB-2 prolongs cytokine-induced signal transducers and activators of transcription activation and suppresses apoptosis following cytokine withdrawal. *Blood* **98**, 1935–1941.

77 Precious, B., Childs, K., Fitzpatrick-Swallow, V., Goodbourn, S. and Randall, R.E. (2005a) Simian virus 5 V protein acts as an adaptor, linking DDB1 to STAT2, to facilitate the ubiquitination of STAT1. *J Virol* **79**, 13434–13441.

78 Precious, B., Young, D.F., andrejeva, L., Goodbourn, S. and Randall, R.E. (2005b) In vitro and in vivo specificity of ubiquitination and degradation of STAT1 and STAT2 by the V proteins of the paramyxoviruses simian virus 5 and human parainfluenza virus type 2. *J Gen Virol* **86**, 151–158.

79 Ulane, C.M., Rodriguez, J.J., Parisien, J.P. and Horvath, C.M. (2003) STAT3 ubiquitylation and degradation by mumps virus suppress cytokine and oncogene signaling. *J Virol* **77**, 6385–6393.

80 Chen, Z.J. (2005) Ubiquitin signaling in the NF-kappaB pathway. *Nat Cell Biol* **7**, 758–765.

81 Zhang, B., Srirangam, A., Potter, D.A. and Roman, A. (2005) HPV16 E5 protein disrupts the c-Cbl-EGFR interaction and EGFR ubiquitination in human foreskin keratinocytes. *Oncogene* **24**, 2585–2588.

82 Liang, Y., Kurakin, A. and Roizman, B. (2005) Herpes simplex virus 1 infected cell protein 0 forms a complex with CIN85 and Cbl and mediates the degradation of EGF receptor from cell surfaces. *Proc Natl Acad Sci USA* **102**, 5838–5843.

83 Diao, L., Zhang, B., Fan, J., Gao, X., Sun, S., Yang, K., Xin, D., Jin, N., Geng, Y. and Wang, C. (2005) Herpes virus proteins ICP0 and BICP0 can activate NF-kappaB by catalyzing IkappaBalpha ubiquitination. *Cell Signal* **17**, 217–229.

84 Jang, K.L., Shackelford, J., Seo, S.Y. and Pagano, J.S. (2005) Up-regulation of beta-catenin by a viral oncogene correlates with inhibition of the seven in absentia homolog 1 in B lymphoma cells. *Proc Natl Acad Sci USA* **102**, 18431–18436.

85 Merchant, M., Caldwell, R.G. and Longnecker, R. (2000) The LMP2A ITAM is essential for providing B cells with development and survival signals in vivo. *J Virol* **74**, 9115–9124.

86 Morrison, J.A. and Raab-Traub, N. (2005) Roles of the ITAM and PY motifs of Epstein-Barr virus latent membrane protein 2A in the inhibition of epithelial cell differentiation and activation of β-catenin signaling. *J Virol* **79**, 2375–2382.

87 Besnard-Guerin, C., Belaidouni, N., Lassot, I., Segeral, E., Jobart, A., Marchal, C. and Benarous, R. (2004) HIV-1 Vpu sequesters beta-transducin repeat-containing protein (betaTrCP) in the cytoplasm and provokes the accumulation of beta-catenin and other SCFbetaTrCP substrates. *J Biol Chem* **279**, 788–795.

88 Pickart, C.M. and Eddins, M.J. (2004) Ubiquitin: structures, functions, mechanisms. *Biochim Biophys Acta* **1695**, 55–72.

89 Tyers, M. and Willems, A.R. (1999) One ring to rule a superfamily of E3 ubiquitin ligases. *Science* **284**, *601*, 603–604.

90 Zachariae, W., Shevchenko, A., andrews, P.D., Ciosk, R., Galova, M., Stark, M.J., Mann, M. and Nasmyth, K. (1999) Mass spectrometric analysis of the anaphase-promoting complex from yeast: identification of a subunit related to cullins. *Science* **279**, 1216–1219.

91 Huibregtse, J.M., Scheffner, M., Beaudenon, S. and Howley, P.M. (1995) A family of proteins structurally and functionally related to the E6-AP ubiquitin-protein ligase. *Proc Natl Acad Sci USA* **92**, 5249.

92 Scheffner, M., Huibregtse, J.M., Vierstra, R.D. and Howley, P.M. (1993) The HPV-16 E6 and E6-AP complex functions as a ubiquitin-protein ligase in the ubiquitination of p53. *Cell* **75**, 495–505.

93 Joazeiro, C.A. and Hunter, T. (2000) Biochemistry. Ubiquitination–more than two to tango. *Science* **289**, 2061–2062.

94 Dodd, R.B., Allen, M.D., Brown, S.E., Sanderson, C.M., Duncan, L.M., Lehner, P.J., Bycroft, M. and Read, R.J. (2004) Solution structure of the Kaposi's sarcoma-associated herpesvirus K3 N-terminal domain reveals a Novel E2-binding C4HC3-type RING domain. *J Biol Chem* **279**, 53840–53847.

95 Lehner, P.J., Hoer, S., Dodd, R. and Duncan, L.M. (2005) Downregulation of cell surface receptors by the K3 family of viral and cellular ubiquitin E3 ligases. *Immunol Rev* **207**, 112–125.

96 Koegl, M., Hoppe, T., Schlenker, S., Ulrich, H.D., Mayer, T.U. and Jentsch, S. (1999) A novel ubiquitination factor, E4, is involved in multiubiquitin chain assembly. *Cell* **96**, 635–644.

97 Chelbi-Alix, M.K. and de The, H. (1999) Herpes virus induced proteasome-dependent degradation of the nuclear bodies-associated PML and Sp100 proteins. *Oncogene* **18**, 935–941.

98 Everett, R.D., Earnshaw, W.C., Findlay, J. and Lomonte, P. (1999) Specific destruction of kinetochore protein CENP-C and disruption of cell division by herpes simplex virus immediate-early protein Vmw110. *EMBO J*, **18**, 1526–1538.

99 Lomonte, P., Sullivan, K.F. and Everett, R.D. (2001) Degradation of nucleosome-associated centromeric histone H3-like protein CENP-A induced by herpes simplex virus type 1 protein ICP0. *J Biol Chem* **276**, 5829–5835.

100 Everett, R.D. (2000) ICP0 induces the accumulation of colocalizing conjugated ubiquitin. *J Virol* **74**, 9994–10005.

101 Parkinson, J. and Everett, R.D. (2000) Alphaherpesvirus proteins related to herpes simplex virus type 1 ICP0 affect cellular structures and proteins. *J Virol* **74**, 10006–10017.

102 Hagglund, R. and Roizman, B. (2002) Characterization of the novel E3 ubiquitin ligase encoded in exon 3 of herpes simplex virus-1-infected cell protein 0. *Proc Natl Acad Sci USA* **99**, 7889–7894.

103 Van Sant, C., Hagglund, R., Lopez, P. and Roizman, B. (2001) The infected cell protein 0 of herpes simplex virus 1 dynamically interacts with proteasomes, binds and activates the cdc34 E2 ubiquitin-conjugating enzyme, and possesses in vitro E3 ubiquitin ligase activity. *Proc Natl Acad Sci USA* **98**, 8815–8820.

104 Boutell, C., Sadis, S. and Everett, R.D. (2002) Herpes simplex virus type 1 immediate-early protein ICP0 and is isolated RING finger domain act as ubiquitin E3 ligases in vitro. *J Virol* **76**, 841–850.

105 Hagglund, R., Van Sant, C., Lopez, P. and Roizman, B. (2002) Herpes simplex virus 1-infected cell protein 0 contains two E3 ubiquitin ligase sites specific for different E2 ubiquitin-conjugating enzymes. *Proc Natl Acad Sci USA* **99**, 631–636.

106 Boutell, C., Orr, A. and Everett, R.D. (2003) PML residue lysine 160 is required for the degradation of PML induced by herpes simplex virus type 1 regulatory protein ICP0. *J Virol* **77**, 8686–8694.

107 Gu, H. and Roizman, B. (2003) The degradation of promyelocytic leukemia and Sp100 proteins by herpes simplex virus 1 is mediated by the ubiquitin-conjugating enzyme UbcH5a. *Proc Natl Acad Sci USA* **100**, 8963–8968.

108 Bailey, D. and O'Hare, P. (2002) Herpes simplex virus 1 ICP0 co-localizes with a SUMO-specific protease. *J Gen Virol* **83**, 2951–2964.

109 Walsh, D. and Mohr, I. (2004) Phosphorylation of eIF4E by Mnk-1 enhances HSV-1 translation and replication in quiescent cells. *Genes Dev* **18**, 660–672.

110 Canning, M., Boutell, C., Parkinson, J. and Everett, R.D. (2004) A RING finger ubiquitin ligase is protected from autocatalyzed ubiquitination and degradation by binding to ubiquitin-specific protease USP7. *J Biol Chem* **279**, 38160–38168.

111 Boutell, C., Canning, M., Orr, A. and Everett, R.D. (2005) Reciprocal activities between herpes simplex virus type 1 regulatory protein ICP0, a ubiquitin E3 ligase, and ubiquitin-specific protease USP7. *J Virol* **79**, 12342–12354.

112 Chang, L.K., Lee, Y.H., Cheng, T.S., Hong, Y.R., Lu, P.J., Wang, J.J., Wang, W.H., Kuo, C.W., Li, S.S. and Liu, S.T. (2004) Post-translational modification of Rta of Epstein-Barr virus by SUMO-1. *J Biol Chem* **279**, 38803–38812.

113 Imai, N., Matsuda, N., Tanaka, K., Nakano, A., Matsumoto, S. and Kang, W. (2003) Ubiquitin ligase activities of Bombyx mori nucleopolyhedrovirus RING finger proteins. *J Virol* **77**, 923–930.

114 Green, M.C., Monser, K.P. and Clem, R.J. (2004) Ubiquitin protein ligase activity of the anti-apoptotic baculovirus protein Op-IAP3. *Virus Res* **105**, 89–96.

115 Wang, Z., Chua, H.K., Gusti, A.A., He, F., Fenner, B., Manopo, I., Wang, H. and Kwang, J. (2005) RING-H2 protein WSSV249 from white spot syndrome virus sequesters a shrimp ubiquitin-conjugating enzyme, PvUbc, for viral pathogenesis. *J Virol* **79**, 8764–8772.

116 He, F., Fenner, B.J., Godwin, A.K. and Kwang, J. (2006) White spot syndrome virus open reading frame 222 encodes a viral E3 ligase and mediates degradation of a host tumor suppressor via ubiquitination. *J Virol* **80**, 3884–3892.

117 Senkevich, T.G., Koonin, E.V. and Buller, R.M. (1994) A poxvirus protein with a RING zinc finger motif is of crucial importance for virulence. *Virology* **198**, 118–128.

118 Bartee, E., Mansouri, M., Hovey Nerenberg, B.T., Gouveia, K. and Fruh, K. (2004) Downregulation of major histocompatibility complex class I by human ubiquitin ligases related to viral immune evasion proteins. *J Virol* **78**, 1109–1120.

119 Stevenson, P.G., Efstathiou, S., Doherty, P.C. and Lehner, P.J. (2000) Inhibition of MHC class I-restricted antigen presentation by gamma 2-herpesviruses. *Proc Natl Acad Sci USA* **97**, 8455–8460.

120 Boname, J.M. and Stevenson, P.G. (2001) MHC class I ubiquitination by a viral PHD/LAP finger protein. *Immunity* **15**, 627–636.

121 Wang, X., Lybarger, L., Connors, R., Harris, M.R. and Hansen, T.H. (2004b) Model for the interaction of gammaherpesvirus 68 RING-CH finger protein mK3 with major histocompatibility complex class I and the peptide-loading complex. *J Virol* **78**, 8673–8686.

122 Yu, Y.Y., Harris, M.R., Lybarger, L., Kimpler, L.A., Myers, N.B., Virgin, H.W.t. and Hansen, T.H. (2002) Physical association of the K3 protein of gamma-2 herpesvirus 68 with major histocompatibility complex class I molecules with impaired peptide and beta(2)-microglobulin assembly. *J Virol* **76**, 2796–2803.

123 Boname, J.M., de Lima, B.D., Lehner, P.J. and Stevenson, P.G. (2004) Viral degradation of the MHC class I peptide loading complex. *Immunity* **20**, 305–317.

124 Boname, J.M., May, J.S. and Stevenson, P.G. (2005) The murine gamma-herpesvirus-68 MK3 protein causes TAP degradation independent of MHC class I heavy chain degradation. *Eur J Immunol* **35**, 171–179.

125 Wang, X., Ye, Y., Lencer, W. and Hansen, T.H. (2006) The viral E3 ubiquitin ligase mK3 uses the Derlin/p97 endoplasmic reticulum-associated degradation pathway to mediate down-regulation of major histocompatibility complex class I proteins. *J Biol Chem* **281**, 8636–8644.

126 Coscoy, L. and Ganem, D. (2000) Kaposi's sarcoma-associated herpesvirus encodes two proteins that block cell surface display of MHC class I chains by enhancing their endocytosis. *Proc Natl Acad Sci USA* **97**, 8051–8056.

127 Ishido, S., Wang, C., Lee, B.S., Cohen, G.B. and Jung, J.U. (2000b) Downregulation of major histocompatibility complex class I molecules by Kaposi's sarcoma-associated herpesvirus K3 and K5 proteins. *J Virol* **74**, 5300–5309.

128 Coscoy, L., Sanchez, D.J. and Ganem, D. (2001) A novel class of herpesvirus-encoded membrane-bound E3 ubiquitin ligases regulates endocytosis of proteins involved in immune recognition. *J Cell Biol* **155**, 1265–1273.

129 Haglund, K., Sigismund, S., Polo, S., Szymkiewicz, I., Di Fiore, P.P. and Dikic, I. (2003) Multiple monoubiquitination of RTKs is sufficient for their endocytosis and degradation. *Nat Cell Biol* **5**, 461–466.

130 Cadwell, K. and Coscoy, L. (2005) Ubiquitination on nonlysine residues by a viral E3 ubiquitin ligase. *Science* **309**, 127–130.

131 Mansouri, M., Douglas, J., Rose, P.P., Gouveia, K., Thomas, G., Means, R.E., Moses, A.V. and Fruh, K. (2006) Kaposi sarcoma herpesvirus K5 removes CD31/PECAM from endothelial cells. *Blood* **108**, 1932–1940.

132 Sanchez, D.J., Gumperz, J.E. and Ganem, D. (2005) Regulation of CD1d expression and function by a herpesvirus infection. *J Clin Invest* **115**, 1369–1378.

133 Ishido, S., Choi, J.K., Lee, B.S., Wang, C., DeMaria, M., Johnson, R.P., Cohen, G.B. and Jung, J.U. (2000a) Inhibition of natural killer cell-mediated cytotoxicity by Kaposi's sarcoma-associated herpesvirus K5 protein. *Immunity* **13**, 365–374.

134 Coscoy, L. and Ganem, D. (2001) A viral protein that selectively downregulates ICAM-1 and B7-2 and modulates T cell costimulation. *J Clin Invest* **107**, 1599–1606.

135 Fruh, K., Bartee, E., Gouveia, K. and Mansouri, M. (2002) Immune evasion by a novel family of viral PHD/LAP-finger proteins of gamma-2 herpesviruses and poxviruses. *Virus Res* **88**, 55–69.

136 Guerin, J.L., Gelfi, J., Boullier, S., Delverdier, M., Bellanger, F.A., Bertagnoli, S., Drexler, I., Sutter, G. and

Messud-Petit, F. (2002) Myxoma virus leukemia-associated protein is responsible for major histocompatibility complex class I and Fas-CD95 down-regulation and defines scrapins, a new group of surface cellular receptor abductor proteins. *J Virol* **76**, 2912–2923.

137 Mansouri, M., Bartee, E., Gouveia, K., Hovey Nerenberg, B.T., Barrett, J., Thomas, L., Thomas, G., McFadden, G. and Fruh, K. (2003) The PHD/LAP-domain protein M153R of myxomavirus is a ubiquitin ligase that induces the rapid internalization and lysosomal destruction of CD4. *J Virol* **77**, 1427–1440.

138 Collin, N., Guerin, J.L., Drexler, I., Blanie, S., Gelfi, J., Boullier, S., Foucras, G., Sutter, G. and Messud-Petit, F. (2005) The poxviral scrapin MV-LAP requires a myxoma viral infection context to efficiently downregulate MHC-I molecules. *Virology* **343**, 171–178.

139 Deshaies, R.J. (1999) SCF and Cullin/Ring H2-based ubiquitin ligases. *Annu Rev Cell Dev Biol* **15**, 435–467.

140 Blanchette, P., Cheng, C.Y., Yan, Q., Ketner, G., Ornelles, D.A., Dobner, T., Conaway, R.C., Conaway, J.W. and Branton, P.E. (2004) Both BC-box motifs of adenovirus protein E4orf6 are required to efficiently assemble an E3 ligase complex that degrades p53. *Mol Cell Biol* **24**, 9619–9629.

141 Harada, J.N., Shevchenko, A., Shevchenko, A., Pallas, D.C. and Berk, A.J. (2002) Analysis of the adenovirus E1B-55K-anchored proteome reveals its link to ubiquitination machinery. *J Virol* **76**, 9194–9206.

142 Ulane, C.M. and Horvath, C.M. (2002) Paramyxoviruses SV5 and HPIV2 assemble STAT protein ubiquitin ligase complexes from cellular components. *Virology* **304**, 160–166.

143 Li, T., Chen, X., Garbutt, K.C., Zhou, P. and Zheng, N. (2006) Structure of DDB1 in complex with a paramyxovirus V protein: viral hijack of a propeller cluster in ubiquitin ligase. *Cell* **124**, 105–117.

144 Bishop, K.N., Holmes, R.K., Sheehy, A.M., Davidson, N.O., Cho, S.J. and Malim, M.H. (2004) Cytidine deamination of retroviral DNA by diverse APOBEC proteins. *Curr Biol* **14**, 1392–1396.

145 Mehle, A., Strack, B., Ancuta, P., Zhang, C., McPike, M. and Gabuzda, D. (2004b) Vif overcomes the innate antiviral activity of APOBEC3G by promoting its degradation in the ubiquitin-proteasome pathway. *J Biol Chem* **279**, 7792–7798.

146 Sheehy, A.M., Gaddis, N.C. and Malim, M.H. (2003) The antiretroviral enzyme APOBEC3G is degraded by the proteasome in response to HIV-1 Vif. *Nat Med* **9**, 1404–1407.

147 Yu, X., Yu, Y., Liu, B., Luo, K., Kong, W., Mao, P. and Yu, X.F. (2003) Induction of APOBEC3G ubiquitination and degradation by an HIV-1 Vif-Cul5-SCF complex. *Science* **302**, 1056–1060.

148 Mehle, A., Goncalves, J., Santa-Marta, M., McPike, M. and Gabuzda, D. (2004a) Phosphorylation of a novel SOCS-box regulates assembly of the HIV-1 Vif-Cul5 complex that promotes APOBEC3G degradation. *Genes Dev* **18**, 2861–2866.

149 Yu, Y., Xiao, Z., Ehrlich, E.S., Yu, X. and Yu, X.F. (2004) Selective assembly of HIV-1 Vif-Cul5-ElonginB-ElonginC E3 ubiquitin ligase complex through a novel SOCS box and upstream cysteines. *Genes Dev* **18**, 2867–2872.

150 Liu, B., Sarkis, P.T., Luo, K., Yu, Y. and Yu, X.F. (2005b) Regulation of Apobec3F and human immunodeficiency virus type 1 Vif by Vif-Cul5-ElonB/C E3 ubiquitin ligase. *J Virol* **79**, 9579–9587.

151 Schrofelbauer, B., Yu, Q., Zeitlin, S.G. and Landau, N.R. (2005) Human immunodeficiency virus type 1 Vpr induces the degradation of the UNG and SMUG uracil-DNA glycosylases. *J Virol* **79**, 10978–10987.

152 Ali, S.H., Kasper, J.S., Arai, T. and DeCaprio, J.A. (2004) Cul7/p185/p193 binding to simian virus 40 large T antigen has a role in cellular transformation. *J Virol* **78**, 2749–2757.

153 Knight, J.S., Sharma, N. and Robertson, E.S. (2005b) SCFSkp2 complex targeted by Epstein-Barr virus essential nuclear antigen. *Mol Cell Biol* **25**, 1749–1763.

154 Tang, W., Pavlish, O.A., Spiegelman, V.S., Parkhitko, A.A. and Fuchs, S.Y. (2003) Interaction of Epstein-Barr virus latent membrane protein 1 with SCFHOS/beta-TrCP E3 ubiquitin ligase regulates extent of NF-kappaB activation. *J Biol Chem* **278**, 48942–48949.

155 Kondo, S., Seo, S.Y., Yoshizaki, T., Wakisaka, N., Furukawa, M., Joab, I., Jang, K.L. and Pagano, J.S. (2006) EBV latent membrane protein 1 up-regulates hypoxia-inducible factor 1alpha through Siah1-mediated down-regulation of prolyl hydroxylases 1 and 3 in nasopharyngeal epithelial cells. *Cancer Res* **66**, 9870–9877.

156 Caldwell, R.G., Wilson, J.B., anderson, S.J. and Longnecker, R. (1999) Epstein-Barr virus LMP2A drives B cell development and survival in the absence of normal B cell receptor signals. *Immunity* **9**, 405–411.

157 Winberg, G., Matskova, L., Chen, F., Plant, P., Rotin, D., Gish, G., Ingham, R., Ernberg, I. and Pawson, T. (2000) Latent membrane protein 2A of Epstein-Barr virus binds WW domain E3 protein-ubiquitin ligases that ubiquitinate B-cell tyrosine kinases. *Mol Cell Biol* **20**, 8526–8535.

158 Ingham, R.J., Raaijmakers, J., Lim, C.S., Mbamalu, G., Gish, G., Chen, F., Matskova, L., Ernberg, I., Winberg, G. and Pawson, T. (2005) The Epstein-Barr virus protein, latent membrane protein 2A, co-opts tyrosine kinases used by the T cell receptor. *J Biol Chem* **280**, 34133–34142.

159 Mulder, L.C., Chakrabarti, L.A. and Muesing, M.A. (2002) Interaction of HIV-1 integrase with DNA repair protein hRad18. *J Biol Chem* **277**, 27489–27493.

160 Barboric, M., Zhang, F., Besenicar, M., Plemenitas, A. and Peterlin, B.M. (2005) Ubiquitylation of Cdk9 by Skp2 facilitates optimal Tat transactivation. *J Virol* **79**, 11135–11141.

161 Margottin, F., Bour, S.P., Durand, H., Selig, L., Benichou, S., Richard, V., Thomas, D., Strebel, K. and Benarous, R. (1999) A novel human WD protein, h-beta TrCp, that interacts with HIV-1 Vpu connects CD4 to the ER degradation pathway through an F-box motif. *Mol Cell* **1**, 565–574.

162 Schubert, U. and Strebel, K. (1994) Differential activities of the human immunodeficiency virus type 1-encoded Vpu protein are regulated by phosphorylation and occur in different cellular compartments. *J Virol* **68**, 2260–2271.

163 Willey, R.L., Maldarelli, F., Martin, M.A. and Strebel, K. (1992) Human immunodeficiency virus type 1 Vpu protein regulates the formation of intracellular gp160-CD4 complexes. *J Virol* **66**, 226–234.

8
The Ubiquitin–Proteasome System in Parkinson's Disease

Kevin St. P. McNaught

8.1
Introduction

Parkinson's disease (PD) is a neurological disorder and is characterized clinically by bradykinesia (slowness of movement), rigidity, tremor, postural instability and gait dysfunction [1, 2]. However, non-motor features (e.g. autonomic dysfunction and dementia) often develop in these patients, especially during the advanced stages of the illness [1, 2]. PD affects both males and females, occurs in all racial/ ethnic groups, and is found worldwide. Reports of incidence and prevalence rates of the illness vary, but most studies show that the occurrence of the disorder increases with aging [3, 4]. For example, Van Den Eeden and colleagues found an overall population incidence rate of PD annually to be 13.4 per 100 000 individuals, but this rate increases to 38.8 and 107.2 per 100 000 individuals in the age range of 60–69 and 70–99 years, respectively [4].

PD is defined pathologically by degeneration of the dopaminergic neurons in the substantia nigra pars compacta (SNc), leading to destruction of the nigrostriatal pathway, and consequently reduction of dopamine levels in the striatum [5]. Neuronal death with depletion of respective neurotransmitters also occur in other areas of the central nervous system (CNS), in particular the noradrenergic neurons in the locus coeruleus (LC), dorsal motor nucleus of the vagus (DMN), cholinergic neurons in the nucleus basalis of Meynert (NBM), and cells in the olfactory system [5–7]. Further, pathology can occur in some regions of the peripheral nervous system (PNS), such as autonomic ganglia (e.g. superior cervical ganglion) and the mesenteric plexus in the wall of the gut [5]. Pathology in the extranigral regions likely plays a role in the development of both motor and non-motor dysfunction in PD patients. Characteristically, neurodegeneration at the various pathological sites is accompanied by protein accumulation, aggregation and the formation of Lewy body inclusions in PD (Figure 8.1) [5, 6].

PD can occur through inheritance or may develop sporadically. Approximately 10–15% of cases are thought to be genetic in origin and specific linkages and gene mutations have been identified in small numbers of familial cases [8]. Most cases

Protein Degradation, Vol. 4: The Ubiquitin-Proteasome System and Disease.
Edited by R. J. Mayer, A. Ciechanover, M. Rechsteiner

ISBN: 978-3-527-31436-2

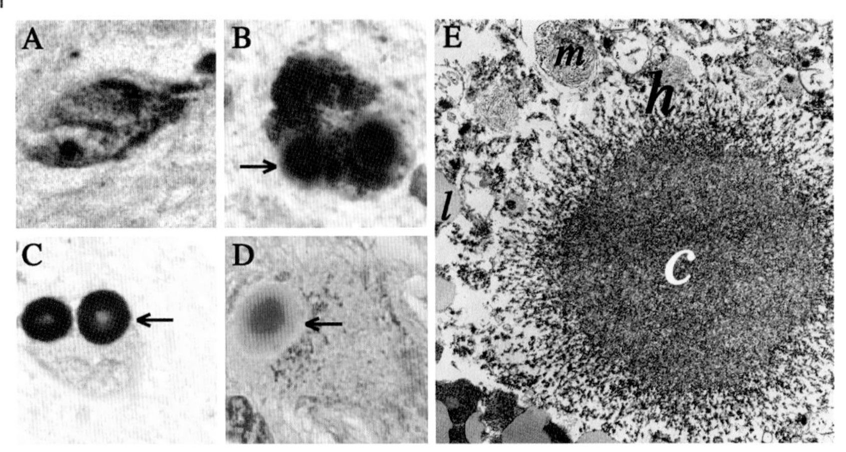

Fig. 8.1. Lewy bodies in the SNc in sporadic PD. (A–C) A standard immunohistochemical protocol; 3,3′-diaminobenzidine was used as a brown chromogen to stain sections of the SNc from normal control (A) and PD subjects (B, C) subjects. This procedure shows the presences of two Lewy bodies (arrows) containing ubiquitin (B) and -synuclein (B) in dopaminergic neurons in PD, but no inclusion body is present in normal controls (A). The dark granular substance in the neurons is neuromelanin. (D) Conventional Hematoxylin (blue) & Eosin (pink) histological staining reveals (arrow) a spherical Lewy body in SNc dopamine neurons with a distinct central core and a peripheral halo. (E) Electron micrograph of a Lewy body reveals that the core (*c*) contains granular material and the outer halo (*h*) is composed of radiating filaments. In panel E, (*l*) represents lipofuscin deposits, and (*m*) indicates a mitochondrion.

(≈90%) of PD are sporadic and of unknown cause. A widely held hypothesis relating to the cause of sporadic PD suggests that exposure to environmental toxins leads to the development of the illness in individuals who are rendered susceptible due to their genetic profile, poor ability to metabolize toxins and/or advancing age [9]. The pathogenic process is unknown, but has been linked to a variety of factors, including oxidative stress [10], mitochondrial dysfunction [11], inflammation [12], excitotoxicity [13] and apoptosis [14]. However, it is not clear as to how these cellular, biochemical and molecular changes relate to each other and to neuronal degeneration in PD.

In recent years, a growing body of genetic, postmortem and experimental evidence have converged to suggest that failure of the ubiquitin–proteasome system (UPS) and altered protein handling play a major role in the etiopathogenesis of sporadic and the various familial forms of PD (Figure 8.2) [15–17]. In this chapter, we will examine the range of defects that occur in the various hereditary and sporadic forms of PD, and consider how they might be linked to the UPS and lead to pathogenesis.

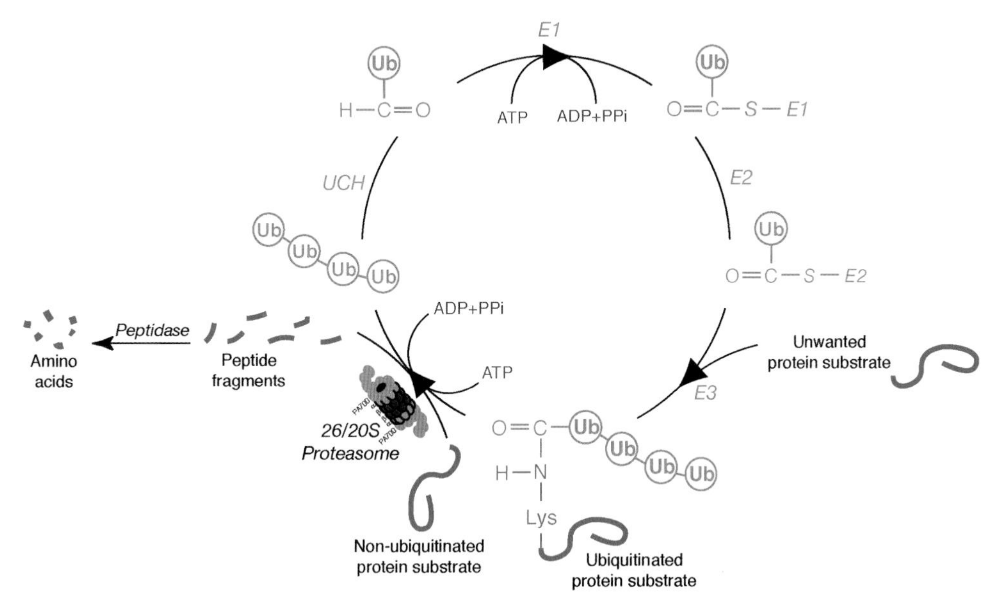

Fig. 8.2. The UPS and Parkinson's disease. Recent genetic, postmortem and experimental evidence suggest that failure of the UPS plays a role in the etiopathogenesis of familial and sporadic PD [15–17]. The UPS comprises two processes that occur consecutively to degrade unwanted proteins that are either abnormal (i.e. incomplete, mutant, misfolded, denatured, oxidized and otherwise damaged proteins) or normal (e.g. turnover of short-lived regulatory proteins). In the first step, a ubiquitin molecule (a 76-amino acid, 8.5-kDa polypeptide) is conjugated to unwanted proteins via a covalent isopeptide bond between the C-terminal Gly residue of ubiquitin and an internal Lys residue of the substrate protein. Thereafter, additional ubiquitin molecules are attached to the previously conjugated ubiquitin (at a Lys residue) in a sequential manner to form a polyubiquitin chain. Ubiquitination is ATP dependent and is mediated by three different enzymes acting in sequence, namely a ubiquitin-activating enzyme (E1) which activates ubiquitin by forming a thioester, followed by a ubiquitin-conjugating enzyme (E2) that carries activated ubiquitin as a thioester, and finally an ubiquitin ligase (E3) which transfers activated ubiquitin to the substrate protein. In mammalian cells, it appears that only one E1 enzyme exists, while 20–40 E2 enzymes have been identified and there are 500–1000 E3 enzymes which can be grouped into distinct families (e.g. HECT domain and RING-finger domain E3s). The selectivity of protein ubiquitination is assured by the fact that each E3 enzyme is specific for one or a limited number of different proteins. Additionally, some proteins require post-translational modification (e.g. phosphorylation of I B) before they can undergo ubiquitination and this provides a further degree of selectivity in the process [132]. In the second step of the UPS, unwanted proteins previously tagged by ubiquitin are unfolded by PA700 to permit entry into the inner chamber of the 26S proteasome complex where they are degraded in an ATP-dependent manner. The degradation products of 26S proteasomes are 2–25-residue peptide fragments that are further hydrolyzed by peptidases to produce their constituent amino acids which are re-used in protein synthesis [203]. Following recognition but before entry into the proteasome, polyubiquitin chains are separated from protein conjugates then disassembled by de-ubiquitination enzymes (ubiquitin C-terminal hydrolases) into monomeric ubiquitin which is re-used in the ubiquitination cycle. It is noteworthy that short peptides and some proteins (e.g. oxidatively damaged proteins) can be degraded by the 20S proteasome (the catalytic core of the 26S proteasome) without prior ubiquitination [24]. A variety of studies indicate that the failure of the UPS at various points as indicated, plays a role in protein accumulation, Lewy body formation and neurodegeneration in familial and sporadic forms of PD [15–17].

8.2 Protein Handling in the CNS

Neuronal activity is associated with the generation of proteins that are abnormal, such as incomplete, mutant, misfolded, denatured, oxidized and otherwise damaged proteins [18, 19]. This is prominent in the CNS due to the relatively high utilization of oxygen and elevated rate of metabolism, and the enzymatic- and auto-oxidation of neurotransmitters such as dopamine, all of which facilitate the production of reactive oxygen species and other free radicals that can induce protein damage [20, 21]. Abnormal proteins have a tendency to misfold, aggregate, interfere with intracellular processes and induce cytotoxicity [22–25]. As such, their production must be limited or they must be rapidly cleared so as to maintain the integrity and viability of cells [18, 19]. Indeed, in the CNS, a tight balance between the production of abnormal proteins and their clearance is crucial since these neurons have a limited ability for repair/regeneration and so are crucially dependent on existing neuronal populations. This equilibrium might be difficult to maintain in the CNS since the long lifespan of neurons is associated with alterations in a variety of intracellular process, such as oxidative stress and mitochondrial dysfunction which cause protein damage and accumulation [20]. Also, in the aging CNS, there is a marked increase in oxidative damage [20, 26, 27]. Finally, proteasomal function declines with aging in various regions of the CNS [20, 26, 28]. Thus, the CNS appears to be crucially dependent on protein clearance systems to maintain its structural and functional integrity.

There are two systems that mediate the majority of protein degradation and clearance within cells [29]. An autophagic process involving cathepsins (cysteine proteases) is responsible for the degradation of membrane and extracellular components following endocytosis into the lysosome [29]. The UPS is primarily responsible for the clearance of abnormal and cytoplasmic proteins and this process occurs in the cytoplasm, nucleus and endoplasmic reticulum (ER) of cells including neurons in the CNS (Figure 8.2) [18, 29–33]. This pathway also plays a significant role in the turnover of short-lived regulatory/functional proteins and is therefore intimately linked with a variety of inter-/intra-cellular processes [18, 29–31].

Molecular chaperones or heat shock proteins (HSPs), such as HSP70 and HSP90, are a highly conserved class of proteins that contribute to protein handling within cells [34, 35]. HSPs act to facilitate the proper folding and localization of proteins, and serve to prevent inappropriate interactions within and between proteins that can otherwise lead to misfolding and aggregation. Additionally, HSPs promote the refolding of proteins that become abnormal. Importantly, HSPs function synergistically with the UPS in several ways, notably in their ability to alter the folding pattern of abnormal proteins to facilitate their recognition and entry into 26/20S proteasomes for degradation [35–38]. HSPs also play a role in the assembly of the 26S proteasome complex [39].

Cells normally maintain a dynamic balance between the generation of abnormal proteins and their clearance by the UPS, HSPs and other proteolytic systems.

Disturbance of this equilibrium, either by the excess production of abnormal proteins or reduced degradation, leads to an adverse state called proteolytic stress [17, 40]. During proteolytic stress, poorly degraded or undegraded proteins tend to accumulate and aggregate with each other and with normal proteins [18, 24, 41]. Such protein aggregates can disrupt intracellular processes and induce cytotoxicity [22–25]. In recent years, several studies have shown that when UPS-mediated degradation fails, cells can activate a secondary and possibly cytoprotective response. This is the transport, sequestration and compartmentalization of poorly degraded/undegraded proteins and aggregates to form aggresomes and inclusion bodies [23, 42–45]. In many neurodegenerative disorders, where proteolytic stress is a key factor, protein aggregates and inclusion bodies can be seen within different compartments of the cell [46]. In PD, these proteinaceous inclusions are referred to as Lewy bodies and they are typically seen in the cytoplasm of neurons at the various pathological sites (Figure 8.1) [40, 47].

8.3 The UPS and Protein Mishandling in PD

During the past 10 years, there have been several discoveries of gene mutations that cause rare familial forms of PD [15, 17, 48]. Although the clinical spectrum and pathology of these illnesses often differ significantly from each other and from typical sporadic PD, it appears that they share similar pathogenic mechanisms, namely protein mishandling and aggregation [15, 17, 48]. This concept relates to the observation that these mutations can affect proteins that have a high propensity to misfold and aggregate, or impair the activity of UPS enzymes and related proteolytic systems [15, 17, 48]. Several studies have demonstrated that in patients with sporadic PD there is a reduction in proteasomal subunits and enzymatic activity in the SNc which degenerates, but an upregulation of proteasomal function in regions of the brain that have been spared [49]. Taken together, these findings suggest that failure of the UPS could be a central and common defect that underlies the pathogenesis of the various familial and sporadic forms of PD (Figure 8.2).

8.4 Parkin

An hereditary form of PD, autosomal recessive juvenile parkinsonism (AR-JP) first described in Japanese families, was linked to chromosome 6q25.2–q27 (PARK 2) in 1997 [50, 51]. This locus was found to host the gene that codes for a 465-amino acid/52-kDa protein called parkin [52, 53]. It is now appreciated that many deletions, point mutations and mutations spanning the entire parkin gene cause familial PD [15]. Estimates suggest that parkin mutations account for approximately 50% of early-onset (<45 years) familial cases of PD [54]. It is noteworthy that parkin mutations have also been associated with late-onset (≥60 years old) hereditary PD [55].

Clinically, AR-JP is similar to common sporadic PD, but there are notable differences. AR-JP can have a very early age of onset, ranging from 7–72 years (average, 30 years), and demonstrates a rather slow rate of progression [54]. Similarly, the neuropathology of AR-JP differs from sporadic PD in that neurodegeneration in the familial disorder is restricted to the SNc and LC, whereas cell death is more widespread in sporadic PD [56]. Further, Lewy bodies or other protein aggregates are largely absent in patients with AR-JP [56]. It is also noteworthy that Lewy bodies have been found in a patient with parkin-linked autosomal dominant parkinsonism and clinical features more typical of sporadic PD [57].

Parkin is expressed in the cytoplasm, nucleus, golgi apparatus and processes of neurons throughout the CNS [58]. Several studies have shown that parkin is an E3 ubiquitin ligase [59–63]. Similar to other E3 ubiquitin ligases, parkin contains a RING-finger domain (comprising two RING-finger motifs separated by an in-between-RING domain) at the C-terminal, a central linker region, and a ubiquitin-like (UBL) domain at the N-terminal. Parkin acts in conjunction with several E2 enzymes, Ubc6, UbcH7 and UbcH8, to ubiquitinate a variety of substrates. These include synphilin-1, CDCrel-1, parkin-associated endothelin receptor-like receptor (Pael-R), an *O*-glycosylated isoform of α-synuclein (αSp22), cyclin E α/β-tubulin, p38 subunit of the aminoacyl-tRNA synthetase complex, and synaptotagmin X1 [15, 59, 61, 62]. Interestingly, parkin may polyubiquitinate proteins with linkages at lysine 48 (K48) or lysine (K63) [64]. Parkin has been shown to interact through its UBL domain with the 26S proteasome Rpn10/S5a subunit that, along with Rpt5/S6′, plays a role in the recognition of ubiquitinated substrates by the PA700 proteasome activator [31, 65]. It was shown that parkin also interacts with a protein complex containing CHIP/HSP70 and which promotes parkin's activity [66, 67]. Parkin also interacts with proteasomal subunits [68].

Precisely how parkin induces pathology in familial PD is not known, but could relate to a loss of E3 ubiquitin ligase and hence impairment of protein ubiquitination. The levels of parkin and its enzymatic activity are decreased in the SNc and LC which degenerate in AR-JP [56, 59, 60, 69]. This defect may thus underlie the accumulation of undegraded parkin substrates, including Pael-R and αSp22, in these brain areas [60, 62]. It has been shown that normal parkin prevents ER dysfunction and unfolded protein-induced cell death following overexpression of Pael-R in cultured cells and *Drosophila* [61, 62, 70]. So, it is reasonable to consider that accumulation of undegraded substrate proteins disrupts intracellular processes leading to neurodegeneration in familial PD.

Mutations of parkin in transgenic mice do not cause nigrostriatal degeneration as seen in AR-JP [71–74]. The frequency of point mutations, deletions and duplications of parkin is similar in AR-JP (3.8%) and normal control (3.1%) subjects [75]. Taken together, these observations raise the possibility that additional factors, for example exposure to environmental substances or other gene alterations, might be necessary to trigger the development of parkinsonism in some individuals carrying mutations in parkin.

8.5
UCH-L1

It was discovered in 1998 that an I93M missense mutation in the gene (4p14; PARK 5) encoding ubiquitin C-terminal L1 (UCH-L1), a 230-amino acid/26-kDa de-ubiquitinating enzyme, caused autosomal dominant PD in two siblings of a European family [76]. The parents were asymptomatic, suggesting that the gene defect causes disease with incomplete penetrance. The affected individuals had clinical features that resemble sporadic PD, including a good response to levodopa, but the age (49 and 51 years) of onset was relatively early. Postmortem analyses on one of the siblings revealed Lewy bodies in the brain [77]. More recent genetic screening studies have failed to detect UCH-L1 mutations in families with PD, suggesting that this mutations is a very rare cause of the illness [78]. Several studies have found that the UCH-L1 gene is a susceptibility locus in sporadic PD and that polymorphisms, such as the S18Y substitution, confer some degree of protection against developing the illness [79]. However, a recent study has failed to find any association between UCH-L1 polymorphisms and PD [80].

UCH-L1 is expressed exclusively in neurons in many areas of the CNS [81], and constitutes 1–2% of the soluble proteins in the brain [81–83]. It is not known how alterations in the UCH-L1 gene alters the UPS, proteolysis and protein levels in PD. Mutations in UCH-L1 causes a reduction in de-ubiquitinating activity *in vitro* and in the brain of transgenic mice with the neurological disorder gracile axonal dystrophy (GAD) [76, 84, 85]. Toxin- or mutation-induced inhibition of UCH-L1's activity leads to a marked decrease in the levels of ubiquitin in cultured cells and in the brain of GAD mice [85, 86]. Impairment of ubiquitin C-terminal hydrolases leads to degeneration of dopaminergic neurons with protein accumulation and formation of Lewy body-like inclusions in rat ventral midbrain cell cultures [86]. A recent study showed that UCH-L1 has E3 ubiquitin ligase activity, but it remains unclear if and how the PD-related mutation alters this function of the protein [87].

Therefore, it is possible that a mutation in UCH-L1 alters UPS function leading to altered proteolysis and ultimately cell death. However, more studies are required to decipher the mechanism by which mutations in UCH-L1 leads to pathogenesis in familial PD.

8.6
α-Synuclein

The first gene, at chromosome 4q21–q23 (PARK 1&4), to be associated with PD was reported in several European families during the 1990s [88–90]. Genetic analyses showed that the defects were A53T and A30P point mutations in the gene that encodes for a previously discovered 140-amino acid/14-kDa protein known as α-synuclein [89, 90]. Subsequently, an E46K mutation in α-synuclein was reported in another European family with autosomal dominant PD (plus features of dementia with Lewy bodies) [91], but no other point mutation has been found. In recent years, duplication and triplication of the normal α-synuclein gene have been found to cause autosomal dominant PD in European and American families [92–97].

Familial PD caused by α-synuclein shares features of common sporadic PD, but there are also significant differences, in particular the relatively early age of onset (mean, between 40 and 50 years) and high occurrence of dementia in patients with α-synuclein mutations. Also, patients with duplication/triplication of the α-synuclein gene tend to have the neurodegenerative disorder, dementia with Lewy bodies, in addition to or instead of parkinsonism [92–97]. Pathological studies have shown that patients with the A53T mutation or multiplication mutation of the α-synuclein gene show a marked increase in α-synuclein levels with protein aggregation in various regions of the brain [95, 98, 99]. However, in patients with the A53T mutation, Lewy bodies are rarely present and there is a marked accumulation of α-synuclein and tau in the cerebral cortex and striatum [98, 99]. Also, patients with triplication of the normal α-synuclein gene have vacuoles in the cortex, neuronal death in the hippocampus and the inclusions bodies in glial cells [95]. These findings show that there are significant differences between the pathology in α-synuclein-linked familial and common sporadic PD.

α-Synuclein is a member of a family of related proteins that also include β- and γ-synucleins [100]. α-Synuclein, so called because of its intracellular localization to synapses and nuclear envelope when first discovered [101, 102], is expressed throughout the CNS [81]. The protein is enriched in presynaptic nerve terminals and associates with lipid membranes and vesicles [100]. The normal function of α-synuclein is unknown, but their is some evidence that it plays a role in synaptic neurotransmission [100, 103]. Since the discovery of α-synuclein-linked familial PD, there has been a great deal of effort aimed at deciphering how mutations in this protein induce neurodegeneration. The dominant mode of inheritance suggests a gain of function. Wild type α-synuclein is monomeric and intrinsically unstructured/natively unfolded at low concentrations, but in high concentration it has a propensity to oligomerize and aggregate into β-pleated sheets [104, 105]. Mutations in the protein increase this potential for misfolding, oligomerization and aggregation [104, 106–110]. Oligomerization of α-synuclein produces intermediary species (protofibrils) that form annular structures with pore-like properties that permeabilize synthetic vesicular membranes *in vitro* [106–108]. It has been suggested that protofibrils are the toxic α-synuclein species that are responsible for cell death [109], but this concept is largely based on studies of the biophysical and conformational properties of α-synuclein *in vitro*. With the discovery that Lewy bodies in patients with sporadic PD stained positively for α-synuclein, it has been considered that α-synuclein might also play an important role in the development of sporadic PD.

It is possible that the cytotoxicity of mutant/excess α-synuclein involves interference with proteolysis. Wild type α-synuclein is a substrate for both the 26S and 20S proteasome and is preferentially degraded in a ubiquitin-independent manner [111–113]. *In vitro* and *in vivo* studies have demonstrated that mutant α-synuclein, which misfolds, oligomerizes and aggregates, is resistant to UPS-mediated degradation and also inhibits this pathway [114–116]. As a result, there is accumulation of a wide range of proteins in addition to α-synuclein in cells expressing mutant α-synuclein [114–116]. As previously discussed, high levels of undegraded or poorly degraded proteins have a tendency to aggregate with each other and other proteins, form inclusion bodies, disrupt intracellular processes, and cause cell death [22]. Recent studies indicate that α-synuclein can also be broken down by the 20S proteasome through endoproteolytic degradation that does not involve the N- or C-terminus [111–113]. This type of degradation yields truncated α-synuclein fragments which are particularly prone to aggregate, promote aggregation of the full length protein as well as other proteins, and cause cytotoxicity [117]. Thus, it is reasonable to consider that alterations in the α-synuclein gene can cause the UPS to fail and this defect may underlie protein aggregation, Lewy body formation and neurodegeneration in hereditary PD.

Numerous studies, employing a variety of approaches, have examined the effects of expressing familial PD-related mutant (and wild-type) α-synuclein in transgenic animals [118]. Expression of mutant (A53T, A30P) or wild-type α-synuclein in transgenic *Drosophila* [119], or the adenoviral-mediated expression of A53T mutant or wild type α-synuclein in the SNc of adult non-human primates (common marmosets) [120], causes dopamine cell degeneration. However, overexpression of A53T, A30P or wild-type α-synuclein causes inclusion body formation but does not cause neurodegeneration in transgenic mice [118]. Interestingly, the non-human forms of α-synuclein normally have a threonine in the alanine position and do not show aggregation as is found with the human mutation [89], possibly because α-synuclein is degraded differently in these species. α-Synuclein can also be degraded by the lysosomal system and there is evidence of impaired chaperone-mediated clearance by autophagy of the mutant form of the protein [121, 122].

The relative roles of the UPS and lysosomal systems in the degradation of wild-type and mutant α-synuclein has not been clearly defined, and it is possible that defects in the lysosomal systems could contribute to the protein accumulation and aggregation found in α-synuclein-linked familial PD. Additionally, the observation that not all carriers of point mutations in α-synuclein develop PD, suggests that additional factors, such as environmental toxins, might be required to trigger the development of PD in these individuals.

8.7 Dardarin/LRRK2

In 2002, a large Japanese family having autosomal dominant PD with incomplete penetrance was linked to a mutation on chromosome 12p11.2–q13.1 (PARK 8) [123]. This linkage has since been found in several families from different countries, and it is estimated that the mutation could account for 5% of familial cases of PD [8]. The gene defects in these patients are several missense mutations in the gene encoding a 2527-amino acid/≈250-kDa protein called dardarin or LRRK2 (leucine-rich repeat kinase 2) [124, 125]. Notably, not all individuals with these mutations develop parkinsonism, suggesting the possible requirement of other etiological factors to act as a trigger for the illness [126].

The clinical spectrum of dardarin/LRRK2-linked PD is similar to sporadic PD and has an age of onset ranging from 32 to 79 years. There are, however, important pathological differences between these two forms of PD [123, 125, 127]. In dardarin/LRRK2-linked familial PD, all subjects have nigrostriatal degeneration, some have Lewy bodies and some do not have these inclusions; others have extensive cortical Lewy bodies consistent with "dementia with Lewy bodies", and some have tau-immunoreactive glial and neuronal inclusions consistent with tauopathies such as progressive supranuclear palsy. Interestingly, some patients with this mutation have a late onset form of PD with no family history and pathology changes characteristic of sporadic PD. It has been estimated that the LRRK2 mutation might account for as many as 7% of familial cases and 1.5–3% of cases of sporadic PD [126, 128, 129].

Dardarin/LRRK2 is expressed throughout the brain [124, 130], but its normal function is unknown. Based on its similarity with other proteins, it has been suggested that dardarin/LRRK2 might be a cytoplasmic kinase [124, 125]. It remains to be determined how mutations in dardarin/LRRK2 alter the structure and function of the protein, but a recent study indicated that PD-related mutations cause an increase in phosphorylation activity [131]. Some proteins, such as I B, require phosphorylation as a prerequisite to their ubiquitination and proteasomal degradation [132]. Indeed, the phopsphorylation status of serine 129 of α-synuclein appears to play a significant role in its ability to induce toxicity in *Drosophila* [133]. It will therefore be interesting to determine whether mutations in dardarin/LRRK2 lead to altered phosphorylation with increased aggregation of target proteins and impairment of the UPS.

8.8 PINK1

Autosomal recessive early-onset PD in several European families was found to result from missense and exon-deletion mutations in a gene located at chromosome 1p35–p36 (PARK 6) which codes for a 581-amino acid/62.8-kDa protein designated PINK1 (PTEN (phosphatase and tensin homolog deleted on chromosome 10)-induced kinase 1) [134–136]. This mutation has since been found in additional families, and may account for 1–2% of early-onset cases of PD [137, 138]. Clinically, this form of PD is characterized by early onset (20–40 years) of symptoms, slow progression and good response to levodopa [134, 138].

PINK1 is localized to mitochondria but additional studies are required to determine its cellular and anatomical distribution [136]. The normal function of PINK1 is unknown. It appears to be a serine/threonine protein kinase which phosphorylates proteins involved in signal transduction pathways [136]. In cell culture studies, wild-type PINK1 prevents proteasome inhibitor-induced mitochondrial dysfunction and apoptosis, but this protection is lost with familial PD-related mutations [136]. These findings raise the possibility that mutations in PINK1 could render neurons susceptible to agents, such as abnormal proteins and toxins, that act on proteasomes to induce cell death.

Thus, the pathogenic process in PINK1-related familial PD might involve alterations in protein handling. It is interesting that familial PD-related mutations in PINK1 have been found in normal control subjects who do not have clinical features of parkinsonism [139], again raising the possibility that multiple factors may be necessary for the development of PD [140].

8.9 DJ-1

Genetic studies in 2001 of several European families with autosomal recessive early-onset parkinsonism found linkage to chromosome 1p36 (PARK 7) [141]. Subsequently, these families were found to have missense and deletion mutations in the gene that encodes a previously known 189-amino acid/20-kDa protein called DJ-1 [142–144]. Since then, no additional mutation in DJ-1 has been reported and it is thought that this defect could account for only <1% of early-onset cases of the illness [145]. Clinically, DJ-1-linked PD is similar to parkin-related PD, namely early onset (20–40 years) of symptoms, slow progression, presence of dystonia, levodopa-responsiveness, and the occurrence of psychiatric disturbance [142, 143]. At present, the neuropathological features in DJ-1 patients are not known.

In the CNS, DJ-1 is more prominent in astrocytes compared to neurons, and is present in the cytosol and nucleus of cells [146, 147]. The normal function of DJ-1 is not known. There is evidence to suggest that it acts as an antioxidant or a sensor of oxidative stress [148, 149]. Additionally, the molecular structure and *in vitro* properties of DJ-1 indicate that it might act as a molecular chaperone and a protease [150–152]. Recently, DJ-1 was found to interact with parkin, CHIP and HSP70, suggesting a link to these proteolytic systems [153].

The mechanism by which mutations in DJ-1 induces pathogenesis is unknown. The recessive pattern of inheritance raises the possibility that the mutations induce a loss of function of the protein. The PD-related mutations (e.g. L166P) destabilize and inactivate the protein, impair its proteolytic activity, and promote its rapid degradation by the proteasome [150, 154]. In cell cultures, overexpression of DJ-1 protects cells from oxidative stress, but knockdown of DJ-1 increases susceptibility to oxidative stress, endoplasmic reticulum stress and proteasomal inhibition [148, 149]. Further, mutations in DJ-1 reduce its ability to inhibit the aggregation of α-synuclein both *in vitro* and *in vivo* [155]. Interestingly, a recent study has shown that deletion of DJ-1 in transgenic mice does not induce neurodegeneration [156], suggesting that other factors might be involved in the pathogenic process in PD.

Thus, it may be speculated that mutations in DJ-1 lead to a loss of its putative antioxidant, chaperone and proteolytic activity. Such defects, if proven to be the case in future studies, would indicate that altered protein handling also plays a role in the pathogenesis of this familial form of PD.

8.10 Proteasomal Dysfunction in Sporadic PD

The majority of PD cases occur sporadically with insidious onset and are of unknown cause. At present, there is no convincing evidence to suggest that a defect in either parkin, UCH-L1, α-synuclein, dardarin/LRRK2, PINK1 or DJ-1, is responsible for sporadic PD. However, these or other genes could be involved as a susceptibility factor in this form of the illness. It is widely believed that gene and/or aging-related-susceptibility coupled with exposure to environmental toxins underlies the etiology of sporadic PD [9]. Thus, it is interesting that variability in the genes encoding α-synuclein [157], parkin [158], UCH-L1 [79], but not PINK1 [159], dardarin/LRRK2 [160], or DJ-1 [161], have been associated with an increased risk of developing sporadic PD. Several etiopathogenic factors, including oxidative stress, mitochondrial dysfunction [11], inflammation [12], excitotoxicity [13] and apoptosis [14], have been linked with the neurodegenerative process. Most recently, failure of the UPS, specifically at the level of the 26/20S proteasome, has been implicated in the pathogenesis of sporadic PD [17].

8.10.1 Altered Proteasomal Function

Over many years of research in PD, there has been indirect, but nevertheless significant, findings which suggested that proteasomal dysfunction plays a role in the vulnerability and degeneration of the SNc and perhaps other regions in PD. The mRNA level and enzymatic activity of 26/20S proteasomes and proteasome activators decrease with advancing age in the midbrain and other areas of the CNS [28, 162, 163]. In comparison with other brain areas, the SNc has a higher level of basal oxidative stress and protein oxidation; these processes are elevated in parallel with aging [27]. Therefore, declining proteasomal activity coupled with increasing oxidative protein damage with advancing age could underlie the age-related increase in susceptibility of the SNc to proteolytic stress and degeneration. Indeed, mild neuronal loss with Lewy bodies are found in the SNc of 10–15% of individuals who die over the age of 65 years without clinical evidence of neurological disorder [6, 164]. This condition, referred to as incidental Lewy body disease (ILBD), occurs with 10 times the frequency of PD and is thought to represent the pre-symptomatic phase of sporadic PD [164].

There is a marked increase in the levels of oxidatively damaged, 4-hydroxynonenal-conjugated, nitrated, phosphorylated and ubiquitinated proteins in the SNc and other areas of the brain in PD [47]. Indeed, protein aggregates and Lewy body inclusions containing a wide variety of proteins, including α-synuclein and ubiquitin, can be seen at the various pathological sites in patients with PD [47]. These observations suggest that the UPS may be inhibited and/or saturated, resulting in protein accumulation and aggregation in the disorder. The accumulation of both ubiquitinated and non-ubiquitinated proteins (e.g. oxidized proteins and α-synuclein) [24, 111–113] in the brain indicates that a defect in proteolysis at a central and common point, i.e. the 20S proteasome core, is likely since both groups of proteins accumulate in the illness.

Studies of proteasomal function have shown that the chymotrypsin-like, trypsin-like and PGPH enzymatic activities of the 26/20S proteasomes were reduced by approximately 44–55% in the SNc in PD compared to age-matched controls [165–168]. In contrast, the three proteolytic activities of the proteasome were unchanged in regions that do not degenerate in PD, namely the frontal cortex, striatum, hippocampus, pons and cerebellum [165–168]. Interestingly, Tofaris and colleagues showed proteasomal impairment in PD cases with relatively mild neuropathology, suggesting that altered proteasomal function occurs early in the pathogenic process [167]. There is a 40% reduction in the content of proteasome α-subunits, but not β-subunits, in the SNc in PD compared to age-matched controls [165]. In contrast, the levels of α-subunits were increased by 9% in the cerebral cortex and by 29% in the striatum in PD. Immunohistochemical staining demonstrated reduced levels of 26/20S proteasomal α-subunits, but not β-subunits, within dopaminergic neurons in the SNc of PD subjects compared to age-matched controls [165]. As discussed previously, the PA700 proteasome activator is a complex of over 20+ different subunits with varying molecular weights [31]. In PD, there was either no change (42, 46 and 95 kDa bands) or up to a 33% loss (52.5, 75 and 81 kDa bands) of PA700 subunits in the SNc [165]. In contrast, there was a marked increase in the levels of subunits at the 81, 75, 52.5 and 42 kDa bands in the frontal cortex and/or the striatum of PD subjects compared to controls. This observation is consistent with other studies showing a significant upregulation of proteasomal function in cortical areas in patients with PD [168]. In normal control subjects, the levels of the PA28 proteasome activator were very low in the SNc compared to the frontal cortex and striatum [165]. In PD brains, PA28 immunoreactivity was almost undetectable in the SNc and levels were reduced in the frontal cortex (24%) and striatum (16%) in comparison to controls [165]. These findings indicate that, in sporadic PD, there is inhibition of proteasomal function in regions that degenerate while there is upregulation, perhaps a compensatory and cytoprotective response, in areas that do not degenerate. Interestingly, a recent study showed that proteasomal function is impaired in lymphocytes of PD but not Alzheimer's disease patients, although this defect might be related to drug therapy [169].

8.10.2
Role of Proteasomal Dysfunction in the Neurodegenerative Process

The involvement of altered proteasomal function in the pathogenesis in PD is supported by several observations. Proteasomes not only play a critical role in

Table 8.1. Alterations in UPS-linked cellular processes in Parkinson's disease.

Cellular processes linked to the UPS	Alterations in Parkinson's disease
Degradation and clearance of abnormal proteins [18]	Yes: failure of the UPS and protein aggregation
Antioxidant defense mechanisms [171, 176]	Yes: oxidative stress [10]
Mitochondrial function [172, 188, 200, 204]	Yes: complex I activity impaired [11]
Inflammatory response [174, 177]	Yes: microglial activation and gliosis [12, 205]
Immune processes [206]	Yes: complement activation [12]
Apoptotic signaling [172, 175]	Yes: apoptotic cell death [14]
Synaptic function and neurotransmission [207]	Yes: altered basal ganglia function [208, 209]
Signal transduction [210]	Yes: altered neuronal activity [208, 209]
Protein transport/trafficking [211]	Yes: inclusion body formation [47]
Gene transcription [212]	Yes: altered expression of a variety of proteins [186]
Development and differentiation [207]	–
Regulation of cell cycle and division [213]	–

The ubiquitin–proteasome system (UPS) controls the levels of short-lived regulatory/functional proteins that mediate a wide variety of cellular processes. Thus, failure of the UPS to degrade proteins not only causes protein accumulation and aggregation, but it also alters cellular functions. Many of these cellular and biochemical defects occur in Parkinson's disease and likely play role in the neurodegenerative process.

the degradation and clearance of unwanted proteins, but they also play a major role in controlling the levels of short-lived regulatory/functional proteins and are intimately linked with a variety of cellular processes (Table 8.1). Indeed, proteasomes are linked with antioxidant defense mechanisms [170, 171], mitochondrial activity [172, 173], inflammatory responses [174], and anti-apoptotic pathways [175] (Table 8.1). Thus, inhibition of proteasomal function disrupts these processes and causes oxidative stress [176], mitochondrial dysfunction [176], pro-inflammatory reactions [177] and apoptotic cell death [172]. Most of these proteasome-linked cellular processed have been found to be altered in PD, further supporting the concept that proteasomal dysfunction plays a role in the pathogenic process (Table 8.1).

Typically, inhibition of proteasomal function causes protein accumulation and the formation of aggresomes which are intracytoplasmic proteinaceous inclusions formed at the centrosome in response to inadequate protein degradation [23, 43, 178]. The demonstration that Lewy bodies contain the centrosome-related marker γ-tubulin, UPS components and HSP, and share other compositional and organization features of aggresomes, suggests that they might form in an aggresome-related manner as a result of proteolytic stress PD [40, 47]. Indeed, this concept raises the possibility that Lewy bodies might represent a cytoprotective response aimed at promoting the survival of the neurons in which they are formed in PD [47, 179].

Impairment of proteasomal function typically induces cell death and this often occurs via an apoptotic mechanism. It has been shown that application of proteasome inhibitors to cultured cells, or injection of these agents into the brain of rats, induces preferential degeneration of dopaminergic neurons in the SNc [86, 180–184]. In these model systems, dopaminergic cell death is accompanied by an accumulation of protein and the formation of α-synuclein/ubiquitin-immunoreactive inclusions that resemble Lewy bodies [86, 180–184]. Recently, it was shown that systemic exposure of rats to PSI (Z-Ile-Glu(OtBu)-Ala-Leu-al, a peptide aldehyde) or epoxomicin (Ac(Me)-Ile-Ile-Thr-Leu-EX, a peptide α′,β′-epoxyketone), which are synthetic and bacterial proteasome inhibitors respectively, induces a model that closely recapitulated many features of PD [185]. Proteasome inhibitor-treated rats developed progressive, PD-like, motor dysfunctions that could be improved with the administration of dopaminergic agents (i.e. levodopa and apomorphine). Positron emission tomography imaging (PET) demonstrated a gradual loss of dopaminergic nerve terminals in the striatum, and postmortem analyses showed striatal dopamine depletion and progressive neurodegeneration with apoptosis and inflammation in the SNc. Also, neuronal death occurred in the LC, DMN and NMB. At the various pathological sites in the rats treated with proteasome inhibitors, there was a 43–82% inhibition of proteasomal function, accumulation of proteins and the formation of intraneuronal α-synuclein/ubiquitin-positive inclusions which resemble Lewy bodies. Thus, this model based on inhibition of proteasomal function more closely recapitulates the behavioral, imaging, pathological and biochemical features of sporadic PD than any other model of the disorder described to date.

Taken together, the above observations suggest that altered proteasomal function could play a key role in protein accumulation, Lewy body formation and neurodegeneration in the SNc and perhaps other brain regions in sporadic PD.

8.10.3 The Cause of Proteasomal Dysfunction

The question arises as to what causes proteasomal dysfunction in sporadic PD. This is not known, but there are several possibilities. It could result from undiscovered gene mutations. Interestingly, DNA microarray analyses were recently used to demonstrate a reduction in the mRNA levels of 20S proteasome α-subunits (PSAM2, PSMA3 and PSMA5) and a non-ATPase subunit (PSMD8/Rpn12), and an ATPase subunit (PSMC4/Rpt3) of PA700, in the SNc in PD [186]. Proteasomal dysfunction may develop secondary to the other biochemical defects that occur in PD, such as oxidative stress or mitochondrial dysfunction [49, 187–189]. Indeed, there could be a close relationship and interplay between proteasomal dysfunction and the various cellular, biochemical and molecular changes that have been detected in PD [49]. An alternative hypothesis is that proteasomal dysfunction in PD could be the result of exposure to toxic substances. Inhibitors of the proteasome are widely distributed in the environment [190]. They are produced by bacteria (e.g. actinomycetes which infect the below-ground portion of crops) [191,

192], fungi (e.g. *Apiospora montagne* which infests wheat/flour) [193], plants [194–196], and the chemical/pharmaceutical industry [190, 197]. Indeed, lactacystin and epoxomicin, which are among the most potent proteasome inhibitors known, are naturally produced by actinomycete (*Streptomyces*) bacteria [198, 199]. These microbes are found globally in the soil and aquatic habitats of gardens and farmland, and are well known for infecting root vegetables and potatoes (causing "scab" formation) [198, 199]. Also, structurally-related analogs and the active pharmacophore of natural and synthetic compounds known to potently inhibit the proteasome, such as PSI, are also present in the environment [190]. Notably, agrochemicals such as the fungicide maneb (specifically its active metabolite) [197] and pesticides including rotenone, have been shown to impair proteasomal function [172, 200, 201]. Thus, humans could be exposed to proteasome inhibitors via the food chain. Indeed, the distribution of proteasome inhibitors in agrarian environments could underlie the finding that rural living and drinking well water are both associated with a high risk of developing PD [202].

8.11 Conclusion

The etiology of PD has fascinated researchers since the illness was discovered. Over the years, a variety of factors have been considered and it remains unclear if or how they contribute to development of the disorder. In recent years, there has been increasing evidence to suggest that failure of the UPS and altered protein handling as key, and perhaps common, are defects that might underlie the various familial and sporadic forms of PD. The evidence for UPS dysfunction in some forms of PD is substantial and convincing. In other types of PD, however, the evidence implicating UPS failure is speculative but is nevertheless intriguing. A determination that the UPS plays a role in the etiopathogenesis of disorder is not merely academic as this could reveal novel targets that can be exploited to develop neuroprotective medicines and possibly a diagnostic biomarker that are currently not available to patients with PD.

References

1 Lang, A.E. and Lozano, A.M. (1998) Parkinson's disease. First of two parts. *N Engl J Med* **339**, 1044–1053.

2 Lang, A.E. and Lozano, A.M. (1998) Parkinson's disease. Second of two parts. *N Engl J Med* **339**, 1130–1143.

3 Marras, C. and Tanner, C.T. (2000) Epidemiology of Parkinson's disease, in *Movement Disorders: Neurologic Principles and Practice*, Vol. 2nd edn. (eds R.L. Watts, and W.C Koller), The McGraw-Hull Companies, Inc., 2004, pp. 177–195.

4 Van Den Eeden, S.K. et al. (2003) Incidence of Parkinson's disease: variation by age, gender, and race/ethnicity. *Am J Epidemiol* **157**, 1015–1022.

5 Forno, L.S. (1996) Neuropathology of Parkinson's disease. *J Neuropathol Exp Neurol* **55**, 259–272.

6 Braak, H. et al. (2003) Staging of brain pathology related to sporadic Parkinson's disease. *Neurobiol Aging* **24**, 197–211.

7 Zarow, C., Lyness, S.A., Mortimer, J.A. and Chui, H.C. (2003) Neuronal loss is greater in the locus coeruleus than nucleus basalis and substantia nigra in Alzheimer and Parkinson diseases. *Arch Neurol* **60**, 337–341.

8 Farrer, M.J. (2006) Genetics of Parkinson disease: paradigm shifts and future prospects. *Nat Rev Genet* **7**, 306–318.

9 Tanner, C.M. (2003) Is the cause of Parkinson's disease environmental or hereditary? Evidence from twin studies. *Adv Neurol* **91**, 133–142.

10 Jenner, P. (2003) Oxidative stress in Parkinson's disease. *Ann Neurol* **53**, S26–S36; discussion S36–S38.

11 Orth, M. and Schapira, A.H. (2002) Mitochondrial involvement in Parkinson's disease. *Neurochem Int* **40**, 533–541.

12 McGeer, P.L. and McGeer, E.G. (2004) Inflammation and neurodegeneration in Parkinson's disease. *Parkinsonism Relat Disord* **10** Suppl 1, S3–S7.

13 Beal, M.F. (1998) Excitotoxicity and nitric oxide in Parkinson's disease pathogenesis. *Ann Neurol* **44**, S110–S114.

14 Tatton, W.G., Chalmers-Redman, R., Brown, D. and Tatton, N. (2003) Apoptosis in Parkinson's disease: signals for neuronal degradation. *Ann Neurol* **53** Suppl 3, S61–S70; discussion S70–S72.

15 Hattori, N. and Mizuno, Y. (2004) Pathogenetic mechanisms of parkin in Parkinson's disease. *Lancet* **364**, 722–724.

16 Moore, D.J., West, A.B., Dawson, V.L. and Dawson, T.M. (2005) Molecular pathophysiology of Parkinson's disease. *Ann Rev Neurosci* **25**, 55–84.

17 McNaught, K.S. and Olanow, C.W. (2003) Proteolytic stress: A unifying concept in the etiopathogenesis of familial and sporadic Parkinson's disease. *Ann Neurol* **53**(3 Suppl 1), S73–S86.

18 Goldberg, A.L. (2003) Protein degradation and protection against misfolded or damaged proteins. *Nature* **426**, 895–899.

19 Sherman, M.Y. and Goldberg, A.L. (2001) Cellular defenses against unfolded proteins: a cell biologist thinks about neurodegenerative diseases. *Neuron* **29**, 15–32.

20 Keller, J.N. et al. (2004) Autophagy, proteasomes, lipofuscin, and oxidative stress in the aging brain. *Int J Biochem Cell Biol* **36**, 2376–2391.

21 Tse, D.C., McCreery, R.L. and Adams, R.N. (1976) Potential oxidative pathways of brain catecholamines. *J Med Chem* **19**, 37–40.

22 Bence, N.F., Sampat, R.M. and Kopito, R.R. (2001) Impairment of the ubiquitin–proteasome system by protein aggregation. *Science* **292**, 1552–1555.

23 Kopito, R.R. (2000) Aggresomes, inclusion bodies and protein aggregation. *Trends Cell Biol* **10**, 524–530.

24 Grune, T., Jung, T., Merker, K. and Davies, K.J. (2004) Decreased proteolysis caused by protein aggregates, inclusion bodies, plaques, lipofuscin, ceroid, and "aggresomes" during oxidative stress, aging, and disease. *Int J Biochem Cell Biol* **36**, 2519–2530.

25 Bennett, E.J., Bence, N.F., Jayakumar, R. and Kopito, R.R. (2005) Global impairment of the ubiquitin–proteasome system by nuclear or cytoplasmic protein aggregates precedes inclusion body formation. *Mol Cell* **17**, 351–365.

26 Keller, J.N., Hanni, K.B. and Markesbery, W.R. (2000) Possible involvement of proteasome inhibition in aging: implications for oxidative stress. *Mech Ageing Dev* **113**, 61–70.

27 Floor, E. and Wetzel, M.G. (1998) Increased protein oxidation in human substantia nigra pars compacta in comparison with basal ganglia and prefrontal cortex measured with an improved dinitrophenylhydrazine assay. *J Neurochem* **70**, 268–275.

28 Zeng, B.Y., Medhurst, A.D., Jackson, M., Rose, S. and Jenner, P. (2005) Proteasomal activity in brain differs

between species and brain regions and changes with age. *Mech Ageing Dev* **126**, 760–766.

29 Ciechanover, A. (2005) Proteolysis: from the lysosome to ubiquitin and the proteasome. *Nat Rev Mol Cell Biol* **6**, 79–87.

30 Pickart, C.M. (2001) Mechanisms Underlying Ubiquitination. *Annu Rev Biochem* **70**, 503–533.

31 Pickart, C.M. and Cohen, R.E. (2004) Proteasomes and their kin: proteases in the machine age. *Nat Rev Mol Cell Biol* **5**, 177–187.

32 Adori, C. et al. (2006) Subcellular distribution of components of the ubiquitin–proteasome system in non-diseased human and rat brain. *J Histochem Cytochem* **54**, 263–267.

33 Mengual, E., Arizti, P., Rodrigo, J., Gimenez-Amaya, J.M. and Castano, J.G. (1996) Immunohistochemical distribution and electron microscopic subcellular localization of the proteasome in the rat CNS. *J Neurosci* **16**, 6331–6341.

34 Hartl, F.U. and Hayer-Hartl, M. (2002) Molecular chaperones in the cytosol: from nascent chain to folded protein. *Science* **295**, 1852–1858.

35 Muchowski, P.J. and Wacker, J.L. (2005) Modulation of neurodegeneration by molecular chaperones. *Nat Rev Neurosci* **6**, 11–22.

36 Luders, J., Demand, J. and Hohfeld, J. (2000) The ubiquitin-related BAG-1 provides a link between the molecular chaperones Hsc70/Hsp70 and the proteasome. *J Biol Chem* **275**, 4613–4617.

37 Conconi, M. et al. (1998) Protection from oxidative inactivation of the 20S proteasome by heat-shock protein 90. *Biochem J* **333** (Pt 2), 407–415.

38 Imai, J., Yashiroda, H., Maruya, M., Yahara, I. and Tanaka, K. (2003) Proteasomes and molecular chaperones: cellular machinery responsible for folding and destruction of unfolded proteins. *Cell Cycle* **2**, 585–590.

39 Imai, J., Maruya, M., Yashiroda, H., Yahara, I. and Tanaka, K. (2003) The molecular chaperone Hsp90 plays a role in the assembly and maintenance of the 26S proteasome. *EMBO J* **22**, 3557–3567.

40 McNaught, K.S., Shashidharan, P., Perl, D.P., Jenner, P. and Olanow, C.W. (2002) Aggresome-related biogenesis of Lewy bodies. *Eur J Neurosci* **16**, 2136–2148.

41 Rajan, R.S., Illing, M.E., Bence, N.F. and Kopito, R.R. (2001) Specificity in intracellular protein aggregation and inclusion body formation. *Proc Natl Acad Sci U S A* **98**, 13060–13065.

42 Johnston, J.A., Illing, M.E. and Kopito, R.R. (2002) Cytoplasmic dynein/dynactin mediates the assembly of aggresomes. *Cell Motil Cytoskeleton* **53**, 26–38.

43 Johnston, J.A., Ward, C.L. and Kopito, R.R. (1998) Aggresomes: a cellular response to misfolded proteins. *J Cell Biol* **143**, 1883–1898.

44 Kawaguchi, Y. et al. (2003) The deacetylase HDAC6 regulates aggresome formation and cell viability in response to misfolded protein stress. *Cell* **115**, 727–738.

45 Arrasate, M., Mitra, S., Schweitzer, E.S., Segal, M.R. and Finkbeiner, S. (2004) Inclusion body formation reduces levels of mutant huntingtin and the risk of neuronal death. *Nature* **431**, 805–810.

46 Ciechanover, A. and Brundin, P. (2003) The ubiquitin proteasome system in neurodegenerative diseases. Sometimes the chicken, sometimes the egg. *Neuron* **40**, 427–446.

47 Olanow, C.W., Perl, D.P., DeMartino, G.N. and McNaught, K.S. (2004) Lewy-body formation is an aggresome-related process: a hypothesis. *Lancet Neurol* **3**, 496–503.

48 Petrucelli, L. and Dawson, T.M. (2004) Mechanism of neurodegenerative disease: role of the ubiquitin proteasome system. *Ann Med* **36**, 315–320.

49 McNaught, K.S. and Olanow, C.W. (2006) Protein aggregation in the pathogenesis of familial and sporadic Parkinson's disease. *Neurobiol Aging* **25**, 530–545.

50 Matsumine, H. et al. (1997) Localization of a gene for an autosomal recessive form of juvenile Parkinsonism to chromosome 6q25.2–27. *Am J Hum Genet* **60**, 588–596.

51 Yamamura, Y., Sobue, I., ando, K., Iida, M. and Yanagi, T. (1973) Paralysis agitans of early onset with marked diurnal fluctuation of symptoms. *Neurology* **23**, 239–244.
52 Kitada, T. et al. (1998) Mutations in the parkin gene cause autosomal recessive juvenile parkinsonism. *Nature* **392**, 605–608.
53 Mizuno, Y., Hattori, N., Mori, H., Suzuki, T. and Tanaka, K. (2001) Parkin and Parkinson's disease. *Curr Opin Neurol* **14**, 477–482.
54 Lucking, C.B. et al. (2000) Association between early-onset Parkinson's disease and mutations in the parkin gene. French Parkinson's Disease Genetics Study Group. *N Engl J Med* **342**, 1560–1567.
55 Foroud, T. et al. (2003) Heterozygosity for a mutation in the parkin gene leads to later onset Parkinson disease. *Neurology* **60**, 796–801.
56 Mori, H. et al. (1998) Pathologic and biochemical studies of juvenile parkinsonism linked to chromosome 6q. *Neurology* **51**, 890–892.
57 Farrer, M. et al. (2001) Lewy bodies and parkinsonism in families with parkin mutations. *Ann Neurol* **50**, 293–300.
58 Horowitz, J.M. et al. (2001) Immunodetection of Parkin protein in vertebrate and invertebrate brains: a comparative study using specific antibodies. *J Chem Neuroanat* **21**, 75–93.
59 Shimura, H. et al. (2000) Familial Parkinson disease gene product, parkin, is a ubiquitin-protein ligase. *Nat Genet* **25**, 302–305.
60 Shimura, H. et al. (2001) Ubiquitination of a New Form of {alpha}-Synuclein by Parkin from Human Brain: Implications for Parkinson's Disease. *Science* **293**, 263–269.
61 Imai, Y., Soda, M. and Takahashi, R. (2000) Parkin suppresses unfolded protein stress-induced cell death through its E3 ubiquitin-protein ligase activity. *J Biol Chem* **275**, 35661–35664.
62 Imai, Y. et al. (2001) An Unfolded Putative Transmembrane Polypeptide, which Can Lead to Endoplasmic Reticulum Stress, Is a Substrate of Parkin. *Cell* **105**, 891–902.
63 Zhang, Y. et al. (2000) Parkin functions as an E2-dependent ubiquitin-protein ligase and promotes the degradation of the synaptic vesicle-associated protein, CDCrel-1. *Proc Natl Acad Sci U S A* **97**, 13354–13359.
64 Lim, K.L. et al. (2005) Parkin mediates nonclassical, proteasomal-independent ubiquitination of synphilin-1: implications for Lewy body formation. *J Neurosci* **25**, 2002–2009.
65 Sakata, E. et al. (2003) Parkin binds the Rpn10 subunit of 26S proteasomes through its ubiquitin-like domain. *EMBO Rep* **4**, 301–306.
66 Imai, Y. et al. (2002) CHIP is associated with Parkin, a gene responsible for familial Parkinson's disease, and enhances its ubiquitin ligase activity. *Mol Cell* **10**, 55–67.
67 Cyr, D.M., Hohfeld, J. and Patterson, C. (2002) Protein quality control: U-box-containing E3 ubiquitin ligases join the fold. *Trends Biochem Sci* **27**, 368–375.
68 Dachsel, J.C. et al. (2005) Parkin interacts with the proteasome subunit alpha4. *FEBS Lett* **579**, 3913–3919.
69 Shimura, H. et al. (1999) Immunohistochemical and subcellular localization of Parkin protein: absence of protein in autosomal recessive juvenile parkinsonism patients. *Ann Neurol* **45**, 668–672.
70 Yang, Y., Nishimura, I., Imai, Y., Takahashi, R. and Lu, B. (2003) Parkin suppresses dopaminergic neuron-selective neurotoxicity induced by Pael-R in Drosophila. *Neuron* **37**, 911–924.
71 Itier, J.M. et al. (2003) Parkin gene inactivation alters behaviour and dopamine neurotransmission in the mouse. *Hum Mol Genet* **12**, 2277–2291.
72 Goldberg, M.S. et al. (2003) Parkin-deficient mice exhibit nigrostriatal deficits but not loss of dopaminergic neurons. *J Biol Chem* **278**, 43628–4335.
73 Von Coelln, R. et al. (2004) Loss of locus coeruleus neurons and reduced startle in parkin null mice. *Proc Natl Acad Sci USA* **101**, 10744–10749.

74 Perez, F.A. and Palmiter, R.D. (2005) Parkin-deficient mice are not a robust model of parkinsonism. *Proc Natl Acad Sci USA* **102**, 2174–2179.

75 Lincoln, S.J. et al. (2003) Parkin variants in North American Parkinson's disease: cases and controls. *Mov Disord* **18**, 1306–1311.

76 Leroy, E. et al. (1998) The ubiquitin pathway in Parkinson's disease. *Nature* **395**, 451–452.

77 Auberger, G. et al. (2005) Is the PARK5 I93M mutation a cause of Parkinson's disease with cognitive deficits and cortical Lewy pathology? 16th International Congress on Parkinson's Disease and Related Disorders Berlin, PT042.

78 Wintermeyer, P. et al. (2000) Mutation analysis and association studies of the UCHL1 gene in German Parkinson's disease patients. *Neuroreport* **11**, 2079–2082.

79 Maraganore, D.M. et al. (2004) UCHL1 is a Parkinson's disease susceptibility gene. *Ann Neuron* **55**, 512–521.

80 Healy, D.G. et al. (2006) UCHL-1 is not a Parkinson's disease susceptibility gene. *Ann Neurol* **59**, 627–633.

81 Solano, S.M., Miller, D.W., Augood, S.J., Young, A.B. and Penney, J.B., Jr. (2000) Expression of alpha-synuclein, parkin, and ubiquitin carboxy-terminal hydrolase L1 mRNA in human brain: genes associated with familial Parkinson's disease. *Ann Neurol* **47**, 201–210.

82 Wilkinson, K.D., Deshpande, S. and Larsen, C.N. (1992) Comparisons of neuronal (PGP 9.5) and non-neuronal ubiquitin C-terminal hydrolases. *Biochem Soc Trans* **20**, 631–637.

83 Wilkinson, K.D. et al. (1989) The neuron-specific protein PGP 9.5 is a ubiquitin carboxyl-terminal hydrolase. *Science* **246**, 670–673.

84 Nishikawa, K. et al. (2003) Alterations of structure and hydrolase activity of parkinsonism-associated human ubiquitin carboxyl-terminal hydrolase L1 variants. *Biochem Biophys Res Commun* **304**, 176–183.

85 Osaka, H. et al. (2003) Ubiquitin carboxy-terminal hydrolase L1 binds to and stabilizes monoubiquitin in neuron. *Hum Mol Genet* **12**, 1945–1958.

86 McNaught, K.S.P. et al. (2002) Impairment of the ubiquitin–proteasome system causes dopaminergic cell death and inclusion body formation in ventral mesencephalic cultures. *J Neurochem* **81**, 301–306.

87 Liu, Y., Fallon, L., Lashuel, H.A., Liu, Z. and Lansbury, P.T., Jr. (2002) The UCH-L1 gene encodes two opposing enzymatic activities that affect alpha-synuclein degradation and Parkinson's disease susceptibility. *Cell* **111**, 209–218.

88 Polymeropoulos, M.H. et al. (1996) Mapping of a gene for Parkinson's disease to chromosome 4q21–q23. *Science* **274**, 1197–1199.

89 Polymeropoulos, M.H. et al. (1997) Mutation in the alpha-synuclein gene identified in families with Parkinson's disease. *Science* **276**, 2045–2047.

90 Kruger, R. et al. (1998) Ala30Pro mutation in the gene encoding alpha-synuclein in Parkinson's disease. *Nat Genet* **18**, 106–108.

91 Zarranz, J.J. et al. (2004) The new mutation, E46K, of alpha-synuclein causes Parkinson and Lewy body dementia. *Ann Neurol* **55**, 164–173.

92 Chartier-Harlin, M.C. et al. (2004) Alpha-synuclein locus duplication as a cause of familial Parkinson's disease. *Lancet* **364**, 1167–119.

93 Ibanez, P. et al. (2004) Causal relation between alpha-synuclein gene duplication and familial Parkinson's disease. *Lancet* **364**, 1169–1171.

94 Singleton, A.B. et al. (2003) alpha-Synuclein locus triplication causes Parkinson's disease. *Science* **302**, 841.

95 Muenter, M.D. et al. (1998) Hereditary form of parkinsonism – dementia. *Ann Neurol* **43**, 768–781.

96 Miller, D.W. et al. (2004) Alpha-synuclein in blood and brain from familial Parkinson disease with SNCA locus triplication. *Neurology* **62**, 1835–1838.

97 Farrer, M. et al. (2004) Comparison of kindreds with parkinsonism and alpha-synuclein genomic multiplications. *Ann Neurol* **55**, 174–179.

98 Kotzbauer, P.T. et al. (2004) Fibrillization of alpha-synuclein and tau in familial Parkinson's disease caused by the A53T alpha-synuclein mutation. *Exp Neurol* **187**, 279–288.

99 Duda, J.E. et al. (2002) Concurrence of alpha-synuclein and tau brain pathology in the Contursi kindred. *Acta Neuropathol (Berl)* **104**, 7–11.

100 Goedert, M. (2001) Alpha-synuclein and neurodegenerative diseases. *Nat Rev Neurosci* **2**, 492–501.

101 Maroteaux, L., Campanelli, J.T. and Scheller, R.H. (1988) Synuclein: a neuron-specific protein localized to the nucleus and presynaptic nerve terminal. *J Neurosci* **8**, 2804–2815.

102 Jakes, R., Spillantini, M.G. and Goedert, M. (1994) Identification of two distinct synucleins from human brain. *FEBS Lett* **345**, 27–32.

103 Abeliovich, A. et al. (2000) Mice lacking alpha-synuclein display functional deficits in the nigrostriatal dopamine system. *Neuron* **25**, 239–252.

104 Conway, K.A., Harper, J.D. and Lansbury, P.T. (1998) Accelerated in vitro fibril formation by a mutant alpha-synuclein linked to early-onset Parkinson disease. *Nat Med* **4**, 1318–1320.

105 Weinreb, P.H., Zhen, W., Poon, A.W., Conway, K.A. and Lansbury, P.T., Jr. (1996) NACP, a protein implicated in Alzheimer's disease and learning, is natively unfolded. *Biochemistry* **35**, 13709–13715.

106 Conway, K.A. et al. (2000) Acceleration of oligomerization, not fibrillization, is a shared property of both alpha-synuclein mutations linked to early-onset Parkinson's disease: implications for pathogenesis and therapy. *Proc Natl Acad Sci USA* **97**, 571–576.

107 Li, J., Uversky, V.N. and Fink, A.L. (2001) Effect of familial Parkinson's disease point mutations A30P and A53T on the structural properties, aggregation, and fibrillation of human alpha-synuclein. *Biochemistry* **40**, 11604–11613.

108 Lashuel, H.A. et al. (2002) Alpha-synuclein, especially the Parkinson's disease-associated mutants, forms pore-like annular and tubular protofibrils. *J Mol Biol* **322**, 1089–1102.

109 Caughey, B. and Lansbury, P.T. (2003) Protofibrils, pores, fibrils, and neurodegeneration: separating the responsible protein aggregates from the innocent bystanders. *Annu Rev Neurosci* **26**, 267–298.

110 Pandey, N., Schmidt, R.E. and Galvin, J.E. (2006) The alpha-synuclein mutation E46K promotes aggregation in cultured cells. *Exp Neurol* **197**, 515–520.

111 Bennett, M.C. et al. (1999) Degradation of alpha-synuclein by proteasome. *J Biol Chem* **274**, 33855–33858.

112 Tofaris, G.K., Layfield, R. and Spillantini, M.G. (2001) alpha-Synuclein metabolism and aggregation is linked to ubiquitin-independent degradation by the proteasome. *FEBS Lett* **25504**, 1–5.

113 Liu, C.W., Corboy, M.J., DeMartino, G.N. and Thomas, P.J. (2003) Endoproteolytic activity of the proteasome. *Science* **299**, 408–411.

114 Tanaka, Y. et al. (2001) Inducible expression of mutant alpha-synuclein decreases proteasome activity and increases sensitivity to mitochondria-dependent apoptosis. *Hum Mol Genet* **10**, 919–926.

115 Stefanis, L., Larsen, K.E., Rideout, H.J., Sulzer, D. and Greene, L.A. (2001) Expression of A53T mutant but not wild-type alpha-synuclein in PC12 cells induces alterations of the ubiquitin-dependent degradation system, loss of dopamine release, and autophagic cell death. *J Neurosci* **21**, 9549–9560.

116 Snyder, H. et al. (2003) Aggregated and monomeric alpha-synuclein bind to the S6' proteasomal protein and inhibit proteasomal function. *J Biol Chem* **278**, 11753–11759.

117 Liu, C.W. et al. (2005) A precipitating role for truncated alpha-synuclein and the proteasome in alpha-synuclein aggregation: implications for pathogenesis of Parkinson's disease. *J Biol Chem* **280**, 22670–22678.

118 Fernagut, P.O. and Chesselet, M.F. (2004) Alpha-synuclein and transgenic mouse models. *Neurobiol Dis* **17**, 123–130.

119 Feany, M.B. and Bender, W.W. (2000) A Drosophila model of Parkinson's disease. *Nature* **404**, 394–398.

120 Kirik, D. et al. (2003) Nigrostriatal alpha-synucleinopathy induced by viral vector-mediated overexpression of human alpha-synuclein: a new primate model of Parkinson's disease. *Proc Natl Acad Sci USA* **100**, 2884–2889.

121 Lee, H.J., Khoshaghideh, F., Patel, S. and Lee, S.J. (2004) Clearance of alpha-synuclein oligomeric intermediates via the lysosomal degradation pathway. *J Neurosci* **24**, 1888–1896.

122 Cuervo, A.M., Stefanis, L., Fredenburg, R., Lansbury, P.T. and Sulzer, D. (2004) Impaired degradation of mutant alpha-synuclein by chaperone-mediated autophagy. *Science* **305**, 1292–1295.

123 Funayama, M. et al. (2002) A new locus for Parkinson's disease (PARK8) maps to chromosome 12p11.2–q13.1. *Ann Neurol* **51**, 296–301.

124 Paisan-Ruiz, C. et al. (2004) Cloning of the gene containing mutations that cause PARK8-linked Parkinson's disease. *Neuron* **44**, 595–600.

125 Zimprich, A. et al. (2004) Mutations in LRRK2 cause autosomal-dominant parkinsonism with pleomorphic pathology. *Neuron* **44**, 601–607.

126 Di Fonzo, A. et al. (2005) A frequent LRRK2 gene mutation associated with autosomal dominant Parkinson's disease. *Lancet* **365**, 412–415.

127 Wszolek, Z.K. et al. (2004) Autosomal dominant parkinsonism associated with variable synuclein and tau pathology. *Neurology* **62**, 1619–1622.

128 Nichols, W.C. et al. (2005) Genetic screening for a single common LRRK2 mutation in familial Parkinson's disease. *Lancet* **365**, 410–412.

129 Gilks, W.P. et al. (2005) A common LRRK2 mutation in idiopathic Parkinson's disease. *Lancet* **365**, 415–416.

130 Simon-Sanchez, J., Herranz-Perez, V., Olucha-Bordonau, F. and Perez-Tur, J. (2006) LRRK2 is expressed in areas affected by Parkinson's disease in the adult mouse brain. Eur *J Neurosci* **23**, 659–666.

131 West, A.B. et al. (2005) Parkinson's disease-associated mutations in leucine-rich repeat kinase 2 augment kinase activity. *Proc Natl Acad Sci USA* **102**, 16842–16847.

132 DiDonato, J. et al. (1996) Mapping of the inducible IkappaB phosphorylation sites that signal its ubiquitination and degradation. *Mol Cell Biol* **16**, 1295–1304.

133 Chen, L. and Feany, M.B. (2005) Alpha-synuclein phosphorylation controls neurotoxicity and inclusion formation in a Drosophila model of Parkinson disease. *Nat Neurosci* **8**, 657–663.

134 Valente, E.M. et al. (2001) Localization of a novel locus for autosomal recessive early-onset parkinsonism, PARK6, on human chromosome 1p35–p36. Am *J Hum Genet* **68**, 895–900.

135 Valente, E.M. et al. (2002) PARK6-linked parkinsonism occurs in several European families. *Ann Neurol* **51**, 14–18.

136 Valente, E.M. et al. (2004) Hereditary early-onset Parkinson's disease caused by mutations in PINK1. *Science* **304**, 1158–1160.

137 Hatano, Y. et al. (2004) Novel PINK1 mutations in early-onset parkinsonism. *Ann Neurol* **56**, 424–427.

138 Healy, D.G., Abou-Sleiman, P.M. and Wood, N.W. (2004) PINK, PANK, or PARK? A clinicians' guide to familial parkinsonism. *Lancet Neurol* **3**, 652–662.

139 Rogaeva, E. et al. (2004) Analysis of the PINK1 gene in a large cohort of cases with Parkinson disease. *Arch Neurol* **61**, 1898–1904.

140 Bonifati, V. et al. (2005) Early-onset parkinsonism associated with PINK1 mutations: frequency, genotypes, and phenotypes. *Neurology* **65**, 87–95.

141 van Duijn, C.M. et al. (2001) Park7, a novel locus for autosomal recessive early-onset parkinsonism, on chromosome 1p36. *Am J Hum Genet* **69**, 629–634.

142 Bonifati, V. et al. (2003) Mutations in the DJ-1 gene associated with autosomal recessive early-onset parkinsonism. *Science* **299**, 256–259.

143 Bonifati, V., Oostra, B.A. and Heutink, P. (2004) Linking DJ-1 to neurodegeneration offers novel insights for understanding the pathogenesis of Parkinson's disease. *J Mol Med* **82**, 163–174.

144 Nagakubo, D. et al. (1997) DJ-1, a novel oncogene which transforms mouse NIH3T3 cells in cooperation with ras. *Biochem Biophys Res Commun* **231**, 509–513.

145 Lockhart, P.J. et al. (2004) DJ-1 mutations are a rare cause of recessively inherited early onset parkinsonism mediated by loss of protein function. *J Med Genet* **41**, e22.

146 Shang, H., Lang, D., Jean-Marc, B. and Kaelin-Lang, A. (2004) Localization of DJ-1 mRNA in the mouse brain. *Neurosci Lett* **367**, 273–277.

147 Bandopadhyay, R. et al. (2004) The expression of DJ-1 (PARK7) in normal human CNS and idiopathic Parkinson's disease. *Brain* **127**, 420–430.

148 Yokota, T. et al. (2003) Down regulation of DJ-1 enhances cell death by oxidative stress, ER stress, and proteasome inhibition. *Biochem Biophys Res Commun* **312**, 1342–1348.

149 Taira, T. et al. (2004) DJ-1 has a role in antioxidative stress to prevent cell death. *EMBO Rep* **5**, 213–218.

150 Olzmann, J.A. et al. (2004) Familial Parkinson's disease-associated L166P mutation disrupts DJ-1 protein folding and function. *J Biol Chem* **279**, 8506–8515.

151 Lee, S.J. et al. (2003) Crystal structures of human DJ-1 and *Escherichia coli* Hsp31, which share an evolutionarily conserved domain. *J Biol Chem* **278**, 44552–44559.

152 Wilson, M.A., St Amour, C.V., Collins, J.L., Ringe, D. and Petsko, G.A. (2004) The 1.8-A resolution crystal structure of YDR533Cp from *Saccharomyces cerevisiae*: a member of the DJ-1/ThiJ/PfpI superfamily. *Proc Natl Acad Sci USA* **101**, 1531–1536.

153 Moore, D.J. et al. (2005) Association of DJ-1 and parkin mediated by pathogenic DJ-1 mutations and oxidative stress. *Hum Mol Genet* **14**, 71–84.

154 Moore, D.J., Zhang, L., Dawson, T.M. and Dawson, V.L. (2003) A missense mutation (L166P) in DJ-1, linked to familial Parkinson's disease, confers reduced protein stability and impairs homo-oligomerization. *J Neurochem* **87**, 1558–1567.

155 Shendelman, S., Jonason, A., Martinat, C., Leete, T. and Abeliovich, A. (2004) DJ-1 is a redox-dependent molecular chaperone that inhibits alpha-synuclein aggregate formation. *PLoS Biol* **2**, e362.

156 Goldberg, M.S. et al. (2005) Nigrostriatal dopaminergic deficits and hypokinesia caused by inactivation of the familial parkinsonism-linked gene DJ-1. *Neuron* **45**, 489–496.

157 Mellick, G.D., Maraganore, D.M. and Silburn, P.A. (2005) Australian data and meta-analysis lend support for alpha-synuclein (NACP-Rep1) as a risk factor for Parkinson's disease. *Neurosci Lett* **375**, 112–116.

158 Oliveira, S.A. et al. (2003) Parkin mutations and susceptibility alleles in late-onset Parkinson's disease. *Ann Neurol* **53**, 624–629.

159 Healy, D.G. et al. (2004) The gene responsible for PARK6 Parkinson's disease, PINK1, does not influence common forms of parkinsonism. *Ann Neurol* **56**, 329–335.

160 Paisan-Ruiz, C. et al. (2006) Testing association between LRRK2 and Parkinson's disease and investigating linkage disequilibrium. *J Med Genet* **43**, e9.

161 Morris, C.M., O'Brien, K.K., Gibson, A.M., Hardy, J.A. and Singleton, A.B. (2003) Polymorphism in the human DJ-1 gene is not associated with sporadic dementia with Lewy bodies or Parkinson's disease. *Neurosci Lett* **352**, 151–153.

162 El-Khodor, B.F., Kholodilov, N.G., Yarygina, O. and Burke, R.E. (2001) The expression of mRNAs for the proteasome complex is developmentally regulated in the rat mesencephalon. *Brain Res Dev Brain Res* **129**, 47–56.

163 Gaczynska, M., Osmulski, P.A. and Ward, W.F. (2001) Caretaker or undertaker? The role of the proteasome in aging. *Mech Ageing Dev* **122**, 235–254.

164 Gibb, W.R. and Lees, A.J. (1988) The relevance of the Lewy body to the pathogenesis of idiopathic Parkinson's disease. *J Neurol Neurosurg Psychiatry* **51**, 745–752.

165 McNaught, K.S., Belizaire, R., Isacson, O., Jenner, P. and Olanow, C.W. (2003) Altered Proteasomal Function in Sporadic Parkinson's Disease. *Exp Neurol* **179**, 38–45.

166 McNaught, K.S. and Jenner, P. (2001) Proteasomal function is impaired in substantia nigra in Parkinson's disease. *Neurosci Lett* **297**, 191–194.

167 Tofaris, G.K., Razzaq, A., Ghetti, B., Lilley, K. and Spillantini, M.G. (2003) Ubiquitination of alpha-synuclein in Lewy bodies is a pathological event not associated with impairment of proteasome function. *J Biol Chem* **278**, 44405–444411.

168 Furukawa, Y. et al. (2002) Brain proteasomal function in sporadic Parkinson's disease and related disorders. *Ann Neurol* **51**, 779–782.

169 Blandini, F. et al. (2006) Peripheral proteasome and caspase activity in Parkinson disease and Alzheimer disease. *Neurology* **66**, 529–534.

170 Atlante, A., Bobba, A., Calissano, P., Passarella, S. and Marra, E. (2003) The apoptosis/necrosis transition in cerebellar granule cells depends on the mutual relationship of the antioxidant and the proteolytic systems which regulate ROS production and cytochrome c release en route to death. *J Neurochem* **84**, 960–971.

171 Jha, N., Kumar, M.J., Boonplueang, R. and andersen, J.K. (2002) Glutathione decreases in dopaminergic PC12 cells interfere with the ubiquitin protein degradation pathway: relevance for Parkinson's disease? *J Neurochem* **80**, 555–561.

172 Hoglinger, G.U. et al. (2003) Dysfunction of mitochondrial complex I and the proteasome: interactions between two biochemical deficits in a cellular model of Parkinson's disease. *J Neurochem* **86**, 1297–1307.

173 Lee, H.J., Shin, S.Y., Choi, C., Lee, Y.H. and Lee, S.J. (2001) Formation and removal of alpha-synuclein aggregates in cells exposed to mitochondrial inhibitors. *J Biol Chem* **27**, 27.

174 Li, Z., Jansen, M., Pierre, S.R. and Figueiredo-Pereira, M.E. (2003) Neurodegeneration: linking ubiquitin/proteasome pathway impairment with inflammation. *Int J Biochem Cell Biol* **35**, 547–552.

175 Jesenberger, V. and Jentsch, S. (2002) Deadly encounter: ubiquitin meets apoptosis. *Nat Rev Mol Cell Biol* **3**, 112–121.

176 Kikuchi, S. et al. (2003) Effect of proteasome inhibitor on cultured mesencephalic dopaminergic neurons. *Brain Res* **964**, 228–236.

177 Rockwell, P., Yuan, H., Magnusson, R. and Figueiredo-Pereira, M.E. (2000) Proteasome inhibition in neuronal cells induces a proinflammatory response manifested by upregulation of cyclooxygenase-2, its accumulation as ubiquitin conjugates, and production of the prostaglandin PGE(2). *Arch Biochem Biophys* **374**, 325–333.

178 Junn, E., Lee, S.S., Suhr, U.T. and Mouradian, M.M. (2002) Parkin accumulation in aggresomes due to proteasome impairment. *J Biol Chem* **277**, 47870–47877.

179 Bodner, R.A. et al. (2006) From the Cover: Pharmacological promotion of inclusion formation: A therapeutic approach for Huntington's and Parkinson's diseases. *Proc Natl Acad Sci USA* **103**, 4246–4251.

180 Petrucelli, L. et al. (2002) Parkin protects against the toxicity associated with mutant alpha-synuclein: proteasome dysfunction selectively affects catecholaminergic neurons. *Neuron* **36**, 1007–1019.

181 Rideout, H.J., Lang-Rollin, I.C., Savalle, M. and Stefanis, L. (2005) Dopaminergic neurons in rat ventral midbrain cultures undergo selective apoptosis and form inclusions, but do not up-regulate iHSP70, following proteasomal inhibition. *J Neurochem* **93**, 1304–1313.

182 McNaught, K.S.P., Bjorklund, L.M., Belizaire, R., Jenner, P. and Olanow, C.W. (2002) Proteasome Inhibition Causes Nigral Degeneration with Inclusion Bodies in Rats. *NeuroReport* **13**, 1437–1441.

183 Fornai, F. et al. (2003) Fine structure and biochemical mechanisms underlying nigrostriatal inclusions and cell death after proteasome inhibition. *J Neurosci* **23**, 8955–8966.

184 Miwa, H., Kubo, T., Suzuki, A., Nishi, K. and Kondo, T. (2005) Retrograde dopaminergic neuron degeneration following intrastriatal proteasome inhibition. *Neurosci Lett* **380**, 93–98.

185 McNaught, K.S.P., Perl, D.P., Brownell, A.L. and Olanow, C.W. (2004) Systemic exposure to proteasome inhibitors causes a progressive model of Parkinson's disease. *Ann Neurol* **56**, 149–162.

186 Grunblatt, E. et al. (2004) Gene expression profiling of parkinsonian substantia nigra pars compacta; alterations in ubiquitin–proteasome, heat shock protein, iron and oxidative stress regulated proteins, cell adhesion/ cellular matrix and vesicle trafficking genes. *J Neural Transm* **111**, 1543–1573.

187 Bulteau, A.L. et al. (1999) Oxidative modification and inactivation of the proteasome during coronary occlusion/ reperfusion. *J Biol Chem* **276**, 30057–30063.

188 Hendil, K.B., Hartmann-Petersen, R. and Tanaka, K. (2002) 26S Proteasomes Function as Stable Entities. *J Mol Biol* **315**, 627–636.

189 Jenner, P. and Olanow, C.W. (1998) Understanding cell death in Parkinson's disease. *Ann Neurol* **44**, S72–S84.

190 Kisselev, A.F. and Goldberg, A.L. (2001) Proteasome inhibitors: from research tools to drug candidates. *Chem Biol* **8**, 739–758.

191 Fenteany, G. and Schreiber, S.L. (1998) Lactacystin, proteasome function, and cell fate. *J Biol Chem* **273**, 8545–8548.

192 Sin, N. et al. (1999) Total synthesis of the potent proteasome inhibitor epoxomicin: a useful tool for understanding proteasome biology. *Bioorg Med Chem Lett* **9**, 2283–2288.

193 Koguchi, Y. et al. (2000) TMC-95A, B, C, and D, novel proteasome inhibitors produced by Apiospora montagnei Sacc. TC 1093. Taxonomy, production, isolation, and biological activities. *J Antibiot (Tokyo)* **53**, 105–109.

194 Nam, S., Smith, D.M. and Dou, Q.P. (2001) Ester bond-containing tea polyphenols potently inhibit proteasome activity in vitro and in vivo. *J Biol Chem* **276**, 13322–13330.

195 Kazi, A. et al. (2003) A natural musaceas plant extract inhibits proteasome activity and induces apoptosis selectively in human tumor and transformed, but not normal and non-transformed, cells. *Int J Mol Med* **12**, 879–887.

196 Jana, N.R., Dikshit, P., Goswami, A. and Nukina, N. (2004) Inhibition of proteasomal function by curcumin induces apoptosis through mitochondrial pathway. *J Biol Chem* **279**, 11680–11685.

197 Zhou, Y., Shie, F.S., Piccardo, P., Montine, T.J. and Zhang, J. (2004) Proteasomal inhibition induced by manganese ethylene-bis-dithiocarbamate: relevance to Parkinson's disease. *Neuroscience* **128**, 281–291.

198 Ensign, J.C., Normand, P., Burden, J.P. and Yallop, C.A. (1993) Physiology of some actinomycete genera. *Res Microbiol* **144**, 657–660.

199 Cross, T. (1981) Aquatic actinomycetes: a critical survey of the occurrence, growth and role of actinomycetes in aquatic habitats. *J Appl Bacteriol* **50**, 397–423.

200 Shamoto-Nagai, M. et al. (2003) An inhibitor of mitochondrial complex I, rotenone, inactivates proteasome by oxidative modification and induces aggregation of oxidized proteins in SH-SY5Y cells. *J Neurosci Res* **74**, 589–597.

201 Wang, X.F. and Li, J.M. (2004) Effects of pesticides on proteasomal activity and cell viability in HEK and SH-SY5Y cells. *Mov Disord* **10** (Suppl 9), P553.

202 Priyadarshi, A., Khuder, S.A., Schaub, E.A. and Priyadarshi, S.S. (2001) Environmental risk factors and

Parkinson's disease: a metaanalysis. *Environ Res* **86**, 122–127.

203 Saric, T., Graef, C.I. and Goldberg, A.L. (2004) Pathway for degradation of peptides generated by proteasomes: A key role for thimet oligopeptidase and other metallopeptidases. *J Biol Chem* **279**, 46723–46732.

204 Sullivan, P.G. et al. (2004) Proteasome inhibition alters neural mitochondrial homeostasis and mitochondria turnover. *J Biol Chem* **279**, 20699–20707.

205 Hunot, S. and Hirsch, E.C. (2003) Neuroinflammatory processes in Parkinson's disease. *Ann Neurol* **53** Suppl 3, S49–S58; discussion S58–S60.

206 Goldberg, A.L., Cascio, P., Saric, T. and Rock, K.L. (2002) The importance of the proteasome and subsequent proteolytic steps in the generation of antigenic peptides. *Mol Immunol* **39**, 147–164.

207 Hegde, A.N. and DiAntonio, A. (2002) Ubiquitin and the synapse. *Nat Rev Neurosci* **3**, 854–861.

208 Bezard, E., Gross, C.E. and Brotchie, J.M. (2003) Presymptomatic compensation in Parkinson's disease is not dopamine-mediated. *Trends Neurosci* **26**, 215–221.

209 Obeso, J.A., Rodriguez-Oroz, M.C., Lanciego, J.L. and Rodriguez Diaz, M. (2004) How does Parkinson's disease begin? The role of compensatory mechanisms. *Trends Neurosci* **27**, 125–127; author reply 127–128.

210 Wilkinson, K.D. (1999) Ubiquitin-dependent signaling: the role of ubiquitination in the response of cells to their environment. *J Nutr* **129**, 1933–1936.

211 Aguilar, R.C. and Wendland, B. (2003) Ubiquitin: not just for proteasomes anymore. *Curr Opin Cell Biol* **15**, 184–190.

212 Muratani, M. and Tansey, W.P. (2003) How the ubiquitin–proteasome system controls transcription. *Nat Rev Mol Cell Biol* **4**, 192–201.

213 Adams, J. (2004) The proteasome: a suitable antineoplastic target. *Nat Rev Cancer* **4**, 349–360.

9
The Molecular Pathway to Neurodegeneration in Parkin-Related Parkinsonism

Ryosuke Takahashi

9.1
Introduction

Parkinson's disease (PD) is the most common neurodegenerative disease of the motor system amongst elderly people. The prevalence of PD is approximately 1% of people by the age of 70 years [1]. PD is characterized by a progressive loss of dopaminergic neurons in the pars compacta of the substantia nigra accompanied by the formation of Lewy bodies. Lewy bodies are intra-neuronal fibrillary inclusions mainly composed of α-synuclein [2]. They are regarded as the hallmark of idiopathic PD. Loss of neurons within the pars compacta of the substantia nigra causes progressive motor disturbances, classically tremor, rigidity, bradykinesia and postural instability. To date, there is no known effective therapy to prevent or retard neurodegeneration as a result of PD [1, 3].

Most cases of PD develop sporadically, however, fewer than 10% of cases are familial and presumably inherited [4]. Autosomal recessive juvenile parkinsonism (AR-JP) accounts for approximately 50% of cases of early-onset familial PD in affected European families [5]. It is characterized by several unique features, including young age of onset (usually under 40 years of age), dystonia, and a marked response to dopamine. The neuropathological hallmark of AR-JP is selective degeneration of dopaminergic neurons in the substantia nigra zona compacta, similar to that observed in the idiopathic form of PD. However, AR-JP is not usually associated with Lewy bodies [6, 7].

Mutations in the parkin gene are responsible for AR-JP [8]. In this chapter, the role of parkin in the ubiquitin–proteasome system will be focused and discussed in light of recent findings.

Protein Degradation, Vol. 4: The Ubiquitin-Proteasome System and Disease.
Edited by R. J. Mayer, A. Ciechanover, M. Rechsteiner

ISBN: 978-3-527-31436-2

9.2 Parkin is an E3 Ubiquitin Ligase

9.2.1 Parkin and the Ubiquitin-Proteasome System

Parkin is a 465-amino acid protein characterized by a ubiquitin-like domain at its NH_2-terminus, as well as two RING-finger motifs and an IBR (in-between RING fingers) motif at its COOH terminus (RING-IBR-RING or RBR domain) [9]. The RING domain has been shown to be a feature of ubiquitin ligase involved in the ubiquitination reaction [10]. Polyubiquitination involves a sequence of reactions performed by ubiquitin-activating (E1), ubiquitin-conjugating (E2) and ubiquitin ligating (E3) enzymes. E3 interacts with specific substrate(s) and facilitates the formation of covalent bonds between the COOH terminus of ubiquitin and ε-lysine, either on a target protein or on the last ubiquitin of a protein-bound polyubiquitin chain in concert with its partner E2s. Yeast protein UFD2 is a multi-ubiquitin chain elongation factor, also called E4, required for efficient multi-ubiquitination of a substrate [11]. Polyubiquitin chains are thought to be potent targeting signals for the degradation of proteins within 26S proteasomes.

Several groups have shown that wild-type parkin is an E3 ubiquitin ligase [12–14] (Figure 9.1). Parkin ubiquitinates substrate proteins or itself in concert with E2s, such as UbcH7, UbcH8, Ubc6 and Ubc7 [12–14]. Moreover, several AR-JP-related missense mutations have been identified in the ubiquitin-like

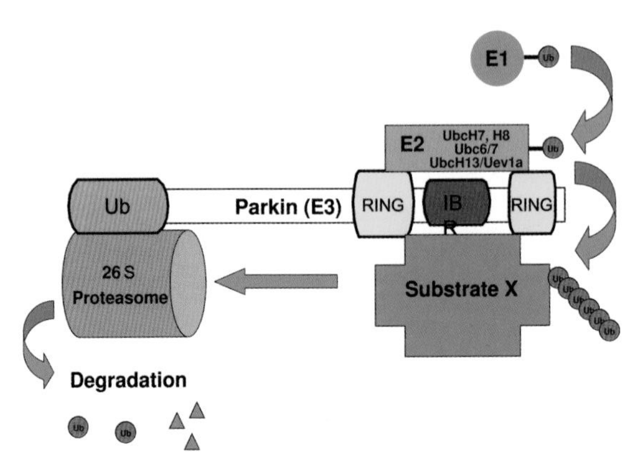

Fig. 9.1. Function of parkin in the ubiquitin proteasomal pathway. Parkin is an E3 ubiquitin ligase that recognizes substrate X and promotes ubiquitination in adjunct with two other ubiquitination enzymes, E1 and E2. Polyubiquitinated substrate X is recognized and degraded by the 26S proteasome. The N-terminal ubiquitin-like domain and the C-terminal RING-IBR-RING domain of parkin serve as recruitment domains for 26S proteasome and E2 enzymes, respectively. Some of the known substrates of parkin associate with its RING-IBR-RING domain.

domain of the parkin gene [15]. Furthermore, an NMR analysis has revealed binding between the ubiquitin-like domain of parkin and the Rpn 10 subunit of the 26S proteasome [16], strongly suggesting the link between Parkin and the UPS (Figure 9.1).

9.2.2
Proteasome-independent Role of Parkin

Polyubiquitin chains are formed through distinct types of linkages using one of the seven internal lysine residues (K6, K11, K27, K29, K33, K48 and K63) within the previous ubiquitin molecule [17].

Recently, parkin was shown to be a dual function ubiqutin ligase that mediates both K48- and K63-linked polyubiquitination [18]. K48-linked polyubiquitin chain, the best characterized form of polyubiquitin, leads the proteins to degradation via 26S proteasomes, constituting the ubiquitin–proteasome system as mentioned before. On the other hand, K63-linked chains act as proteasome-independent signals in several different cellular pathways [17]. Dual specificity seems to be determined by the E2 enzymes that parkin recruits. In the assembly of a K63-linked polyubiquitin chain, parkin interacts with the UbcH13/Uev1a heterodimer [18]. Parkin mediates K63-linked, proteasome-independent ubiquitination of its substrate synphilin 1 [19]. Although K63-linked ubiqitination is implicated in inclusion body formation, further study is required to clarify its physiological relevance [20].

9.2.3
Multiple Monoubiquitination is Mediated by Parkin

Surprisingly, two recent reports have shown that Parkin mediates multiple monoubiquitination *in vitro* [21, 22]. Both reports have demonstrated that the second RING finger is responsible for E3 activity in an *in-vitro* ubiquitination assay, where bacterially-produced recombinant Parkin was used. Previous findings that mutations in regions other than the second RING finger showed reduced E3 activity *in vivo* might be ascribed to their insolubility and sequestration [23–26]. Parkin itself as well as maltose binding protein (MBP) connected to Parkin as a pseudosubstrate and p38 as a substrate, have been shown to be monoubiquitinated *in vitro* and *in vivo* respectively. In contrast to a previous report that Parkin accelerates polyubiquitin chain formation [18], Parkin has been shown to mediate monoubiquitination in concert with Ubc13 as well as Ubc7 or Ubc H7 under pure *in vitro* conditions [21, 22]. These results suggest that Parkin may mediate monoubiquitination regardless of its partner E2s.

A recent report showed that Parkin mediates monoubiquitination of an adaptor protein Eps15 with two ubiquitin-interacting motifs (UIMs) [27]. Eps15 interacts with and positively regulates the endocytosis of ubiquitinated epithelial growth factor receptor (EGFR). Parkin-mediated ubiquitination of Eps15 inhibits its ability to bind with and promote endocytosis of EGFR, resulting in suppression of EGFR

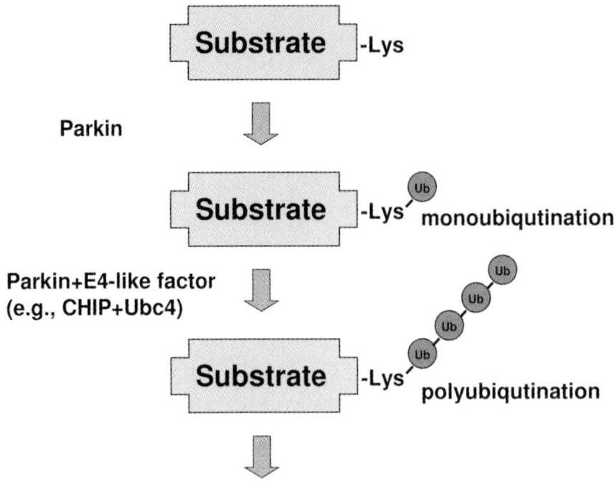

Fig. 9.2. Hypothetical two-step ubiquitination of Parkin substrates. At the first step, Parkin monoubiqutinates its substrate. Then, E4-like factors promote the elongation of polyubiquitin chain on the substrate molecules, thereby targeting the substrates to 26S proteasome. CHIP and its partner E2 Ubc4 are candidates for such E4-like molecules [28].

internalization and degradation, and promoting phosphoinositide 3-kinase (PI(3)K)-Akt signaling. Since Akt plays an important role in neuronal survival, this proteasome-independent function of Parkin may explain some aspects of neurodegeneration.

On the other hand, monoubiquitination may lead to proteasomal degradation. A previous report showed that the carboxy-terminus of Hsc70-interacting protein (CHIP), a U-box motif containing E3 protein, together with Ubc4, serves as an E4-like protein and cooperates with Parkin to form polyubiquitin chains [28]. Given the presence of E4-like factor, monoubiquitination catalyzed by Parkin may eventually target the proteins to degradation via the 26S proteasome (Figure 9.2). Whether Parkin can mediate both monoubiquitination and polyubiquitination should be re-examined and clarified in light of the recent findings.

9.2.4 Modulators of Parkin E3 Activity

Parkin is a component of a high molecular weight complex located in cells and the function of parkin seems to be modulated by its binding partners [28, 29]. Two parkin-associated proteins have been shown to promote the elimination of Pael-R by parkin: CHIP and Hsp70 [28].

CHIP contains a U-box motif, which is structurally similar to the RING-finger motif and exhibits U box-dependent E3 activity [30–32]. On the other hand, CHIP has been shown to downregulate chaperone ATPase activity [33]. Moreover, CHIP has been shown to ubiquitinate improperly-folded protein in a chaperone-dependent manner [32]. When bound to parkin however, CHIP markedly enhances parkin-mediated ubiquitination of Pael-R *in vitro* [28]. Consistent with this observation, overexpression of CHIP accelerates Pael-R degradation in cultured cells, leading to a marked reduction in the steady-state level of Pael-R protein.

In contrast to CHIP, Hsp70 has been observed to inhibit ubiquitination of Pael-R *in vitro* and to increase levels of the soluble form of Pael-R *in vivo*, presumably by facilitating the proper folding of Pael-R. Moreover, Hsp70 inhibits CHIP-mediated degradation of soluble and probably functional Pael-R, so that only insoluble aggregates of the receptor are removed.

It has also been shown that bcl-2-associated athanogene 5 (BAG5), a BAG-family member, directly interacts with parkin and the chaperone Hsp70. BAG5, similar to CHIP, downregulates chaperone ATPase activity. Within this complex, BAG5 inhibits both parkin E3 ubiquitin ligase activity and Hsp70-mediated refolding of misfolded proteins. BAG5 enhances parkin sequestration within protein aggregates and attenuates parkin-dependent preservation of proteasome function [34].

Two binding partners of Parkin, 14-3-3 eta and Nrdp1/FLRF are also found to be negative regulators of Parkin E3 activity [35, 36].

9.3 Substrates of Parkin

9.3.1 Parkin Substrates and their Recognition Mechanisms

Although Parkin may mediate diverse forms of ubiquitination, it is likely that some, but not all, are involved in the UPS. Given that Parkin targets its ubiquitinated substrates to the 26S proteasome, Parkin dysfunction should lead to the accumulation of its substrate. Accumulation of toxic substrate(s) of Parkin (substrate-X) due to loss of parkin E3 activity or disruption of the parkin–proteasomal interaction in AR-JP patients with a genetic defect of parkin, should result in the development of dopaminergic neurodegeneration.

Based on this hypothesis, the identification of such toxic substrate(s) is the key to understanding the molecular mechanisms underlying AR-JP.

To date, 13 proteins have been identified as substrates of parkin [37]: CDCrel-1 , synaptotagmin XI [38], synphilin-1 [39], glycosylated α-synuclein [40], α/β-tubulin [41], the p38 subunit of an aminoacyl-tRNA synthetase (ARS) complex [42], *P*arkin-*a*ssociated *e*ndothelin receptor-like *r*eceptor (Pael-R) [43], the expanded form of polyglutamine [44], and cyclin E [45], SEPT5_v2/CDCrel-2 [46], misfolded dopamine transporter [47], far upstream element-binding protein 1 [48], RanBP2 [49] and Eps15 [27] (Table 9.1). It has been shown that the unmodified form of α-synuclein, a major component of Lewy body, is not a substrate for Parkin [39].

Although there are no apparent common properties among Parkin substrates, it has been noted that several different substrates are found within Lewy bodies.

Table 9.1. The reported substrates of parkin

Protein	Biological function	Lewy body
O-glycosylated α-synuclein	Septin family protein with unknown function	–
CDCrel-1	Isoform of α-synuclein with unknown function	N.D.
(Misfolded) Pae1 receptor	Orphan G-protein coupled receptor	+
p38 subunit of the aminoacyl-tRNA synthetase	Role in protein biosynthesis	+
Synaptotagmin XI	Regulation of exocytosis of neurotransmitters	+
Expanded polyglutamine(polyQ) proteins	Aberrant proteins responsible for polyQ diseases	–
α/β-Tubulins	Microtubule proteins	+
Synphilin-1	α-Synuclein-binding protein	+
Cyclin E	Cell cycle regulation of mitotic cells; unknown function in neurons	N.D.
SEPT5_v2/CDCrel-2	SEPT5_v2 is highly homologous with CDCrel-1	N.D.
Misfolded dopamine transporter	Regulation of dopamine uptake	N.D.
Far upstream element-binding protein-1	A binding partner of p38	N.D.
RanBP2	Small ubiquitin-related modifier (SUMO) E3 ligase family protein	N.D.
EPS15	Adaptor protein with ubiquitin-interacting motifs (UIMs)	N.D.

N.D., not detected.

Considering that the components of Lewy bodies consist of misfolded proteins, molecular chaperones and proteasome subunits, it is likely that Parkin ubiquitinates a subset of misfolded proteins. Consistent with this idea, an expanded form of polyglutamine, which is a causative agent of polyglutamine diseases such as Huntington's disease, has been identified as a parkin substrate [44]. Moreover, there is evidence to suggest that binding between parkin and polyglutamine is mediated by Hsp70, which is known to be a binding partner of Parkin. Hsp70-mediated substrate recognition explains the diversity of substrate specificity observed with parkin (Figure 9.3).

Among the various substrate molecules discussed above, the Pael receptor (Pael-R), CDC-rel1, cyclin E, synphilin-1, and the p38 subunit of aminoacyl tRNA synthetase have either been shown or suggested to promote cell death under certain conditions, and so represent the proteins which are most likely relevant to neurodegeneration in AR-JP.

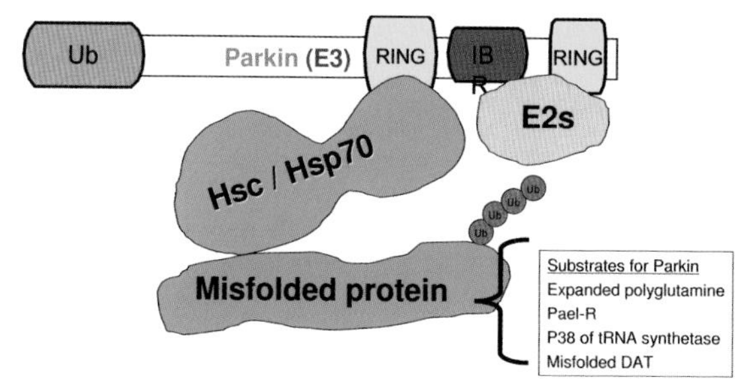

Fig. 9.3. Parkin may recognize a subset of misfolded proteins through Hsc/Hsp70. Parkin interacts with Hsc/Hsp70 through its first RING domain [28]. Parkin may recognize misfolded proteins including Pael-R, and p38 by using Hsc/Hsp70 as a substrate recognition subunit.

9.3.2 The Link between Substrate Accumulation and Cell Death: Pael-R

9.3.2.1 Pael-R and Endoplasmic Reticulum Stress

Pael-R is a putative G-protein-coupled orphan receptor, which is highly expressed in the central nervous system, especially in the substantia nigra [43, 50, 51]. Although the physiological function of Pael-R is implicated in dopamine metabolism, its ligand has yet to be identified [52]. It has been shown that misfolded Pael-R was ubiquitinated by parkin at the level of the endoplasmic reticulum and the disturbance of Pael-R degradation leads to ER stress-induced cell death.

The endoplasmic reticulum (ER) functions as a quality control regulator of membrane and secretory proteins [53]. Newly synthesized secretory proteins are transported to the ER. Inside its lumen, ER chaperones such as BiP/GRP78 bind to these newly synthesized proteins to facilitate their proper folding. After this, proteins enter the conventional secretory pathway. Proteins that are not properly folded are transported back to the cytosol where they are degraded via ubiquitin-proteasomal degradation, a process known as endoplasmic reticulum-associated degradation (ERAD) [54] (Figure 9.4). It has been shown that parkin is an ERAD-related E3 and that Pael-R is a substrate. When insoluble misfolded Pael-R is accumulated in the cells by the inhibition of the proteasome, Pael-R is first accumulated in the ER and then forms a special type of aggregate, known as an aggresome, in the cytoplasm [55]. As these aggresomes form, the cells undergo apoptosis, demonstrating cell death due to the accumulation of Pael-R.

Pael-R-induced cell death was assumed to be mediated by ER stress. Abnormal accumulation of unfolded protein in the ER is a major threat to cell viability, a phenomenon known as ER stress or unfolded protein stress. Cells attempt to adapt to ER stress in several different ways, including transcriptional upregulation of ER chaperones, and suppression of translation. These cellular responses are

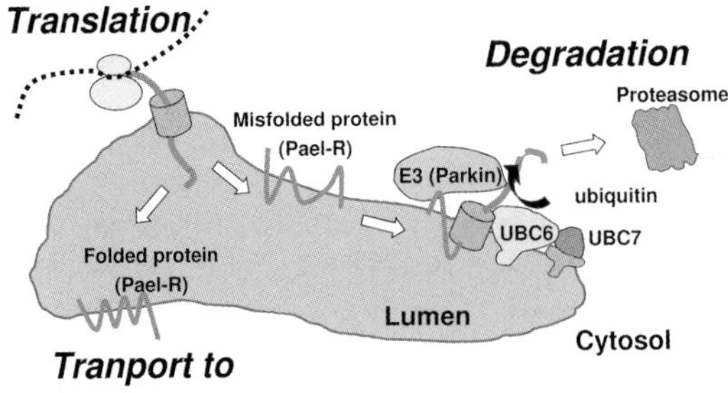

Fig. 9.4. Endoplasmic reticulum-associated degradation (ERAD). ERAD is a protein degradation system for unfolded secretory and membrane proteins. Improperly folded Pael-R is subject to ERAD, and parkin is an E3 involved in ERAD.

collectively known as unfolded protein responses (UPR) [53]. However, when the burden of accumulated protein exceeds these protective mechanisms, cells undergo a death process accompanied by the activation of JNK and caspases as well as upregulation of CHOP [56, 67].

Consistent with the idea that accumulation of misfolded Pael-R contributes to the pathogenesis of AR-JP, the level of detergent-insoluble Pael-R was elevated in the brains of AR-JP patients [43].

9.3.2.2 Pael-R Overexpressing Animals and Dopaminergic Neurodegeneration

The Drosophila model for AR-JP was created by overexpression of Pael-R [58]. When Pael-R was expressed in dopaminergic neurons in Drosophila, the number of dopaminergic neurons observed within the dorsomedial cluster fell to about 50% of that observed in control flies at 40 days of age. Equal numbers of dopaminergic neurons were observed in younger Pael-R and control flies, indicating that the observed cell loss was due to neurodegeneration occurring after birth. Moreover, even when Pael-R expression was driven by a pan-neuronal promoter, only dopaminergic neurons underwent degeneration. This suggests that dopaminergic neurons are selectively vulnerable to Pael-R toxicity.

A recent report showed that Pael-R overexpression in the substantia nigra of mouse brain through adenoviral vectors, resulted in induction of ER stress followed by dopaminergic neuronal death [59]. Pael-R-induced cell death was greatly enhanced in the parkin-deficient mouse and was suppressed by the overexpression of an ER chaperone, ORP150. Moreover, when the animal was pretreated with dopamine synthesis inhibitor, dopaminergic neuronal death was significantly attenuated, indicating that dopamine enhances Pael-R toxicity. It has been reported that dopamine covalently modifies and functionally inactivates Parkin [60]. Although the relationship between Pael-R toxicity and dopamine is still obscure, cellular protective mechanisms against Pael-R toxicity other than Parkin, might also be inactivated by dopamine.

9.3.3
The Link between Substrate Accumulation and Cell Death: CDC-rel1, Synphilin-1, Cyclin E and p38

When CDC-rel1 was introduced into the striatum and the substantia nigra of rat brain by using adeno-associated viral vectors, only dopaminergic cells in the substantia nigra underwent cell death [61]. Since the reduction of dopamine levels by pharmacological treatment alleviated nigral cell death and CDC-rel1 overexpression in PC12 cells decreased the extracellular dopamine level, the accumulation of dopamine by CDC-rel1-mediated exocytosis inhibition is thought to contribute to dopaminergic neuron-selective cell death.

Synphilin-1 is an α-synuclein-interacting protein that promotes the formation of Lewy body-like inclusions in cultured cells [62]. Parkin mediates K63-linked polyubiquitinanation of synphilin-1, apparently contributing to inclusion formation by α-synuclein and synphilin-1 in cultured cells [19]. Moreover, parkin can protect against the toxicity induced by α-synuclein plus synphilin-1 overexpression following proteasome inhibition [39].

Cyclin E has been implicated in glutamate-induced neuronal death, since it is accumulated in primary neuronal cultures in response to glutamanergic excitotoxin kainate. Interestingly, parkin overexpression inhibits the accumulation of cyclin E and cell death induced by kainate treatment, whereas RNAi-mediated parkin downregulation showed the opposite effects. The mechanism underlying cyclin E-induced cell death is not clear.

The p38 subunit plays an essential role in the *in vivo* assembly of the ARS complex [63]. When overexpressed in dopaminergic neuroblastoma cells, it forms aggresomes and induces cell death by unknown mechanisms. Parkin promotes the formation of ubiquitinated p38-positive inclusion bodies and suppresses the p38-induced cell toxicity [42]. It has been noted that only p38 is shown to be upregulated by 15% in the ventral midbrain of the parkin-null mouse among all the substrates identified [48]. The role of p38 in dopaminergic neuronal death should be validated in animal models in the future.

9.4
The Animal Models of AR-JP

To establish animal models of AR-JP, parkin gene deletion mutants for Drosophila and mouse were created [64–69]. However, the phenotypes of parkin-null mutant animals are very different from those of AR-JP patients.

9.4.1
Drosophila Model of AR-JP

The parkin gene deletion mutant Drosophila are small in size, have a short life span and become vulnerable to oxidative stress [64, 65]. The most remarkable phenotypes of the mutant fly are apoptotic muscle degeneration and disturbances in spermatogenesis, which result in locomotor dysfunction and male sterility respectively [64, 65]. Ultrastructural analysis revealed abnormal mitochondrial morphology in both muscle and sperm [64]. On the other hand, the number of dopaminergic neurons was not reduced, although shrinkage of the cell bodies and decreased tyrosine hydroxylase immunostaining in proximal dendrites of dopaminergic neurons were observed [64].

9.4.2 Parkin-null Drosophila and Drosophila

Mutations in the PTEN-induced putative kinase 1(PINK1) are responsible for the autosomal recessive form of familial Parkinson's disease termed PARK6 [70]. PINK1 is a putative mitochondrial protein kinase, whose function is totally unknown. Three recent reports have shown that deletion of Drosophila PINK1 leads to an almost identical phenotype to that of Parkin-deficient Drosophila, i.e. it produces mitochondrial dysfunction resulting in male sterility, apoptotic muscle degeneration, and moderate loss of dopaminergic neurons [71–73]. Interestingly, the disease phenotype of PINK1-deleted Drosophila was rescued by Parkin overexpression, but not vice versa, suggesting that Parkin functions downstream of PINK1.

9.4.3 Mouse Model of AR-JP

Parkin knockout mice, in which exon 3 or 7 is deleted, have been described by three different groups [66–68]. Dopaminergic neuronal loss was not observed in any of the reports, even in aged mice. However, regarding the parkin exon-3 deletion mutant mouse, the dopamine level in the limbic system was elevated and the level of dopamine transporters was lowered according to one report, while the extracellular dopamine concentration in the striatum was increased according to a second report [66, 67]. These changes are accompanied by behavioral or electrophysiological alterations. In addition, in the mutant mouse reported by Goldberg et al., mild mitochondrial dysfunction and mild increase of oxidative stress were observed [74].

On the other hand, in exon-7-deleted parkin mutant mice, abnormalities in the nigrostriatal dopaminergic system were not detected [68]. However, noradrenergic neurons in the locus coeruleus were decreased by 20% in 70% of the total number in mice. These mild noradrenergic neuronal losses were detected as early as 2 months after birth and do not appear to progress with further aging. Consistent with this neuronal loss, the level of noradrenalin in the brain and spinal cord was reduced, accompanied by a significant reduction in the noradrenalin-dependent startle response.

In contrast to these reports, a recent extensive analysis of parkin exon-2 deletion mutant mouse revealed that the behavioral profile and cathecholamine levels in the brain were not different from those of control mice [69]. Moreover these mutant mice were not more sensitive to 6-hydroxydopamine or methamphetamine neurotoxicity, indicating that Parkin-deficient mice are not a robust model of parkinsonism [75].

9.4.4 The Problems with Animal Models of AR-JP

The differences between parkin-deficient fly and mouse models may be explained by the difference in the endogenous substrates or the presence of redundant pathways dealing with parkin substrates in mice. Drosophila and human parkin shows a similar cell protective effect against human Pael-R- and alpha-synuclein-mediated toxicities, suggesting that the substrates are conserved to some extent [58, 76, 77]. It is particularly important to investigate whether the relationship between Parkin and PINK1 is conserved in mice and humans.

The reason why dopaminergic cell loss does not occur in the parkin knockout mouse may be due to the existence of redundant ubiquitination pathways in mice. For example, Pael-R is known to be ubiquitinated by an ER-resident E3, Hrd1 as well as by Parkin [78]. In addition, the absence of dopaminergic cell loss can also probably be attributed to the relatively short lifespan of mice (2–3 years), which would not provide enough time for the toxic substrates to accumulate in concentrations sufficient to cause cell death.

It has been noted that disturbances of the nigrostriatal system, which may represent the early signs of neurodegeneration, are suggested to occur in two exon 3-deletion mutant mice. However, the individual key findings in these papers are not in accord and no dopaminergic phenotype was detected in the exon 7- and exon 2-deletion mutant mice with respect to parkin [68, 69]. Some of the discrepancies in the detection of mild phenotypes might be caused by the different techniques employed or differences in the genetic backgrounds of the mice. Taking these possibilities into consideration, detailed and careful comparison of the phenotypes of these different parkin knockout mice should be carried to identify the real and reproducible phenotype.

9.5 Future Directions

Seven years have passed since parkin was identified as a ubiquitin ligase, and since then 13 different molecules have been isolated as parkin substrates. Some of the substrate molecules appear to explain the pathogenetic mechanisms underlying AR-JP. However, proof of accumulation of known substrates in the parkin knockout mouse brain has not been obtained except for p38, probably because of the relatively short lifespan of the mouse. So, what then is the next step?

One of the potentially promising approaches is to examine whether the nigral dopaminergic neurons in parkin-deficient mice are vulnerable to a specific stress

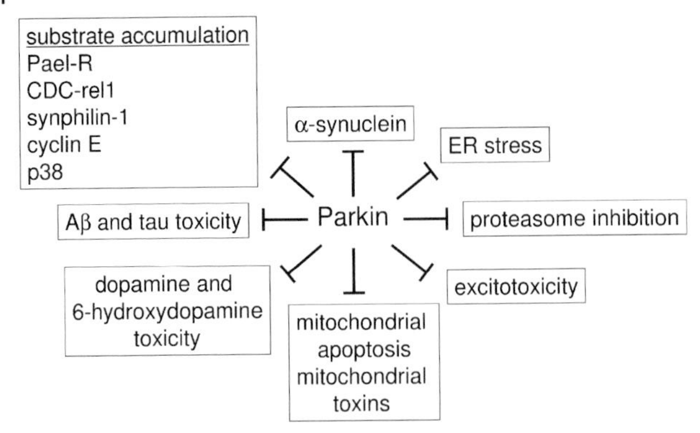

Fig. 9.5. Parkin protects cells from various stresses.

or overexpression of a specific substrate using virus vector or transgenic approaches, as has been applied to Pael-R [59]. On one hand, it is possible that the disease-causing substrate(s) has not been identified and further efforts to identify such a substrate(s) will also be important.

On the other hand, parkin appears to have cell-protective functions against various stresses (Figure 9.5). According to the reports to date, Parkin protects cells against ER stress [13], proteasomal inhibition [77], excitotoxicity [45], ceramide-induced mitochondrial apoptosis [79], mitochondrial toxins [80], intracellular Aβ [80], tau [81, 82], dopamine or 6-hydoxydopamine toxicity [47, 83] and α-synuclein-induced cell death [58, 76, 77, 84]. It is intriguing to ask whether clearance/sequestration of certain parkin substrate(s) contributes to such cell protective effects. Whether the pathways to neurodegeneration caused by parkin mutations are multiple or not should be clarified in the future.

References

1 Savitt, J.M., Dawson, V.L. and Dawson, T.M. (2006). Diagnosis and treatment of Parkinson disease: molecules to medicine. *J Clin Invest* **116**, 1744–1754.

2 Eriksen, J.L., et al. (2003). Caught in the act: alpha-synuclein is the culprit in Parkinson's disease. *Neuron* **40**, 453–456.

3 Dauer, W. and Przedborski, S. (2003). Parkinson's disease: mechanisms and models. *Neuron* **39**, 889–909.

4 Dawson, T.M. and Dawson, V.L. (2003). Rare genetic mutations shed light on the pathogenesis of Parkinson disease. *J Clin Invest* **111**, 145–151.

5 Lucking, C.B., et al. (2000). Association between early-onset Parkinson's disease and mutations in the parkin gene. *N Engl J Med* **342**, 1560–1567.

6 Ishikawa, A. and Tsuji, S. (1996). Clinical analysis of 17 patients in 12 Japanese families with autosomal-recessive type juvenile parkinsonism. *Neurology* **47**, 160–166.

7 Takahashi, H., et al. (1994). Familial juvenile parkinsonism: clinical and pathologic study in a family. *Neurology* **44**, 437–441.

8 Kitada, T., et al. (1998). Mutations in the parkin gene cause autosomal recessive

juvenile parkinsonism. *Nature* **392**, 605–608.
9 Marin, I., et al. (2004). Parkin and relatives: the RBR family of ubiquitin ligases. *Physiol Genomics* **17**, 253–263.
10 Joazeiro, C.A. and Weissman, A.M. (2000). RING finger proteins: mediators of ubiquitin ligase activity. *Cell* **102**, 549–552.
11 Koegl, M., et al. (1999). A novel ubiquitination factor, E4, is involved in multiubiquitin chain assembly. *Cell* **96**, 635–644.
12 Shimura, H., et al. (2000). Familial Parkinson disease gene product, parkin, is a ubiquitin-protein ligase. *Nat Genet* **25**, 302–305.
13 Imai, Y., Soda, M. and Takahashi, R. (2000). Parkin suppresses unfolded protein stress-induced cell death through its E3 ubiquitin-protein ligase activity. *J Biol Chem* **275**, 35661–35664.
14 Zhang, Y., et al. (2000). Parkin functions as an E2-dependent ubiquitin-protein ligase and promotes the degradation of the synaptic vesicle-associated protein, CDCrel-1. *Proc Natl Acad Sci USA* **97**, 13354–13359.
15 Mata, I.F., Lockhart, P.J. and Farrer, M.J. (2004). Parkin genetics: one model for Parkinson's disease. *Hum Mol Genet* **13** Spec No 1: R127–R133.
16 Sakata, E., et al. 2003. Parkin binds the Rpn10 subunit of 26S proteasomes through its ubiquitin-like domain. *EMBO Rep* **4**, 301–306.
17 Pickart, C.M. and Fushman, D. (2004). Polyubiquitin chains: polymeric protein signals. *Curr Opin Chem Biol* **8**, 610–616.
18 Doss-Pepe, E.W., Chen, L. and Madura, K. (2005). Alpha-synuclein and parkin contribute to the assembly of ubiquitin lysine 63-linked multiubiquitin chains. *J Biol Chem* **280**, 16619–16624.
19 Lim, K.L., et al. (2005). Parkin mediates nonclassical, proteasomal-independent ubiquitination of synphilin-1: implications for Lewy body formation. *J Neurosci* **25**, 2002–2009.
20 Lim, K.L., Dawson, V.L. and Dawson, T.M. (2006). Parkin-mediated lysine 63-linked polyubiquitination: a link to protein inclusions formation in Parkinson's and other conformational diseases? *Neurobiol Aging* **27**, 524–529.
21 Matsuda, N., et al. (2006). Diverse effects of pathogenic mutations of Parkin that catalyze multiple monoubiquitylation *in vitro*. *J Biol Chem* **281**, 3204–3209.
22 Hampe, C., et al. (2006). Biochemical analysis of Parkinson's disease-causing variants of Parkin, an E3 ubiquitin-protein ligase with monoubiquitylation capacity. *Hum Mol Genet* **15**, 2059–2075.
23 Gu, W.J., et al. (2003). The C289G and C418R missense mutations cause rapid sequestration of human Parkin into insoluble aggregates. *Neurobiol Dis* **14**, 357–364.
24 Henn, I.H., et al. (2005). Pathogenic mutations inactivate parkin by distinct mechanisms. *J Neurochem* **92**, 114–122.
25 Wang, C., et al. (2005). Alterations in the solubility and intracellular localization of parkin by several familial Parkinson's disease-linked point mutations. *J Neurochem* **93**, 422–431.
26 Sriram, S.R., et al. (2005). Familial-associated mutations differentially disrupt the solubility, localization, binding and ubiquitination properties of parkin. *Hum Mol Genet* **14**, 2571–2586.
27 Fallon, L., et al. (2006). A regulated interaction with the UIM protein Eps15 implicates parkin in EGF receptor trafficking and PI(3)K-Akt signalling. *Nat Cell Biol* **8**, 834–842.
28 Imai, Y., et al. (2002). CHIP is associated with Parkin, a gene responsible for familial Parkinson's disease, and enhances its ubiquitin ligase activity. *Mol Cell* **10**, 55–67.
29 Winklhofer, K.F., et al. (2003). Inactivation of parkin by oxidative stress and C-terminal truncations: a protective role of molecular chaperones. *J Biol Chem* **278**, 47199–47208.
30 Hatakeyama, S., et al. (2001). U box proteins as a new family of ubiquitin-protein ligases. *J Biol Chem* **276**, 33111–33120.
31 Jiang, J., et al. (2001). CHIP is a U-box-dependent E3 ubiquitin ligase: identification of Hsc70 as a target for ubiquitylation. *J Biol Chem* **276**, 42938–42944.

32 Murata, S., et al. (2001). CHIP is a chaperone-dependent E3 ligase that ubiquitylates unfolded protein. *EMBO Rep* **2**, 1133–1138.

33 Ballinger, C.A., et al. (1999). Identification of CHIP, a novel tetratricopeptide repeat-containing protein that interacts with heat shock proteins and negatively regulates chaperone functions. *Mol Cell Biol* **19**, 4535–4545.

34 Kalia, S.K., et al. (2004). BAG5 inhibits parkin and enhances dopaminergic neuron degeneration. *Neuron* **44**, 931–945.

35 Sato, S., et al. (2006). 14-3-3eta is a novel regulator of parkin ubiquitin ligase. *EMBO J* **25**, 211–221.

36 Zhong, L., et al. (2005). RING finger ubiquitin-protein isopeptide ligase Nrdp1/FLRF regulates parkin stability and activity. *J Biol Chem* **280**, 9425–9430.

37 Imai, Y. and Takahashi, R. (2004). How do Parkin mutations result in neurodegeneration? *Curr Opin Neurobiol* **14**, 384–389.

38 Huynh, D.P., et al. (2003). The autosomal recessive juvenile Parkinson disease gene product, parkin, interacts with and ubiquitinates synaptotagmin XI. *Hum Mol Genet* **12**, 2587–2597.

39 Chung, K.K., et al. (2001). Parkin ubiquitinates the alpha-synuclein-interacting protein, synphilin-1: implications for Lewy-body formation in Parkinson disease. *Nat Med* **7**, 1144–1150.

40 Shimura, H., et al. (2001). Ubiquitination of a new form of alpha-synuclein by parkin from human brain: implications for Parkinson's disease. *Science* **293**, 263–269.

41 Ren, Y., Zhao, J. and Feng, J. (2003). Parkin binds to alpha/beta tubulin and increases their ubiquitination and degradation. *J Neurosci* **23**, 3316–3324.

42 Corti, O., et al. (2003). The p38 subunit of the aminoacyl-tRNA synthetase complex is a Parkin substrate: linking protein biosynthesis and neurodegeneration. *Hum Mol Genet* **12**, 1427–1437.

43 Imai, Y., et al. (2001). An unfolded putative transmembrane polypeptide, which can lead to endoplasmic reticulum stress, is a substrate of Parkin. *Cell* **105**, 891–902.

44 Tsai, Y.C., et al. (2003). Parkin facilitates the elimination of expanded polyglutamine proteins and leads to preservation of proteasome function. *J Biol Chem* **278**, 22044–22055.

45 Staropoli, J.F., et al. (2003). Parkin is a component of an SCF-like ubiquitin ligase complex and protects postmitotic neurons from kainate excitotoxicity. *Neuron* **37**, 735–749.

46 Choi, P., et al. (2003). SEPT5_v2 is a parkin-binding protein. *Brain Res Mol Brain Res* **117**, 179–189.

47 Jiang, H., et al. (2004). Parkin protects human dopaminergic neuroblastoma cells against dopamine-induced apoptosis. *Hum Mol Genet* **13**, 1745–1754.

48 Ko, H.S., et al. (2006). Identification of far upstream element-binding protein-1 as an authentic Parkin substrate. *J Biol Chem* **281**, 16193–16196.

49 Um, J.W., et al. (2006). Parkin ubiquitinates and promotes the degradation of RanBP2. *J Biol Chem* **281**, 3595–3603.

50 Zeng, Z., et al. (1997). A novel endothelin receptor type-B-like gene enriched in the brain. *Biochem Biophys Res Commun* **233**, 559–567.

51 Donohue, P.J., et al. (1998). A human gene encodes a putative G protein-coupled receptor highly expressed in the central nervous system. *Brain Res Mol Brain Res* **54**, 152–160.

52 Marazziti, D., et al. (2004). Altered dopamine signaling and MPTP resistance in mice lacking the Parkinson's disease-associated GPR37/parkin-associated endothelin-like receptor. *Proc Natl Acad Sci USA* **101**, 10189–10194.

53 Mori, K. (2000). Tripartite management of unfolded proteins in the endoplasmic reticulum. *Cell* **101**, 451–454.

54 Meusser, B., et al. (2005). ERAD: the long road to destruction. *Nat Cell Biol* **7**, 766–772.

55 Johnston, J.A., Ward, C.L. and Kopito, R.R. (1998). Aggresomes: a cellular response to

misfolded proteins. *J Cell Biol* **143**, 1883–1898.
56 Nakagawa, T., et al. (2000). Caspase-12 mediates endoplasmic-reticulum-specific apoptosis and cytotoxicity by amyloid-beta. *Nature* **403**, 98–103.
57 Urano, F., et al. (2000). Coupling of stress in the ER to activation of JNK protein kinases by transmembrane protein kinase IRE1. *Science* **287**, 664–666.
58 Yang, Y., et al. (2003). Parkin suppresses dopaminergic neuron-selective neurotoxicity induced by Pael-R in Drosophila. *Neuron* **37**, 911–924.
59 Kitao, Y., et al. (2007). Pael receptor induces death of dopaminergic neurons in the substantia nigra via endoplasmic reticulum stress and dopamine toxicity, which is enhanced under condition of Parkin inactivation. *Hum Mol Genet.* **16**, 50–60.
60 LaVoie, M.J., et al. (2005). Dopamine covalently modifies and functionally inactivates parkin. *Nat Med* **11**, 1214–1221.
61 Dong, Z., et al. (2003). Dopamine-dependent neurodegeneration in rats induced by viral vector-mediated overexpression of the parkin target protein, CDCrel-1. *Proc Natl Acad Sci USA* **100**, 12438–12443.
62 Engelender, S., et al. (1999). Synphilin-1 associates with alpha-synuclein and promotes the formation of cytosolic inclusions. *Nat Genet* **22**, 110–114.
63 Quevillon, S., et al. (1999). Macromolecular assemblage of aminoacyl-tRNA synthetases: identification of protein–protein interactions and characterization of a core protein. *J Mol Biol* **285**, 183–195.
64 Greene, J.C., et al. (2003). Mitochondrial pathology and apoptotic muscle degeneration in Drosophila parkin mutants. *Proc Natl Acad Sci USA* **100**, 4078–4083.
65 Pesah, Y., et al. (2004). Drosophila parkin mutants have decreased mass and cell size and increased sensitivity to oxygen radical stress. *Development* **131**, 2183–2194.
66 Goldberg, M.S., et al. (2003). Parkin-deficient mice exhibit nigrostriatal deficits but not loss of dopaminergic neurons. *J Biol Chem* **278**, 43628–43635.
67 Itier, J.M., et al. (2003). Parkin gene inactivation alters behaviour and dopamine neurotransmission in the mouse. *Hum Mol Genet* **12**, 2277–2291.
68 Von Coelln, R., et al. (2004). Loss of locus coeruleus neurons and reduced startle in parkin null mice. *Proc Natl Acad Sci USA* **101**, 10744–10749.
69 Perez, F.A. and Palmiter, R.D. (2005). Parkin-deficient mice are not a robust model of parkinsonism. *Proc Natl Acad Sci USA* **102**, 2174–2179.
70 Valente, E.M., et al. (2004). Hereditary early-onset Parkinson's disease caused by mutations in PINK1. *Science* **304**, 1158–1160.
71 Clark, I.E., et al. (2006). Drosophila pink1 is required for mitochondrial function and interacts genetically with parkin. *Nature* **441**, 1162–1166.
72 Park, J., et al. (2006). Mitochondrial dysfunction in Drosophila PINK1 mutants is complemented by parkin. *Nature* **441**, 1157–1161.
73 Yang, Y., et al. (2006). Mitochondrial pathology and muscle and dopaminergic neuron degeneration caused by inactivation of Drosophila Pink1 is rescued by Parkin. *Proc Natl Acad Sci USA* **103**, 10793–10798.
74 Palacino, J.J., et al. (2004). Mitochondrial dysfunction and oxidative damage in parkin-deficient mice. *J Biol Chem* **279**, 18614–18622.
75 Perez, F.A., Curtis, W.R. and Palmiter, R.D. (2005). Parkin-deficient mice are not more sensitive to 6-hydroxydopamine or methamphetamine neurotoxicity. *BMC Neurosci* **6**, 71.
76 Haywood, A.F. and Staveley, B.E. (2004). Parkin counteracts symptoms in a Drosophila model of Parkinson's disease. *BMC Neurosci* **5**, 14.
77 Petrucelli, L., et al. (2002). Parkin protects against the toxicity associated with mutant alpha-synuclein: proteasome dysfunction selectively affects catecholaminergic neurons. *Neuron* **36**, 1007–1019.
78 Omura, T., et al. (2006). A ubiquitin ligase HRD1 promotes the degradation of Pael

receptor, a substrate of Parkin. *J Neurochem.* **99**, 1456–1469.

79 Darios, F., et al. (2003). Parkin prevents mitochondrial swelling and cytochrome c release in mitochondria-dependent cell death. *Hum Mol Genet* **12**, 517–526.

80 Rosen, K.M., et al. (2006). Parkin protects against mitochondrial toxins and beta-amyloid accumulation in skeletal muscle cells. *J Biol Chem* **281**, 12809–12816.

81 Klein, R.L., et al. (2006). Parkin is protective for substantia nigra dopamine neurons in a tau gene transfer neurodegeneration model. *Neurosci Lett* **401**, 130–135.

82 Menendez, J., et al. (2006). Suppression of Parkin enhances nigrostriatal and motor neuron lesion in mice over-expressing human-mutated tau protein. *Hum Mol Genet* **15**, 2045–2058.

83 Vercammen, L., et al. (2006). Parkin protects against neurotoxicity in the 6-hydroxydopamine rat model for Parkinson's disease. *Mol Ther* **14**, 716–723.

84 Lo Bianco, C., et al. (2004). Lentiviral vector delivery of parkin prevents dopaminergic degeneration in an alpha-synuclein rat model of Parkinson's disease. *Proc Natl Acad Sci USA* **101**, 17510–17515.

10
Parkin and Neurodegeneration

Sathya R. Sriram, Valina L. Dawson and Ted M. Dawson

10.1
Introduction

Parkinson's disease (PD) was first described by James Parkinson in his 1817 publication titled "An Essay on the Shaking Palsy" [1], although descriptions of a similar disease are found in ancient Ayurvedic literature in India [2]. PD is estimated to affect about 1–2% of the population over the age of 65 years, with age as the most consistent risk factor [3]. The prevalence of PD amongst different ethnic groups is variable, and the disease has a slightly higher incidence rate in men than women [4]. With a growing aging population and a significantly high mortality rate [5], examining the pathophysiology of this second most common neurodegenerative disorder has received considerable attention.

Clinically, PD is characterized by parkinsonism, which consists of a group of symptoms such as tremor, bradykinesia (slowness of movement), rigidity, and postural instability (difficulty with balance). However, parkinsonism is observed in other brain disorders, making PD diagnosis a challenge in the clinic [3]. In addition, some patients also show signs of elevated anxiety, depression, and dementia. A hallmark feature of PD is the progressive death of selected but heterogeneous populations of neurons throughout the brain, including the substantia nigra pars compacta, coerulus–subcoeruleus complex, brain stem nuclei, nucleus basalis of Meynert, parts of the hypothalamus and cortex, as well as the olfactory bulb [3, 6]. Deficiency of dopamine in the nigrostriatal pathway of the brain is thought to be the major cause of motor dysfunction observed in PD. It is estimated that a loss of 60–70% of dopaminergic neurons in the substantia nigra precedes the onset of symptoms [7]. Functional imaging studies have also been used to follow the nigrostriatal degeneration observed in PD [8]. Another key pathological feature that has classically distinguished PD from other parkinsonism disorders is the presence of dystrophic neurites (Lewy neurites) and eosinophilic cytoplasmic proteinaceous inclusions, called Lewy bodies, in surviving neurons [9].

Primarily a sporadic disease, the etiology of PD is largely unknown. Several environmental factors, including exposure to toxins are associated with an

Protein Degradation, Vol. 4: The Ubiquitin-Proteasome System and Disease.
Edited by R. J. Mayer, A. Ciechanover, M. Rechsteiner

ISBN: 978-3-527-31436-2

Table 10.1. Loci and genes that have been associated with PD.

Locus	Chromosomal location	Gene	Mode of inheritance	Reference
PARK1/PARK4	4q21–q23	*α-synuclein*	AD	20, 26
PARK2	6q25.2–q27	*parkin*	usually AR	21
PARK3	2p13	*unknown*	AD	27
PARK5	4p14	*UCH-L1*	unclear	28
PARK6	1p35–p36	*PINK1*	AR	23
PARK7	1p36	*DJ-1*	AR	22
PARK8	12p11.2–q13.1	*LRRK2*	AD	24, 25
PARK10	1p32	*unknown*	unclear	29
PARK11	2q36–q37	*unknown*	unclear	30

increased risk of PD. Despite some early studies describing monozygotic twins with PD [10, 11], subsequent studies to determine the relative contribution of genetics to the onset of PD concluded that heredity played a negligible role [12–14]. The identification of neurotoxins that selectively damaged dopaminergic neurons and caused parkinsonism symptoms strengthened this theory [15, 16]. However, a later study with a large sample size concluded that while there was little genetic contribution to the development of PD in twins after 50 years of age, genetics contributed significantly to early-onset PD cases [17]. In addition, imaging studies revealed that concordance for nigral pathology may be higher in PD twins than previously described [18]. Over the last decade, genetic susceptibility has been definitively established with the identification of several distinct loci that are strongly associated with familial forms of PD [19–25] (Table 10.1).

The link between rare Mendelian PD and the more widespread sporadic PD is yet to be unequivocally established. However, since clinical and pathological findings overlap noticeably, the assumption that these two forms of the same disease share common causative and pathogenic pathways has encouraged rigorous research in this field [31]. This chapter explores the role of one Mendelian gene, *parkin*, in the pathophysiology of familial and sporadic PD.

10.2 AR JP and Parkin

10.2.1 ARJP: Introduction

Autosomal recessive juvenile parkinsonism (ARJP) is an early-onset, recessively inherited variant of PD with levodopa-responsive classic parkinson's symptoms in addition to some atypical features such as sleep benefit, dystonia (involuntary muscle contractions) and abnormal gait [32–34]. Pathologically, ARJP is characterized by the lack of Lewy bodies with neuronal loss and gliosis restricted largely to the substantia nigra pars compacta and locus coerulus [35]. Linkage analysis in 13 Japanese families with ARJP resulted in the discovery of a locus on chromosome 6q25 that strongly associated with this familial form of PD [36]. Shortly after, chromosomal deletions in five Japanese patients with ARJP were analyzed to identify the causative gene, whose protein product was denoted “parkin”.

Ongoing controversy with regard to whether ARJP is similar to sporadic PD or is its own clinical entity stems from conflicting clinical, imaging and pathological studies in parkin-positive patients. Parkin-linked PD is characterized by significant heterogeneity in clinical symptoms, drug response, ethnicities of patients, age at onset, and progression of disease, with little correlation to the type of mutations identified [37–39]. The wide variation in age at onset, ranging from 7 to 72 years, not only between unrelated patients but also within a single family with the same mutation is remarkable [37, 39–41]. Several positron emission tomography (PET) studies have found little difference in striatal uptake of 18-fluorodopa between patients with sporadic PD and those with mutations in *parkin* [42–44]. However, one PET study suggests that parkin-positive patients with severe clinical manifestations tend to show significant differences in 18-fluorodopa uptake compared to sporadic PD patients [45]. Even more confounding is the observation that two of seven known parkin-linked post-mortem cases have eosinophilic Lewy bodies [46, 47], which is uncharacteristic in ARJP. In addition, one parkin-positive case was reported to have basophilic inclusions that were positive for proteins normally present in Lewy bodies [48]. Due to the slow progression of parkin-linked PD [32], there are limited numbers of cases available for imaging and post-mortem analysis, making conclusive deductions about the pathophysiology of this disease a sizeable challenge.

10.2.2
PARKIN: The Gene

Parkin is one of the largest known genes, spanning over 1.4 Mb and comprising 12 exons encoding a short 4.5-kb mRNA transcript that is expressed in several human tissues, including the brain [21]. The gene is extensively conserved among vertebrates and invertebrates, including rodents, fruit flies, birds, frogs and pufferfish [49, 50]. Mutations in *parkin* account for about 50% of early-onset familial PD with recessive inheritance and about 10–20% of early-onset PD cases with no family history [41]. Over 100 different mutations in *parkin* have been described, including point mutations, intra-exonic deletions, single base pair deletions, mutiple exon deletions, exon multiplications, intronic splice site and promoter region variants [51–53] (Figure 10.1). There may be some evidence that recurrent point mutations in *parkin* arise from common founders, but those involving whole exons may be independent events [54]. The majority of the mutations in *parkin* cause premature termination due to a frameshift or nonsense mutations, resulting in non-functional translation products as demonstrated by the exon 4 deletion in

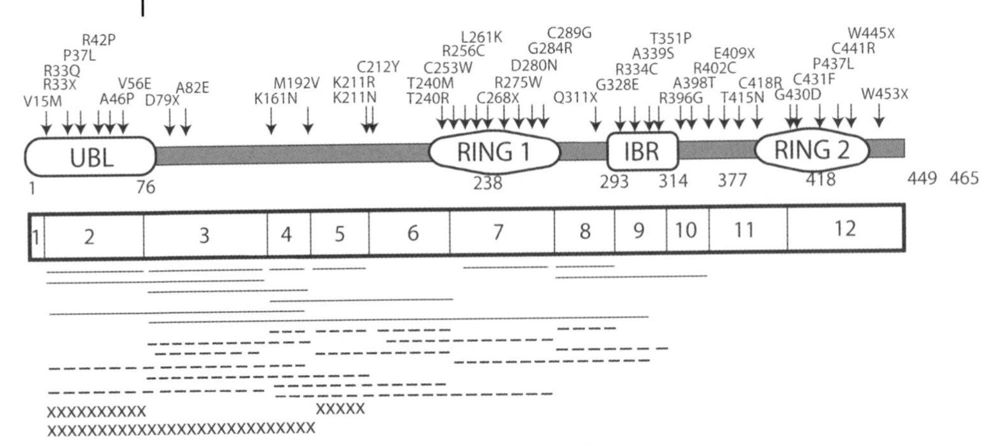

Fig. 10.1. Mutations in Parkin A schematic of pathogenic point mutations (arrows), exon deletions (solid lines), duplications (dashed lines) and triplications (crosses) identified in parkin.

an ARJP patient [55, 56]. On the other hand, missense mutations are hypothesized to destabilize the parkin protein or disable its normal function, resulting in loss-of-function [55]. Polymorphisms in *parkin* were thought to be associated with increased or decreased susceptibility in sporadic PD, but varied results across numerous studies have been inconclusive [57–61].

Mutations in *parkin* mostly show homozygous or compound heterozygous (i.e. different mutations on each allele) inheritance patterns, confirming the autosomal recessive nature of ARJP. However, several published cases were heterozygous for mutations in *parkin* and have forced a paradigm shift in understanding the disease transmission of parkin-linked PD. Some heterozygous mutations have also been associated with increased risk for late-onset PD [62, 63]. More conventional explanations for the presence of heterozygous mutations in *parkin* include haploinsufficiency as a risk factor [47, 53] or that a second mutation was missed either due to the large size and complexity of *parkin* or incomplete screening techniques [53]. Yet another potential explanation is that a single mutation in *parkin* could confer toxic gain-of-function or a dominant-negative function. This may be unlikely, although not improbable, since the described heterozygous mutations have also been found in homozygous or compound heterozygous states, and in some cases, patients with the mutation are asymptomatic [52]. Experimental data from imaging, molecular and biochemical analyses favor the haploinsufficiency model. PET studies in asymptomatic carriers of a single *parkin* mutation show reduced striatal 18-fluorodopa uptake compared to controls [42]. Further, there is evidence for reduced expression due to a single nucleotide polymorphism (–258T/G) in the promoter region of *parkin* in two separate studies involving different ethnicities [64, 65]. Finally, the identification of stress-induced modified parkin with reduced function in sporadic PD cases further supports the haploinsufficiency model [66–68].

10.2.3
PARKIN: Localization and Regulation

In situ hybridization studies show widespread expression of parkin in the rat brain [69]. In dopaminergic neurons of human substantia nigra, parkin mRNA is robustly expressed in a similar pattern as α-synuclein, another PD-linked gene [70]. Although the amount of parkin mRNA is comparable in various regions of the human brain, there seems to be a relative abundance of parkin in the substantia nigra [56]. The parkin protein is also widely expressed in neurons and glia of rodents [71, 72]. Predominantly localized to the cytosol, parkin may be associated with cellular organelles and structures, including the Golgi complex, endoplasmic reticulum, neurites, cytoplasmic and synaptic vesicles [56, 71, 73]. Parkin has also been shown to localize to lipid rafts and postsynaptic densities in rat brain, suggesting a role in synaptic transmission and plasticity [74]. Another study describes parkin immunoreactivity around synaptic vesicles in presynaptic elements of some axons, further supporting a role for parkin in normal synaptic function [75]. Contradictory reports about the presence of parkin in Lewy bodies of sporadic PD patients using different antibodies to parkin, have made it difficult to ascertain a role for parkin in the formation of these inclusion bodies [56, 76–81]. However, the specificity of parkin antibodies has recently been questioned as the majority of parkin antibodies recognize a non-specific protein of the same molecular weight in *parkin*-null mice. Thus definitive localization of parkin will require re-assessment with specific antibodies.

Regulation of parkin levels occurs primarily at the transcriptional level, although there is some evidence of control at the protein level by degradation. Characterization of the *parkin* promoter region led to the identification of a CpG island that is involved in bi-directional transcriptional regulation of parkin and a second gene, PACRG [82]. This shared promoter contains a conserved binding motif for myc-like proteins, through which N-myc has been shown to repress transcription at the *parkin* promoter, plausibly regulating parkin expression during development [83]. The *parkin* promoter variant (–258T/G) that affects gene transcription also affects the physiologic response of the promoter to various cell stressors [64]. *In vitro* luciferase assays to assess the transcriptional activity of the wild-type *parkin* promoter shows significant upregulation under oxidative and proteasomal stress, unlike the –258G variant [64]. Parkin expression is largely absent during embryogenesis, but becomes apparent towards the later stages of development and is predominant in adult cells [49, 83–86]. Levels of exogenous parkin protein may be tightly regulated via rapid degradation by the ubiquitin–proteasome system (UPS) [81]. In addition, there is *in vitro* evidence that the ubiquitin-protein isopeptidase ligase, Nrdp1, may promote parkin turnover [87]. Exogenously expressed truncations were used to demonstrate that its first six residues are involved in regulating cellular levels of parkin [88]. Tunicamycin-induced unfolded protein stress (UPS), but not other types of stress such as H_2O_2, heat shock or ultraviolet light, cause elevated parkin mRNA and protein levels in SH-SY5Y neuroblastoma cells [89]. In addition, tunicamycin treatment in rat primary astrocyte, but not cortical neuronal, cultures results in a modest increase in parkin protein levels [90]. However, another study with SH-SY5Y cells did not show any change in parkin mRNA and protein levels after tunicamycin treatment [91], resulting in some confusion about the association between parkin and the UPS. Parkin mRNA expression increased several fold in response to rotenone, iron and paraquat treatments and parkin becomes insoluble in response to these stressors [92]. The propensity of parkin to become more insoluble and therefore functionally unavailable in aged human tissue, under conditions of stress or mutations supports the loss-of-function and haploinsufficiency hypothesis in parkin-linked PD [76, 92, 93].

10.3 Parkin in the Ubiquitin–Proteasome Pathway

The parkin protein contains an N-terminal ubiquitin-like domain (UBL) and a C-terminal RBR domain, comprising of two RING fingers separated by an in-between-RING (IBR) domain [21]. The RING motif is common to several ubiquitin E3 ligase enzymes that catalyze the conjugation of activated ubiquitin to target substrates [94]. Soon after its discovery, several studies identified parkin as an E2-dependent RING-type E3 ubiquitin ligase in the ubiquitin–proteasome pathway (UPP) [89, 95, 96]. Early studies proposed that there were mutational hotspots, particularly in the functionally significant exon 7 that translate into RING domains [51, 55]. However, compiling all known mutations to date suggests that while there may be some clustering of missense mutations in the functional domains of parkin, deletion mutations are not limited to any one region.

10.3.1 The Ubiquitin–Proteasome Pathway

The UPP is a temporally-regulated highly specific intracellular process, which rapidly catalyzes the turnover of proteins through ubiquitination and proteasome-mediated degradation [97]. Ubiquitin, a highly conserved 76-amino acid protein, acts as functionally distinct signals for proteasomal and lysosomal proteolysis as well as a non-proteolytic signal in protein trafficking and DNA repair [98]. Conjugation of ubiquitin to protein substrates involves three sequential steps: an ATP-dependent activation step catalyzed by the ubiquitin-activating enzyme (E1) that forms a thio-ester bond between ubiquitin and E1; an intermediate step, in which the activated ubiquitin is covalently linked to a ubiquitin-conjugating enzyme (E2) via a similar thio-ester linkage; and finally transferring the ubiquitin molecule to a lysine residue of the substrate in a reaction mediated by a ubiquitin ligase (E3). In addition, an ubiquitin elongation factor (E4) may be recruited to catalyze multiple cycles so four or more ubiquitin molecules can be linked together in a polyubiquitin chain on the substrate. The ubiquitin molecule contains seven lysine (K) residues at amino acid positions 6, 11, 27, 29, 33, 48 and 63. Polyubiquitin chains are synthesized by covalently linking one ubiquitin monomer to the next via iso-peptide bonds between the C-terminus glycine residue of each ubiquitin unit and a specific lysine residue of the previous ubiquitin. Thus, substrates can be modified by mono-ubiquitination, multiple mono-ubiquitinations, or polyubiquitination and the different chains, depending on size and linkage type, provide distinct intracellular signals [99]. In humans, a single E1 enzyme activates ubiquitin for the entire cascade of downstream E2s; about 50 E2s serve multiple E3s to execute distinct biological functions; and hundreds to thousands of E3 enzymes confer specificity to the UPP process by recognizing a limited set of substrates. Regulating E3–substrate or E3–E2 interactions through motif recognition, conformational or covalent modifications provides ample opportunity to tightly control the rate and timing of proteolysis in the cell [100]. Elucidating the mechanism by which an E3 ligase selectively recognizes a particular substrate is a subject of intense research in the field.

Based on the type of conserved domains they contain, the known E3s can be categorized into one of three ubiquitin ligase families: Homologous to E6AP Carboxy Terminus (HECT), Really Interesting New Gene (RING), and UFD2-homology (U-box) proteins [101]. The HECT-type E3s typically contain a domain that is capable of binding activated ubiquitin via a thio-ester bond and serves as a direct intermediate in the transfer of ubiquitin from the E2 to the substrate [101]. The RING-type and U-box-type E3s serve as "bridging" molecules that act as scaffolds to facilitate the transfer of ubiquitin by bringing a ubiquitin-conjugated E2 into close proximity with the target substrate. The RING domain consists of a short stretch of amino acids that is rich in cysteine and histidine residues, and the RING-type E3s are further classified into three sub-families, based on the number and spacing of these conserved residues [101]. Also RING-type E3s may function as single subunit enzymes or co-exist with combinations of other proteins to form a multi-subunit enzyme with more opportunity to dictate substrate specificity [98].

10.3.2 *PARKIN*: An E3 Ubiquitin Ligase

Parkin contains the characteristic RING-IBR-RING (RBR) domain and has been shown to exist as both a single subunit ligase, and in a multi-subunit Skp1-Cullin-F-box (SCF) complex as well [102]. A number of E2s have been shown to associate with parkin, with UbcH7 and UbcH8 being the most common under physiological conditions, although sufficient debate persists on which E2 is preferred conjugating enzyme for parkin [95, 96]. Under conditions of unfolded protein stress, parkin is served by the endoplasmic reticulum membrane (ER)-associated E2s, Ubc6 and Ubc7 [103]. Under these conditions, parkin interacts with Hsp70 and the U-box E3 ligase, CHIP, which modulate the E3 ligase activity of parkin [104]. A similar complex of parkin with Hsp70 and expanded poly-glutamine proteins has also been reported [105]. Further, parkin interacts in a complex with Hsp70 and BAG5, a protein that is upregulated during dopaminergic neuron injury [106]. Parkin localization at post-synaptic densities (PSD) prompted additional investigation, which suggests that parkin interacts with a large multimeric protein complex, implicated in NMDA trafficking, scaffolding, and signaling at the PSD [74]. It remains unclear whether parkin interacts with these complexes preferentially under varying physiological or stress conditions and how these different proteins may modulate its substrate specificity.

10.3.3 Parkin and Lewy Bodies

An increasing number of human diseases are being discovered that are caused by a dysfunctional ubiquitination system. The UPP, along with chaperones, are thought to maintain cell survival and homeostasis by preventing the accumulation of abnormal or toxic proteins that are misfolded or damaged. The well-characterized K48-linked ubiquitin chain on substrates is known to target them for clearance via the 26S proteasome [107, 108]. Therefore a mutation or post-translational modification inhibiting an E3 enzyme or its substrate results in an excess accumulation of the substrate, which may have deleterious consequences. Inclusion bodies are a pathological hallmark of neurodegenerative diseases, such as Lewy bodies in PD, which contain abnormally folded or aggregated disease-associated proteins as well as components of protein quality control machinery, including ubiquitin, proteasome subunits and chaperones [109]. While the mechanism for formation of Lewy bodies still remains unknown 93 years after they were first discovered, there is intense debate about whether these protein-sequestering bodies are neurotoxic or neuroprotective [110]. The discovery that parkin is an E3 ubiquitin ligase, combined with the lack of Lewy bodies in all but two cases of ARJP, implicates a strong role for anomalies in protein homeostasis and UPP in the pathogenesis of parkin-associated PD.

10.3.4 Parkin Substrates

Identification of parkin substrates that may be neurotoxic at elevated steady-state levels is critical to elucidating the underlying neurodegenerative mechanisms in parkin-linked PD. The first parkin substrate to be identified is parkin itself; when exogenously expressed, the E3 ligase can auto-ubiquitinate and promotes its own degradation [96]. In the same study, a yeast two-hybrid screen with full-length parkin yielded another potential substrate: a synaptic vesicle-enriched GTPase called Cell Division Control-Related protein 1 (CDCrel-1) [96]. Parkin binds, ubiquitinates and promotes the proteasome-dependent degradation of CDCrel-1, while pathogenic parkin mutants were unable to turnover the substrate [96]. A later study identified CDCrel-2a as another putative substrate for parkin and showed increased steady-state levels of CDCrel-1 and CDCrel-2a in brains of ARJP patients [111].

Another synaptic vesicle-associated protein, synaptotagmin XI, whose function is unknown, was also described as a parkin substrate [112]. The implication that the parkin substrates, synaptotagmin XI and septin family proteins may be involved in synaptic vesicle transport, docking, and fusion or recycling in the brain, generated immense interest in the role of parkin at the synapse and in pre-synaptic neurotransmission. In a separate study, another yeast two-hybrid screen using full-length parkin as bait revealed a putative G protein-coupled integral membrane polypeptide, named *P*arkin-*a*ssociated *e*ndothelin-*l*ike *R*eceptor (Pael-R), which is degraded by parkin-mediated ubiquitination [103]. Further, this study provides the earliest evidence that parkin has a cytoprotective function under adverse conditions, specifically unfolded protein stress, since abnormally folded Pael-R causes ER stress [103]. Pael-R is accumulated in the detergent-insoluble fraction of ARJP patient brains, suggesting that parkin is crucial for the turnover of this ER-associated substrate [103]. A follow-up study by the same group showed that during unfolded protein response, CHIP promotes the ubiquitination and degradation of Pael-R by parkin [104]. Next, one group demonstrated that parkin interacts with UbcH7 and ubiquitinates a rare O-glycosylated-α-synuclein variant in human brain [113], but the more prevalent non-glycosylated α-synuclein is not a parkin substrate [114]. However, these data have not been replicated in other systems, seriously questioning the physiological relevance of this finding. The lack of altered α-synuclein steady-state levels, where parkin is overexpressed or deficient further supports the idea that parkin has no effect on α-synuclein metabolism [115–119].

While assessing a role for parkin in the ubiquitination of proteins in the Lewy body, since there is high ubiquitin immunoreactivity in these inclusions, the α-synuclein-interacting protein, synphilin-1, was identified as a parkin substrate [114]. When parkin is co-expressed with synphilin-1 and α-synuclein, ubiquitin-positive cytoplasmic inclusions are formed, but familial-linked parkin mutants disrupt ubiquitination of synphilin-1 and the formation of inclusions [114]. This finding is of immense interest since synphilin-1 is found in Lewy bodies [120]. Subsequent studies revealed that synphilin-1 is ubiquitinated by parkin in a non-classical proteasome-independent manner that involves the formation of K63-linked polyubiquitin chains, without appreciable degradation of synphilin-1 by parkin [121]. The ability of parkin to function as a dual ubiquitin ligase, catalyzing the formation of both K48- and K63-type polyubiquitin chains was further confirmed by a study that showed that parkin, in the presence of α-synuclein, promotes the formation of K63-linked chains [122]. Parkin interacts with the E2 complex, comprising UbcH13 and Uev1a, to mediate K63-linkages, supporting the hypothesis that the fate of substrate targets rests on the specific E3–E2 interactions [122]. The relevance of parkin-mediated K63 ubiquitination remains to be clarified. Since K63-linked chains can interact with the proteasome [123], it is plausible that excessive K63 polyubiquitination may interfere with substrate proteolysis and result in accumulated proteins. On the other hand, K63 polyubiquitination may represent an alternate pathway in cells that are stressed with proteasomal overload, diverting the substrates into aggregates. Consistent with the latter hypothesis is the observation that parkin-mediated K63-linked ubiquitination of synphilin-1 enhances the formation of cytoplasmic inclusions, when parkin is co-expressed with synphilin-1 and α-synuclein [121]. These data strongly suggest that the proteasome-independent K63-linked ubiquitination may play a role in inclusion formation in PD as well as other neurodegenerative disease and warrants further study [124].

The race to discover neurotoxic substrates of parkin led to the identification of cell cycle-regulating cyclin E, cytoskeleton-associated α/β tubulin, aminoacyl-tRNA synthetase complex subunit AIMP2 (p38/JTV-1), neuron-specific dopamine transporter, E3 SUMO ligase RanBP2 and the expanded poly-glutamine ataxin-3 fragment [102, 105, 125–128]. An SCF-like ubiquitin ligase complex, comprising parkin, hSel10 and Cul1, is involved in the ubiquitination of parkin and cyclin E [102]. In addition, this report provided evidence for parkin regulation of cyclin E levels in ARJP and PD brains as well as in kainate-induced neuronal apoptosis by overexpression and knockdown of parkin [102]. Strong binding and co-localization between parkin and microtubules was demonstrated in rat cortical neurons and cell lines, with no localization of parkin to the post-synaptic densities, in contrast to prior reports [74, 125]. Furthermore parkin ubiquitinates and accelerates the degradation of α/β tubulin *in vitro* and *in vivo* [125]. Since PD-linked neurotoxins, such as MPP^+ and rotenone, can induce depolymerization of microtubules, and misfolded tubulin may be toxic in the cell, the ability of parkin to degrade these key structural components may be crucial to neuronal survival [125]. A separate study found that parkin strongly associates with and stabilizes microtubules, a process which was not affected by PD-linked mutations, suggesting that the E3 ligase and microtubule-binding activities of parkin are independent [129]. Parkin may recognize misfolded proteins through its association with Hsp70 and mediate ubiquitination and proteasome-dependent degradation of these proteins, as is seen with expanded poly-glutamine ataxin-3 [105]. Another study reports the ability of parkin to preferentially ubiquitinate and degrade misfolded dopamine transporter (DAT), thus preventing misfolded DAT from oligomerizing with properly folded DAT and ensuring sufficient cell surface expression of native DAT [127]. In yet another yeast two-hybrid screen with full-length parkin, the most recent substrate discovered is nuclear membrane-associated RanBP2, which sumolyates the histone deacetylase, HDAC4 [128]. The significance of this novel substrate and the consequential modulation of HDAC4 levels by parkin to PD pathogenesis are far from established.

The p38 subunit of aminoacyl-tRNA synthetase complex, AIMP2, is a parkin substrate, as demonstrated by two separate groups [118, 126]. This co-factor is a key scaffolding component for assembly of the multi-tRNA synthetase complex. As part of a multi-protein complex with Hsp70 and CHIP, parkin interacts with and ubiquitinates AIMP2, promoting its proteasomal degradation [118]. Excessive AIMP2 is toxic, but the AIMP2-induced toxicity is mitigated by exogenously expressed parkin [118, 126]. Overexpression of AIMP2 results in the formation of cytoplasmic AIMP2- and ubiquitin-positive inclusions that is enhanced in the presence of parkin, suggesting that the non-ubiquitinated non-aggregated form of

AIMP2 is deleterious [126]. Interestingly, AIMP2-positive immunoreactivity was observed in Lewy bodies[118, 126]; only three other parkin substrates – synphilin-1, Pael-R, and synaptotagmin XI – have been shown to co-localize in Lewy bodies [78, 112, 120, 130]. In *parkin*-null mice, since parkin dysfunction impairs its ability to effectively mediate the degradation of substrates, authentic substrates of parkin should accumulate [131]. A comprehensive study to assess the steady-state levels of numerous published substrates of parkin revealed that AIMP2 is upregulated in the midbrain and hindbrain of *parkin*-null mice, while none of the other substrates, including CDCrel-1, synphilin-1, Pael-R, cyclin E, and synaptotagmin XI, were upregulated [118]. Parkin mediates proteasome-independent K63-ubiquitination of synphilin-1 (see above), providing a likely explanation for the unchanged steady-state levels of synphilin-1 in *parkin*-null mice. Accumulation of AIMP2 is observed in ARJP patient brains as well as in sporadic PD brains with functional inactivation of parkin due to oxidative, nitrosative and dopaminergic stress [118]. Although other putative substrates have been reported to be upregulated in AR-JP brains [102, 103, 111, 118], AIMP2 is the only substrate that is consistently upregulated in *parkin*-null mice, AR-JP brains and sporadic PD brains; thus, it appears to be the only pathogenic parkin substrate identified to date.

10.4 Parkin in Neuroprotection

10.4.1 Toxic Parkin Substrates

The fundamental question that is yet to be credibly addressed in neurodegenerative diseases is the mechanism by which a selective subset of neurons becomes susceptible to the pathological effects of a mutant gene product. In parkin-linked PD, it is hypothesized that absence of parkin or mutations that abolish its enzymatic activity result in an accumulation of parkin substrates that lead to the specific degeneration of dopaminergic neurons, the most affected neuronal subtype in PD. The most apparent solution to this conundrum would be the identification of a substrate that is expressed only in these neurons. However, only one of the identified putative substrates is exclusively expressed in dopaminergic neurons. The dopamine transporter (DAT) is only found on the cell surface of dopaminergic neurons and is responsible for rapid re-uptake of released dopamine into the neuron [127]. There is *in vitro* evidence that parkin ubiquitinates primarily misfolded DAT in the presence of tunicamycin, which induces unfolded protein stress [127]. The authors propose that dysfunctional parkin causes increased misfolded DAT that interferes with the cell surface expression of normal DAT and dopamine re-uptake, thus sensitizing neurons to extracellular dopamine stress [127]. However, it is imperative to have *in vivo* data to support this hypothesis before any conclusive inference can be made. Alternatively, it is possible that the toxicity threshold tolerated by dopaminergic neurons under certain stress conditions is significantly lower than that of other cell types. Only a few of the identified substrates have been implicated in dopaminergic neuron death. Overexpression of Pael-R in *Drosophila* causes dopaminergic neuron-specific toxicity, which is exacerbated when the expression of endogenous parkin is inhibited by RNAi interference and alleviated by overexpression of parkin in flies [132]. Interestingly, reduced expression of endogenous *Drosophila* parkin alone is not sufficient to cause pathology [132]; it seems that stress, such as that induced by Pael-R overexpression, coupled with parkin insufficiency is necessary for dopaminergic cell death *in vitro*. Another putative parkin substrate, cyclin E, is elevated in primary midbrain dopaminergic cultures when parkin levels are "knocked down" using parkin-specific siRNA and the cultures are treated with the excitotoxin, kainate [102]. The study found that parkin deficiency preferentially sensitizes midbrain dopaminergic neurons, versus midbrain GABAergic cultures, to kainate excitotoxicity, but not MPP^+ toxicity [102]. However, since cyclin E upregulation in toxicity models is not limited to the brain [133], the molecular mechanism underlying dopamine neuron-specific susceptibility needs to be elucidated.

A third parkin substrate, AIMP2, sensitizes human neuroblastoma cells to TNF-α toxicity, which is alleviated by parkin overexpression [118]. Adenovirus-mediated overexpression of AIMP2 in the substantia nigra of mice induces significant dopaminergic neuron loss *in vivo* [118]. Adeno-associated viral delivery of the parkin substrate, CDCrel-1, in the substantia nigra of rats causes significant degeneration of dopamine neurons, but has no effect on neurons in another brain region, the globus pallidus [134]. This site-specific neurodegeneration is prevented by inhibiting dopamine synthesis, suggesting a role for CDCrel-1 in dopamine biochemistry [134]. While both the afore-mentioned studies on viral-mediated overexpression of parkin substrates in rodents show compelling data on selective degeneration, behavioral analyses as well as the therapeutic potential of increasing parkin expression in these animals have yet to be reported upon. Thus, dopaminergic neurons in the substantia nigra may have an increased susceptibility to substrate accumulation. However, PD pathology in the brain extends beyond the substantia nigra and none of the substrate toxicity studies have addressed this issue.

10.4.2
Stress-mediated Toxicity

The protective function of parkin extends beyond preventing the accumulation of its putative substrates. Modulation of parkin levels during unfolded protein stress [89, 135], and the evidence from biochemical studies that parkin interacts with molecular chaperones to preferentially ubiquitinate and degrade misfolded proteins [103, 105, 127], suggest a strong role for parkin in protection against unfolded protein stress. Proteasomal defects apparent in the substantia nigra of sporadic PD patients [136], combined with the presence of UPP components in Lewy bodies [109], have prompted a number of cell culture studies on the effect of parkin in proteolytic stress. In mouse midbrain cultures, proteasome inhibition with

MG-132 and lactacystin resulted in decreased numbers of TH-positive neurons, which was restored by overexpression of parkin [117]. Further, *in vitro* studies show that proteasome inhibition causes endogenous parkin to be recruited into perinuclear microtubule-dependent aggresome-like structures, primarily localized to the centrosome [78, 137, 138]. Overexpression of parkin reduces the MG-132-induced aggresome-like bodies, while protecting the cells from MG-132-induced toxicity [137]. Aggresomes are large non-toxic inclusions formed at the centrosome that sequester misfolded and/or deleterious proteins and are proposed to be involved in the biogenesis of Lewy bodies [139]. Treating cell cultures with a variety of PD-associated toxins, such as manganese and rotenone, induces the formation of similar parkin-positive perinuclear inclusions, which segregates parkin from its normal cellular localization [92, 140]. Parkin overexpression protects dopaminergic cell lines from manganese-induced toxicity, independent of the proteasome system [140]. On the other hand, rotenone treatment causes dose-dependent impairment in proteasome activity, which is relieved by parkin overexpression [92]. Inhibiting the proteasome also abrogates parkin protection against ceramide-mediated cell death *in vitro* [141], supporting the premise that the protective function of parkin against certain stressors may be modulated through maintaining proteasome function. Another study showed that parkin protection against caspase-dependent cell death induced by dopamine treatment is mediated by its ubiquitination/degradation function, although the molecular mechanism is not known [142]. Reduction of endogenous parkin in a glial-like cell line resulted in increased susceptibility to dopamine-induced caspase-dependent as well as H_2O_2-induced caspase-independent cell death, however parkin overexpression did not protect against stress-induced toxicity. The astonishingly large number and variety of parkin substrates, along with the range of cellular stressors that parkin protects against, suggest a vital versatile neuroprotective role for parkin in the survival of dopaminergic neurons.

10.5
Parkin and Other PD-linked Genes

10.5.1
α-Synuclein

In attempting to connect the first two identified PD-linked genes in a common pathogenic pathway, a number of *in vitro* and *in vivo* studies on the ability of parkin to suppress α-synuclein toxicity have been undertaken. Abnormal accumulation of α-synuclein in Lewy bodies is considered to be a pathological hallmark of PD. Unlike parkin, α-synuclein is associated with an autosomal dominant form of PD with Lewy body pathology [19]. Overexpression of wild-type and mutant α-synuclein induces toxicity in cell lines as well as primary midbrain cultures, which is rescued by parkin overexpression in these cultures [117, 143, 144]. While parkin protection in one study is associated with the appearance of high molecular weight non-ubiquitinated α-synuclein species [143], another study argues that parkin mediates protection in a non-proteasomal manner by enhancing the protease activity of calpain to cleave α-synuclein [144]. Parkin overexpression in *Drosophila* mitigates α-synuclein-induced pathology and toxicity [132]. Lentiviral-mediated co-expression of parkin with a α-synuclein pathogenic mutant in rats reduces the number of α-synuclein-induced dopaminergic neuron losses in the substantia nigra, while increasing the amount of phosphorylated α-synuclein inclusions, which is typically found in Lewy bodies [145]. Interestingly, overexpression of glial cell line-derived neurotrophic factor (GDNF), which represents a promising neuroprotective target for PD, was unable to mitigate α-synuclein toxicity in these rats [145]. This suggests a unique role for parkin in neuroprotection against α-synuclein, potentially linking the two familial-linked proteins in a common pathway. However, we have recently shown that there is no observable synergistic effect of parkin deficiency and overexpression of mutant α-synuclein in mice, suggesting that these two genes may have independent mechanisms of pathogenesis [119].

10.5.2 DJ-1

Mutations in *DJ-1* are associated with autosomal recessive early-onset PD, similar to parkin, although it appears to be a rare cause of familial PD [146]. The versatile DJ-1 protein may possess chaperone and oxidative stress-sensing activities, which is confirmed by *in vitro* studies [146]. Parkin interacts selectively but differentially with pathogenic DJ-1 mutants, as well as with wild-type DJ-1 following oxidative stress in cell culture overexpression studies [146]. However, parkin does not ubiquitinate and augment the degradation of mutant DJ-1 as well as wild-type DJ-1 after oxidative stress [146]. In dissecting the biological relevance of the interaction between parkin and DJ-1, we found that in enhancing DJ-1 stability, parkin is part of a larger complex, comprising of CHIP and Hsp70, which can independently associate with the highly unstable pathogenic DJ-1 mutant. This ubiquitination-independent stabilizing effect of parkin is further supported by the significantly increased levels of DJ-1 in the detergent-insoluble fraction of post-mortem PD cortex brains and the marked reduction in DJ-1 levels in the detergent-insoluble fraction of the parkin-deficient post-mortem ARJP cortex brains [146]. Thus the oxidative stress- and mutation-induced association between parkin and DJ-1 may represent a common molecular pathway in the pathogenesis of PD that warrants further study.

10.5.3 LRRK2

Leucine-rich repeat kinase 2 (*LRRK2*) is the most recent gene to be associated with autosomal-dominant PD. A gigantic protein of 2527 amino acids, this mixed-lineage kinase has several predicted functional domains, with point mutations found in almost all the identified domains. Parkin interacts with LRRK2 via its C-terminal RING2 domain in cell culture overexpression studies, an association that is not altered by pathogenic mutants of LRRK2 [147]. This interaction is specific to parkin as LRRK2 does not bind to other known PD-linked genes, including DJ-1, α-synuclein, and tau [147]. Further parkin co-expression increases the number of ubiquitin-positive cytoplasmic aggregates of LRRK2, although there is no evidence that parkin directly ubiquitinates LRRK2 [147]. As more detailed mechanisms of the degenerative pathways involving PD-linked genes are revealed, the relevance of the interactions between these gene products will become more apparent.

10.6 Mechanisms of Parkin Dysfunction

10.6.1 Pathogenic Mutations

To understand how the variety of mutations in *parkin* contribute to its dysfunction to result in a shared pathogenic outcome, cell culture-based overexpression studies in various cell lines have been conducted. Since parkin is associated with recessive forms of PD, it was hypothesized that mutations in *parkin* result in the loss of E3 ligase activity that would translate into accumulated cytotoxic substrates. However, early studies showed that some parkin mutants retain or may have partially disrupted ubiquitination activity [89, 114, 126, 148]. An extensive investigation of several parkin mutants shows that each mutant may be defective in one or more aspects of the ubiquitination/degradation process or is abnormally localized, which manifests as an apparent loss-of-function [131, 149]. Some mutants have completely abolished enzymatic activity, but other mutants are relatively unaffected or have increased activity, as demonstrated by parkin auto-ubiquitination as well as ubiquitination of two established parkin substrates: synphilin-1 and AIMP2 [131]. Regardless of the alternate mechanisms of disruption in the process, all studied parkin mutants, unlike wild-type parkin, were unable to effectively reduce the steady-state levels of substrates [131]. The function of the N-terminal UBL domain in parkin has not as yet been established; although one NMR study suggested that parkin may bind the proteasome via its UBL domain, an interaction that may be abolished by a known pathogenic mutation in the domain [150]. Interestingly, another study reported that the C-terminal IBR-RING domain is essential in mediating the interaction with the 20S proteasomal subunit α4 [151]. Pathogenic mutations in the UBL domain of parkin result in highly unstable protein products that are rapidly degraded, compared to wild type, resulting in less functional parkin in the cell [152].

A consistent finding across several groups is the altered solubility and localization with overexpressed pathogenic parkin mutants [93, 131, 148, 152, 153]. While some parkin mutants, especially those that are localized to the RING finger, reliably aggregate into cytoplasmic inclusions, slight discrepancies exist amongst other mutants that may be attributed to the various cell lines used. It is thought that sequestering enzymatically-active mutants into these aggresome-like bodies would preclude them from ubiquitinating substrates and therefore manifest as loss-of-function mutants. Exogenous PD-associated stress, including oxidative and dopamine stress, result in similar misfolding and aggregation of wild-type parkin protein, suggesting that this mechanism may account for some parkin dysfunction in sporadic PD, where parkin is not mutated. Although they provide substantial insight into the mechanisms of abnormal parkin function and localization, these studies are limited in their interpretation until they are extended into *in vivo* models. We recently showed a significant increase in detergent-insoluble parkin in mouse brain after treatment with the PD toxin, MPTP [92]. These observations, coupled with the presence of increased age-dependent insoluble parkin in human brains, which is marginally enhanced in PD patients [76, 92], strengthens this proposed mechanism of abrogated neuroprotection by parkin through progressive depletion of functionally available parkin.

10.6.2
Cellular Regulators of Parkin

Recent studies emphasize numerous environmentally-inflicted stress conditions that inactivate the enzymatic activity of parkin, plausibly predisposing heterozygous carriers and non-carriers of parkin mutations to PD [66–68, 106, 154]. In cell culture, treatment with inducers of apoptosis, such as the kinase inhibitor, staurosporine, resulted in proteolytic caspase-dependent truncations of parkin that are predicted to be non-functional [154]. This is in line with reported toxicity experiments using staurosporine, where parkin was unable to exert any protection [141]. The *bcl-2-associated athanogene 5* gene product, BAG5, is a molecular mediator of dopaminergic neurodegeneration localized to Lewy bodies, whose expression is induced after dopaminergic neuron injury [106]. BAG5 interacts directly with Hsp70 and parkin, negatively regulating their respective cellular activities as well as the protective function of parkin [106]. The study further suggests that BAG5, through its inhibition of Hsp70, can promote aggregation of overexpressed parkin and presents evidence for BAG5-mediated dopaminergic neurodegeneration in an *in vivo* model of PD [106]. Another reported molecular regulator of parkin function is 14-3-3η, a member of the 14-3-3 family of proteins that is found in Lewy bodies [155]. 14-3-3η binds and suppresses the ubiquitin ligase activity of parkin, but α-synuclein overexpression abrogates the inhibition of parkin function by 14-3-3η, yet again functionally linking these two PD-linked gene products [155].

10.6.3
Post-translational Regulation of Parkin

Environmental stressors are known to cause intracellular changes that can induce post-translational modifications of key proteins. Neurodegenerative pathologic conditions, such as PD, are associated with high levels nitrosative stress [156], therefore it was logical to assess whether parkin is modified under these conditions. *S*-nitrosylation of parkin on select cysteine residues in the RING domain is evident *in vitro* and *in vivo* [66, 68]. This unique post-translational modification of parkin results in a bi-phasic alteration, with an initial spike followed by a gradual decrease, in its enzymatic activity [66, 68]. The pathological relevance of this finding was demonstrated by the detection of *S*-nitrosylated parkin in MPTP- and rotenone-treated mice brain as well as in post-mortem brain tissue of sporadic PD patients [66, 68]. *S*-nitrosylation of parkin ultimately leads to loss of E3 ligase activity and loss of its protective function [66]. Interestingly, AIMP2 accumulates in PD patients with increased nitrosative stress, providing further support for AIMP2 being an authentic parkin substrate [118]. A recent study reports the discovery of dopamine-mediated covalently modified parkin that has reduced solubility and ubiquitin ligase activity *in vitro* and in post-mortem sporadic PD brains [67]. Although this may represent only a small fraction of dysfunctional parkin, the close ties to dopamine oxidation that is restricted to dopaminergic neurons renders it a possible relevant pathological modification. Finally, a small reduction in parkin enzymatic activity as a consequence of serine-phosphorylation *in vitro* has been reported, though the physiological relevance of this modification is not known [157].

10.7 Animal Models of Parkin Deficiency

10.7.1 Drosophila Models

Since loss of parkin function is strongly associated with PD in humans, studying the effects of parkin deficiency *in vivo* would be expected to provide clues to understanding the pathogenesis of parkin-linked PD. Parkin is evolutionarily conserved across several species, allowing the generation of several animal models. A *Drosophila* model of ARJP with targeted disruption of the *Drosophila parkin* (*dParkin*) ortholog resulted in viable, but short-lived flies, with male sterility, and locomotor defects due to mitochondrial pathology and/or muscle degeneration [86]. However, there was no apparent neurodegeneration, other than a marginal shrinkage of dopaminergic cell bodies [86]. A follow-up study on the same *Drosophila* line, using different quantification techniques, suggests that a subset of dopaminergic neurons degenerates in *dParkin* mutants, which is rescued by overexpression of glutathione-*S*-transferase, a protein implicated in cellular response to oxidative stress [158]. An independent study confirmed the observed phenotype in loss-of-function mutations in *dParkin*, and further suggests that these mutant flies have increased sensitivity to stress [159].

10.7.2
Mouse Models

A number of mouse models have been generated, where different exons of *parkin* are deleted by homologous recombination. A *parkin* knockout mouse model, with deletion of exon 3, shows no pathology or neurodegeneration, but displays behavioral, biochemical and electrophysiological dopamine-related alternations [160]. A notable increase in monoamine oxidase activity, which catalyzes dopamine oxidation, in these *parkin*-null mice may be explained by a recent report on the ubiquitination-independent ability of parkin to reduce monoamine oxidase mRNA levels [161]. In another independent *parkin* knockout mouse line, with an exon 3 deletion, some mild nigrostriatal deficits, but no profound loss of dopaminergic neurons, were observed [115]. Interestingly, the authors describe increased extracellular dopamine in the striatum of these mice, and ascribe it to increased dopamine release from nigral neurons [115]. The subsequent discovery that parkin regulates levels of misfolded DAT and cell surface expression of native DAT to ensure functional dopamine re-uptake may help clarify the presence of the excess extracellular dopamine observed [127]. Further proteomic analyses of these mice showed that *parkin*-null mice exhibited changes in abundance of several proteins, a large majority of which are associated with normal mitochondrial and antioxidant function [162]. A third mouse model, with an exon 7 deletion to ensure removal of the first RING finger domain, shows loss of catecholaminergic neurons in the locus coeruleus, a brain region affected in PD, along with loss of norepinephrine in certain brain regions [163]. Accompanying the loss of locus coeruleus neurons and norepinephrine is the marked reduction of acoustic startle, a non-epinephrine-dependent process in rodents. Consistent reports of minor deficits that are classically associated with AR-JP [164] supports the hypothesis that loss of parkin alone is insufficient to cause disease in rodents; perhaps the interaction of parkin with other PD-linked genes or environmental factors, such as stress, play a more significant role than was previously envisaged. The generation of animal models that closely resemble parkin-related PD is vital to comprehending this elusive protein's role in neurodegeneration.

10.8
Concluding Remarks

In the 8 years since the discovery of the *parkin* gene, we have come a long way in understanding parkin genetics and biochemistry (Figure 10.2); however, these interesting insights have only resulted in more questions that need to be addressed. Is the E3 ligase activity of parkin its only function in the cell? Does it truly mediate neuroprotection *in vivo*? Can we develop animal models to efficiently test parkin-related hypotheses? And most enigmatic of them all: what is the pathogenic mechanism by which parkin deficiency causes selective degeneration in the brain? The drive to find a solution and utilize emergent biotechnology to design effective

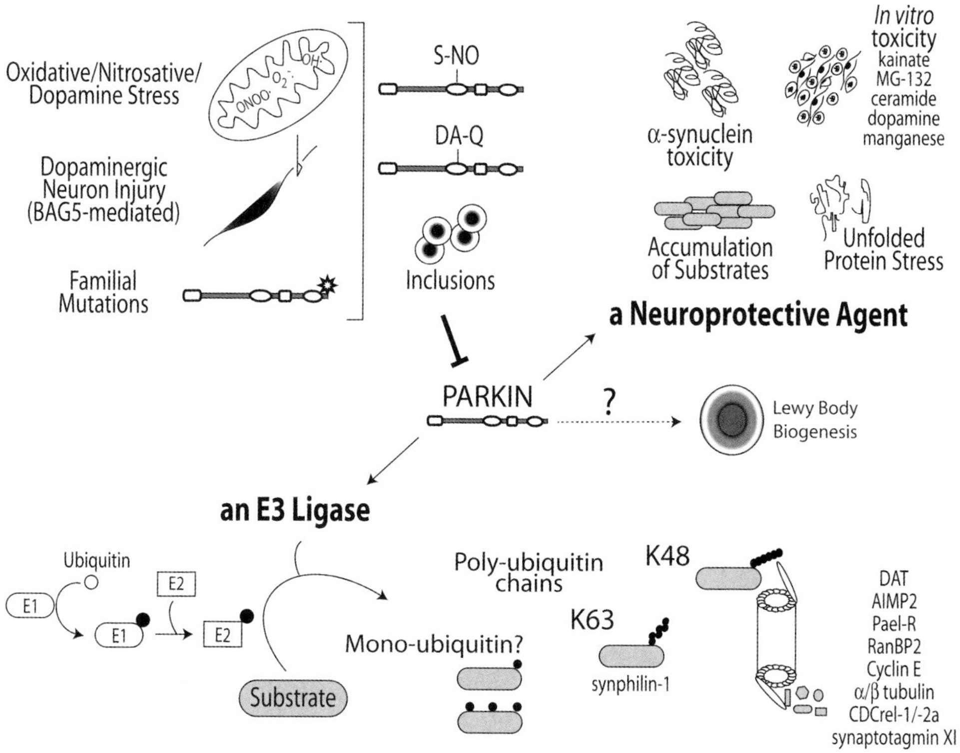

Fig.10.2. Parkin plays a central role as an E3 ligase and a versatile neuroprotective agent. As an E3 ligase, parkin can catalyze K48-linked, K63-linked and mono-ubiquitination of several putative substrates. In addition, parkin protects from toxicity induced by accumulation of some substrates, unfolded protein stress, and overexpression of α-synuclein. A number of *in vitro* studies show that parkin overexpression also protects against other cellular stressors. Mechanisms of parkin dysfunction include familial mutations that disrupt its function by inhibiting its E3 ligase activity or inducing aggregate formation; stress-mediated modifications such as *S*-nitrosylation and dopamine–quinone adduct formation; and negative regulators of parkin activity, such as BAG5, which is induced upon dopaminergic neuron injury.

therapeutics for AR JP, and subsequently for PD, motivates research in this field.

References

1. Parkinson, J. (2002) *J Neuropsychiatry Clin Neurosci* **14**, 223–36; discussion 222.
2. Manyam, B.V. (1990) *Mov Disord* **5**, 47–48.
3. Lang, A.E. and Lozano, A.M. (1998) *N Engl J Med* **339**, 1044–1053.
4. Zhang, Z.X. and Roman, G.C. (1993) *Neuroepidemiology* **12**, 195–208.

5 Morens, D.M., Davis, J.W., Grandinetti, A., Ross, G.W., Popper, J.S. and White, L.R. (1996) *Neurology* **46**, 1044–1050.

6 Braak, H., del Tredici, K., Rub, U., De Vos, R.A., Jansen Steur, E.N. and Braak, E. (2003) *Neurobiol Aging* **24**, 197–211.

7 Fearnley, J.M. and Lees, A.J. (1991) *Brain* **114**(*Pt 5*), 2283–2301.

8 Au, W.L., Adams, J.R., Troiano, A.R. and Stoessl, A.J. (2005) *Brain Res Mol Brain Res* **134**, 24–33.

9 Forno, L.S. (1996) *J Neuropathol Exp Neurol* **55**, 259–272.

10 Koller, W., O'Hara, R., Nutt, J., Young, J. and Rubino, F. (1986) *Ann Neurol* **19**, 402–405.

11 Jankovic, J. and Reches, A. (1986) *Ann Neurol* **19**, 405–408.

12 Ward, C.D., Duvoisin, R.C., Ince, S.E., Nutt, J.D., Eldridge, R. and Calne, D.B. (1983) *Neurology* **33**, 815–824.

13 Marttila, R.J., Kaprio, J., Koskenvuo, M. and Rinne, U.K. (1988) *Neurology* **38**, 1217–1219.

14 Duvoisin, R.C., Eldridge, R., Williams, A., Nutt, J. and Calne, D. (1981) *Neurology* **31**, 77–80.

15 Langston, J.W., Ballard, P., Tetrud, J.W. and Irwin, I. (1983) *Science* **219**, 979–980.

16 Hirsch, E.C., Hoglinger, G., Rousselet, E., Breidert, T., Parain, K., Feger, J., Ruberg, M., Prigent, A., Cohen-Salmon, C. and Launay, J.M. (2003) *J Neural Transm Suppl* 89–100.

17 Tanner, C.M., Ottman, R., Goldman, S.M., Ellenberg, J., Chan, P., Mayeux, R. and Langston, J.W. (1999) *JAMA* **281**, 341–346.

18 Burn, D.J., Mark, M.H., Playford, E.D., Maraganore, D.M., Zimmerman, T.R., Jr., Duvoisin, R.C., Harding, A.E., Marsden, C.D. and Brooks, D.J. (1992) *Neurology* **42**, 1894–1900.

19 Dawson, T.M. and Dawson, V.L. (2003) *J Clin Invest* **111**, 145–151.

20 Polymeropoulos, M.H., Lavedan, C., Leroy, E., Ide, S.E., Dehejia, A., Dutra, A., Pike, B., Root, H., Rubenstein, J., Boyer, R., Stenroos, E.S., Chandrasekharappa, S., Athanassiadou, A., Papapetropoulos, T., Johnson, W.G., Lazzarini, A.M., Duvoisin, R.C., Di Iorio, G., Golbe, L.I. and Nussbaum, R.L. (1997) *Science* **276**, 2045–2047.

21 Kitada, T., Asakawa, S., Hattori, N., Matsumine, H., Yamamura, Y., Minoshima, S., Yokochi, M., Mizuno, Y. and Shimizu, N. (1998) *Nature* **392**, 605–608.

22 Bonifati, V., Rizzu, P., van Baren, M.J., Schaap, O., Breedveld, G.J., Krieger, E., Dekker, M.C., Squitieri, F., Ibanez, P., Joosse, M., Van Dongen, J.W., Vanacore, N., Van Swieten, J.C., Brice, A., Meco, G., Van Duijn, C.M., Oostra, B.A. and Heutink, P. (2003) *Science* **299**, 256–259.

23 Valente, E.M., Abou-Sleiman, P.M., Caputo, V., Muqit, M.M., Harvey, K., Gispert, S., Ali, Z., Del Turco, D., Bentivoglio, A.R., Healy, D.G., Albanese, A., Nussbaum, R., Gonzalez-Maldonado, R., Deller, T., Salvi, S., Cortelli, P., Gilks, W.P., Latchman, D.S., Harvey, R.J., Dallapiccola, B., Auburger, G. and Wood, N.W. (2004) *Science* **304**, 1158–1160.

24 Zimprich, A., Biskup, S., Leitner, P., Lichtner, P., Farrer, M., Lincoln, S., Kachergus, J., Hulihan, M., Uitti, R.J., Calne, D.B., Stoessl, A.J., Pfeiffer, R.F., Patenge, N., Carbajal, I.C., Vieregge, P., Asmus, F., Muller-Myhsok, B., Dickson, D.W., Meitinger, T., Strom, T.M., Wszolek, Z.K. and Gasser, T. (2004) *Neuron* **44**, 601–607.

25 Paisan-Ruiz, C., Jain, S., Evans, E.W., Gilks, W.P., Simon, J., Van der Brug, M., Lopez de Munain, A., Aparicio, S., Gil, A.M., Khan, N., Johnson, J., Martinez, J. R., Nicholl, D., Carrera, I.M., Pena, A.S., De Silva, R., Lees, A., Marti-Masso, J.F., Perez-Tur, J., Wood, N.W. and Singleton, A.B. (2004) *Neuron* **44**, 595–600.

26 Singleton, A.B., Farrer, M., Johnson, J., Singleton, A., Hague, S., Kachergus, J., Hulihan, M., Peuralinna, T., Dutra, A., Nussbaum, R., Lincoln, S., Crawley, A., Hanson, M., Maraganore, D., Adler, C., Cookson, M.R., Muenter, M., Baptista, M., Miller, D., Blancato, J., Hardy, J. and Gwinn-Hardy, K. (2003) *Science* **302**, 841.

27 Gasser, T., Muller-Myhsok, B., Wszolek, Z.K., Oehlmann, R., Calne, D.B.,

Bonifati, V., Bereznai, B., Fabrizio, E., Vieregge, P. and Horstmann, R.D. (1998) *Nat Genet* **18**, 262–265.
28 Leroy, E., Boyer, R., Auburger, G., Leube, B., Ulm, G., Mezey, E., Harta, G., Brownstein, M.J., Jonnalagada, S., Chernova, T., Dehejia, A., Lavedan, C., Gasser, T., Steinbach, P.J., Wilkinson, K.D. and Polymeropoulos, M.H. (1998) *Nature* **395**, 451–452.
29 Hicks, A.A., Petursson, H., Jonsson, T., Stefansson, H., Johannsdottir, H.S., Sainz, J., Frigge, M.L., Kong, A., Gulcher, J.R., Stefansson, K. and Sveinbjornsdottir, S. (2002) *Ann Neurol* **52**, 549–555.
30 Pankratz, N., Nichols, W.C., Uniacke, S.K., Halter, C., Rudolph, A., Shults, C., Conneally, P.M. and Foroud, T. (2003) *Am J Hum Genet* **72**, 1053–1107.
31 Cookson, M.R., Xiromerisiou, G. and Singleton, A. (2005) *Curr Opin Neurol* **18**, 706–711.
32 Ishikawa, A. and Tsuji, S. (1996) *Neurology* **47**, 160–166.
33 Hayashi, S., Wakabayashi, K., Ishikawa, A., Nagai, H., Saito, M., Maruyama, M., Takahashi, T., Ozawa, T., Tsuji, S. and Takahashi, H. (2000) *Mov Disord* **15**, 884–888.
34 van de Warrenburg, B.P., Lammens, M., Lucking, C.B., Denefle, P., Wesseling, P., Booij, J., Praamstra, P., Quinn, N., Brice, A. and Horstink, M. W. (2001) *Neurology* **56**, 555–557.
35 Takahashi, H., Ohama, E., Suzuki, S., Horikawa, Y., Ishikawa, A., Morita, T., Tsuji, S. and Ikuta, F. (1994) *Neurology* **44**, 437–441.
36 Matsumine, H., Saito, M., Shimoda-Matsubayashi, S., Tanaka, H., Ishikawa, A., Nakagawa-Hattori, Y., Yokochi, M., Kobayashi, T., Igarashi, S., Takano, H., Sanpei, K., Koike, R., Mori, H., Kondo, T., Mizutani, Y., Schaffer, A.A., Yamamura, Y., Nakamura, S., Kuzuhara, S., Tsuji, S. and Mizuno, Y. (1997) *Am J Hum Genet* **60**, 588–596.
37 Pramstaller, P.P., Kis, B., Eskelson, C., Hedrich, K., Scherer, M., Schwinger, E., Breakefield, X.O., Kramer, P.L., Ozelius, L.J. and Klein, C. (2002) *Mov Disord* **17**, 424–426.
38 Tassin, J., Durr, A., Bonnet, A.M., Gil, R., Vidailhet, M., Lucking, C.B., Goas, J.Y., Durif, F., Abada, M., Echenne, B., Motte, J., Lagueny, A., Lacomblez, L., Jedynak, P., Bartholome, B., Agid, Y. and Brice, A. (2000) *Brain* **123**, 1112–1121.
39 Klein, C., Pramstaller, P.P., Kis, B., Page, C.C., Kann, M., Leung, J., Woodward, H., Castellan, C.C., Scherer, M., Vieregge, P., Breakefield, X.O., Kramer, P.L. and Ozelius, L.J. (2000) *Ann Neurol* **48**, 65–71.
40 Nichols, W.C., Pankratz, N., Uniacke, S.K., Pauciulo, M.W., Halter, C., Rudolph, A., Conneally, P.M. and Foroud, T. (2002) *J Med Genet* **39**, 489–492.
41 Lucking, C.B., Durr, A., Bonifati, V., Vaughan, J., De Michele, G., Gasser, T., Harhangi, B.S., Meco, G., Denefle, P., Wood, N.W., Agid, Y. and Brice, A. (2000) *N Engl J Med* **342**, 1560–1567.
42 Hilker, R., Klein, C., Ghaemi, M., Kis, B., Strotmann, T., Ozelius, L.J., Lenz, O., Vieregge, P., Herholz, K., Heiss, W.D. and Pramstaller, P.P. (2001) *Ann Neurol* **49**, 367–376.
43 Broussolle, E., Lucking, C.B., Ginovart, N., Pollak, P., Remy, P. and Durr, A. (2000) *Neurology* **55**, 877–879.
44 Khan, N.L., Brooks, D.J., Pavese, N., Sweeney, M.G., Wood, N.W., Lees, A.J. and Piccini, P. (2002) *Brain* **125**, 2248–2256.
45 Portman, A.T., Giladi, N., Leenders, K.L., Maguire, P., Veenma-van Der Duin, L., Swart, J., Pruim, J., Simon, E.S., Hassin-Baer, S. and Korczyn, A.D. (2001) *Neurology* **56**, 1759–1762.
46 Pramstaller, P.P., Schlossmacher, M.G., Jacques, T.S., Scaravilli, F., Eskelson, C., Pepivani, I., Hedrich, K., Adel, S., Gonzales-McNeal, M., Hilker, R., Kramer, P.L. and Klein, C. (2005) *Ann Neurol* **58**, 411–422.
47 Farrer, M., Chan, P., Chen, R., Tan, L., Lincoln, S., Hernandez, D., Forno, L., Gwinn-Hardy, K., Petrucelli, L., Hussey, J., Singleton, A., Tanner, C., Hardy, J. and Langston, J.W. (2001) *Ann Neurol* **50**, 293–300.
48 Sasaki, S., Shirata, A., Yamane, K. and Iwata, M. (2004) *Neurology* **63**, 678–682.

49 Yu, W.P., Tan, J.M., Chew, K.C., Oh, T., Kolatkar, P., Venkatesh, B., Dawson, T.M. and Leong Lim, K. (2005) *Gene* **346**, 97–104.

50 Horowitz, J.M., Vernace, V.A., Myers, J., Stachowiak, M.K., Hanlon, D.W., Fraley, G.S. and Torres, G. (2001) *J Chem Neuroanat* **21**, 75–93.

51 Hedrich, K., Eskelson, C., Wilmot, B., Marder, K., Harris, J., Garrels, J., Meija-Santana, H., Vieregge, P., Jacobs, H., Bressman, S.B., Lang, A.E., Kann, M., Abbruzzese, G., Martinelli, P., Schwinger, E., Ozelius, L.J., Pramstaller, P.P., Klein, C. and Kramer, P. (2004) *Mov Disord* **19**, 1146–1157.

52 West, A.B. and Maidment, N.T. (2004) *Hum Genet* **114**, 327–336.

53 West, A., Periquet, M., Lincoln, S., Lucking, C.B., Nicholl, D., Bonifati, V., Rawal, N., Gasser, T., Lohmann, E., Deleuze, J.F., Maraganore, D., Levey, A., Wood, N., Durr, A., Hardy, J., Brice, A. and Farrer, M. (2002) *Am J Med Genet* **114**, 584–591.

54 Periquet, M., Lucking, C., Vaughan, J., Bonifati, V., Durr, A., De Michele, G., Horstink, M., Farrer, M., Illarioshkin, S. N., Pollak, P., Borg, M., Brefel-Courbon, C., Denefle, P., Meco, G., Gasser, T., Breteler, M.M., Wood, N., Agid, Y. and Brice, A. (2001) *Am J Hum Genet* **68**, 617–626.

55 Kahle, P.J., Leimer, U. and Haass, C. (2000) *Trends Biochem Sci* **25**, 524–527.

56 Shimura, H., Hattori, N., Kubo, S., Yoshikawa, M., Kitada, T., Matsumine, H., Asakawa, S., Minoshima, S., Yamamura, Y., Shimizu, N. and Mizuno, Y. (1999) *Ann Neurol* **45**, 668–672.

57 Hu, C.J., Sung, S.M., Liu, H.C., Lee, C.C., Tsai, C.H. and Chang, J.G. (2000) *Eur Neurol* **44**, 90–93.

58 Oliveri, R.L., Zappia, M., Annesi, G., Bosco, D., Annesi, F., Spadafora, P., Pasqua, A.A., Tomaino, C., Nicoletti, G., Pirritano, D., Labate, A., Gambardella, A., Logroscino, G., Manobianca, G., Epifanio, A., Morgante, L., Savettieri, G. and Quattrone, A. (2001) *Neurology* **57**, 359–362.

59 Satoh, J. and Kuroda, Y. (1999) *Neuroreport* **10**, 2735–2739.

60 Wang, M., Hattori, N., Matsumine, H., Kobayashi, T., Yoshino, H., Morioka, A., Kitada, T., Asakawa, S., Minoshima, S., Shimizu, N. and Mizuno, Y. (1999) *Ann Neurol* **45**, 655–658.

61 Oliveira, S.A., Scott, W.K., Nance, M.A., Watts, R.L., Hubble, J.P., Koller, W.C., Lyons, K.E., Pahwa, R., Stern, M.B., Hiner, B.C., Jankovic, J., Ondo, W.G., Allen, F.H., Jr., Scott, B.L., Goetz, C.G., Small, G.W., Mastaglia, F.L., Stajich, J. M., Zhang, F., Booze, M.W., Reaves, J.A., Middleton, L.T., Haines, J.L., Pericak-Vance, M.A., Vance, J.M. and Martin, E. R. (2003) *Arch Neurol* **60**, 975–980.

62 Oliveira, S.A., Scott, W.K., Martin, E.R., Nance, M.A., Watts, R.L., Hubble, J.P., Koller, W.C., Pahwa, R., Stern, M.B., Hiner, B.C., Ondo, W.G., Allen, F.H., Jr., Scott, B.L., Goetz, C.G., Small, G.W., Mastaglia, F., Stajich, J.M., Zhang, F., Booze, M.W., Winn, M.P., Middleton, L.T., Haines, J.L., Pericak-Vance, M.A. and Vance, J.M. (2003) *Ann Neurol* **53**, 624–629.

63 Foroud, T., Uniacke, S.K., Liu, L., Pankratz, N., Rudolph, A., Halter, C., Shults, C., Marder, K., Conneally, P.M. and Nichols, W.C. (2003) *Neurology* **60**, 796–801.

64 Tan, E.K., Puong, K.Y., Chan, D.K., Yew, K., Fook-Chong, S., Shen, H., Ng, P.W., Woo, J., Yuen, Y., Pavanni, R., Wong, M.C., Puvan, K. and Zhao, Y. (2005) *Hum Genet* **118**, 484–488.

65 West, A.B., Maraganore, D., Crook, J., Lesnick, T., Lockhart, P.J., Wilkes, K.M., Kapatos, G., Hardy, J.A. and Farrer, M.J. (2002) *Hum Mol Genet* **11**, 2787–2792.

66 Chung, K.K., Thomas, B., Li, X., Pletnikova, O., Troncoso, J.C., Marsh, L., Dawson, V.L. and Dawson, T.M. (2004) *Science* **304**, 1328–1331.

67 LaVoie, M.J., Ostaszewski, B.L., Weihofen, A., Schlossmacher, M.G. and Selkoe, D.J. (2005) *Nat Med* **11**, 1214–1221.

68 Yao, D., Gu, Z., Nakamura, T., Shi, Z.Q., Ma, Y., Gaston, B., Palmer, L.A., Rockenstein, E.M., Zhang, Z., Masliah, E., Uehara, T. and Lipton, S.A. (2004)

Proc Natl Acad Sci USA **101**, 10810–10814.

69 D'Agata, V., Zhao, W. and Cavallaro, S. (2000) *Brain Res Mol Brain Res* **75**, 345–349.

70 Solano, S.M., Miller, D.W., Augood, S.J., Young, A.B. and Penney, J.B., Jr. (2000) *Ann Neurol* **47**, 201–210.

71 Huynh, D.P., Dy, M., Nguyen, D., Kiehl, T.R. and Pulst, S.M. (2001) *Brain Res Dev Brain Res* **130**, 173–181.

72 Gu, W.J., Abbas, N., Lagunes, M.Z., Parent, A., Pradier, L., Bohme, G.A., Agid, Y., Hirsch, E.C., Raisman-Vozari, R. and Brice, A. (2000) *J Neurochem* **74**, 1773–1776.

73 Kubo, S.I., Kitami, T., Noda, S., Shimura, H., Uchiyama, Y., Asakawa, S., Minoshima, S., Shimizu, N., Mizuno, Y. and Hattori, N. (2001) *J Neurochem* **78**, 42–54.

74 Fallon, L., Moreau, F., Croft, B.G., Labib, N., Gu, W.J. and Fon, E.A. (2002) *J Biol Chem* **277**, 486–491.

75 Mouatt-Prigent, A., Muriel, M.P., Gu, W.J., El Hachimi, K.H., Lucking, C.B., Brice, A. and Hirsch, E.C. (2004) *J Neural Transm* **111**, 1209–1218.

76 Pawlyk, A.C., Giasson, B.I., Sampathu, D.M., Perez, F.A., Lim, K.L., Dawson, V.L., Dawson, T.M., Palmiter, R.D., Trojanowski, J.Q. and Lee, V.M. (2003) *J Biol Chem* **278**, 48120–48128.

77 Schlossmacher, M.G., Frosch, M.P., Gai, W.P., Medina, M., Sharma, N., Forno, L., Ochiishi, T., Shimura, H., Sharon, R., Hattori, N., Langston, J.W., Mizuno, Y., Hyman, B.T., Selkoe, D.J. and Kosik, K.S. (2002) *Am J Pathol* **160**, 1655–1667.

78 Bandopadhyay, R., Kingsbury, A.E., Muqit, M.M., Harvey, K., Reid, A.R., Kilford, L., Engelender, S., Schlossmacher, M.G., Wood, N.W., Latchman, D.S., Harvey, R.J. and Lees, A.J. (2005) *Neurobiol Dis* **20**, 401–411.

79 Huynh, D.P., Scoles, D.R., Ho, T.H., Del Bigio, M.R. and Pulst, S.M. (2000) *Ann Neurol* **48**, 737–744.

80 Choi, P., Golts, N., Snyder, H., Chong, M., Petrucelli, L., Hardy, J., Sparkman, D., Cochran, E., Lee, J.M. and Wolozin, B. (2001) *Neuroreport* **12**, 2839–2843.

81 Choi, P., Ostrerova-Golts, N., Sparkman, D., Cochran, E., Lee, J.M. and Wolozin, B. (2000) *Neuroreport* **11**, 2635–2638.

82 West, A.B., Lockhart, P.J., O'Farell, C. and Farrer, M.J. (2003) *J Mol Biol* **326**, 11–19.

83 West, A.B., Kapatos, G., O'Farrell, C., Gonzalez-de-Chavez, F., Chiu, K., Farrer, M.J. and Maidment, N.T. (2004) *J Biol Chem* **279**, 28896–28902.

84 Kitada, T., Asakawa, S., Minoshima, S., Mizuno, Y. and Shimizu, N. (2000) *Mamm Genome* **11**, 417–421.

85 Horowitz, J.M., Myers, J., Vernace, V.A., Stachowiak, M.K. and Torres, G. (2001) *Brain Res Dev Brain Res* **126**, 31–41.

86 Greene, J.C., Whitworth, A.J., Kuo, I., andrews, L.A., Feany, M.B. and Pallanck, L.J. (2003) *Proc Natl Acad Sci USA* **100**, 4078–4083.

87 Zhong, L., Tan, Y., Zhou, A., Yu, Q. and Zhou, J. (2005) *J Biol Chem* **280**, 9425–9430.

88 Finney, N., Walther, F., Mantel, P.Y., Stauffer, D., Rovelli, G. and Dev, K.K. (2003) *J Biol Chem* **278**, 16054–16058.

89 Imai, Y., Soda, M. and Takahashi, R. (2000) *J Biol Chem* **275**, 35661–35664.

90 Ledesma, M.D., Galvan, C., Hellias, B., Dotti, C. and Jensen, P.H. (2002) *J Neurochem* **83**, 1431–1440.

91 West, A.B., Gonzalez-de-Chavez, F., Wilkes, K., O'Farrell, C. and Farrer, M.J. (2003) *Neurosci Lett* **341**, 139–142.

92 Wang, C., Ko, H.S., Thomas, B., Tsang, F., Chew, K.C., Tay, S.P., Ho, M.W., Lim, T.M., Soong, T.W., Pletnikova, O., Troncoso, J., Dawson, V.L., Dawson, T.M. and Lim, K.L. (2005) *Hum Mol Genet* **14**, 3885–3897.

93 Wang, C., Tan, J.M., Ho, M.W., Zaiden, N., Wong, S.H., Chew, C.L., Eng, P.W., Lim, T.M., Dawson, T.M. and Lim, K.L. (2005) *J Neurochem* **93**, 422–431.

94 Ciechanover, A. and Brundin, P. (2003) *Neuron* **40**, 427–446.

95 Shimura, H., Hattori, N., Kubo, S., Mizuno, Y., Asakawa, S., Minoshima, S., Shimizu, N., Iwai, K., Chiba, T., Tanaka, K. and Suzuki, T. (2000) *Nat Genet* **25**, 302–305.

96 Zhang, Y., Gao, J., Chung, K.K., Huang, H., Dawson, V.L. and Dawson, T.M.

(2000) *Proc Natl Acad Sci USA* **97**, 13354–13359.
97 Glickman, M.H. and Ciechanover, A. (2002) *Physiol Rev* **82**, 373–428.
98 Pickart, C.M. and Eddins, M.J. (2004) *Biochim Biophys Acta* **1695**, 55–72.
99 Pickart, C.M. (2004) *Cell* **116**, 181–190.
100 Pickart, C.M. (2001) *Annu Rev Biochem* **70**, 503–533.
101 Tanaka, K., Suzuki, T., Hattori, N. and Mizuno, Y. (2004) *Biochim Biophys Acta* **1695**, 235–247.
102 Staropoli, J.F., McDermott, C., Martinat, C., Schulman, B., Demireva, E. and Abeliovich, A. (2003) *Neuron* **37**, 735–749.
103 Imai, Y., Soda, M., Inoue, H., Hattori, N., Mizuno, Y. and Takahashi, R. (2001) *Cell* **105**, 891–902.
104 Imai, Y., Soda, M., Hatakeyama, S., Akagi, T., Hashikawa, T., Nakayama, K.I. and Takahashi, R. (2002) *Mol Cell* **10**, 55–67.
105 Tsai, Y.C., Fishman, P.S., Thakor, N.V. and Oyler, G.A. (2003) *J Biol Chem* **278**, 22044–22055.
106 Kalia, S.K., Lee, S., Smith, P.D., Liu, L., Crocker, S.J., Thorarinsdottir, T.E., Glover, J.R., Fon, E.A., Park, D.S. and Lozano, A.M. (2004) *Neuron* **44**, 931–945.
107 Finley, D., Sadis, S., Monia, B.P., Boucher, P., Ecker, D.J., Crooke, S.T. and Chau, V. (1994) *Mol Cell Biol* **14**, 5501–5509.
108 Chau, V., Tobias, J.W., Bachmair, A., Marriott, D., Ecker, D.J., Gonda, D.K. and Varshavsky, A. (1989) *Science* **243**, 1576–1583.
109 Berke, S.J. and Paulson, H.L. (2003) *Curr Opin Genet Dev* **13**, 253–261.
110 Harrower, T.P., Michell, A.W. and Barker, R.A. (2005) *Exp Neurol* **195**, 1–6.
111 Choi, P., Snyder, H., Petrucelli, L., Theisler, C., Chong, M., Zhang, Y., Lim, K., Chung, K.K., Kehoe, K., D'Adamio, L., Lee, J.M., Cochran, E., Bowser, R., Dawson, T.M. and Wolozin, B. (2003) *Brain Res Mol Brain Res* **117**, 179–189.
112 Huynh, D.P., Scoles, D.R., Nguyen, D. and Pulst, S.M. (2003) *Hum Mol Genet* **12**, 2587–2597.
113 Shimura, H., Schlossmacher, M.G., Hattori, N., Frosch, M.P., Trockenbacher, A., Schneider, R., Mizuno, Y., Kosik, K.S. and Selkoe, D.J. (2001) *Science* **293**, 263–269.
114 Chung, K.K., Zhang, Y., Lim, K.L., Tanaka, Y., Huang, H., Gao, J., Ross, C.A., Dawson, V.L. and Dawson, T.M. (2001) *Nat Med* **7**, 1144–1150.
115 Goldberg, M.S., Fleming, S.M., Palacino, J.J., Cepeda, C., Lam, H.A., Bhatnagar, A., Meloni, E.G., Wu, N., Ackerson, L.C., Klapstein, G.J., Gajendiran, M., Roth, B. L., Chesselet, M.F., Maidment, N.T., Levine, M.S. and Shen, J. (2003) *J Biol Chem* **278**, 43628–43635.
116 Miller, D.W., Crawley, A., Gwinn-Hardy, K., Lopez, G., Nussbaum, R., Cookson, M.R., Singleton, A.B., Hardy, J. and Dogu, O. (2005) *Neurosci Lett* **374**, 189–191.
117 Petrucelli, L., O'Farrell, C., Lockhart, P.J., Baptista, M., Kehoe, K., Vink, L., Choi, P., Wolozin, B., Farrer, M., Hardy, J. and Cookson, M.R. (2002) *Neuron* **36**, 1007–1019.
118 Ko, H.S., von Coelln, R., Sriram, S.R., Kim, S.W., Chung, K.K., Pletnikova, O., Troncoso, J., Johnson, B., Saffary, R., Goh, E.L., Song, H., Park, B.J., Kim, M.J., Kim, S., Dawson, V.L. and Dawson, T.M. (2005) *J Neurosci* **25**, 7968–7978.
119 von Coelln, R., Thomas, B., Andrabi, S.A., Lim, K.L., Savitt, J.M., Saffary, R., Stirling, W., Bruno, K., Hess, E.J., Lee, M.K., Dawson, V.L., Dawson, T.M. (2006) *J Neurosci* **26**, 3685–3696.
120 Wakabayashi, K., Engelender, S., Yoshimoto, M., Tsuji, S., Ross, C.A. and Takahashi, H. (2000) *Ann Neurol* **47**, 521–523.
121 Lim, K.L., Chew, K.C., Tan, J.M., Wang, C., Chung, K.K., Zhang, Y., Tanaka, Y., Smith, W., Engelender, S., Ross, C.A., Dawson, V.L. and Dawson, T.M. (2005) *J Neurosci* **25**, 2002–2009.
122 Doss-Pepe, E.W., Chen, L. and Madura, K. (2005) *J Biol Chem* **280**, 16619–16624.
123 Hofmann, R.M. and Pickart, C.M. (2001) *J Biol Chem* **276**, 27936–27943.
124 Lim, K.L., Dawson, V.L. and Dawson, T.M. (2006) *Neurobiol Aging* **27**, 524–529.

125 Ren, Y., Zhao, J. and Feng, J. (2003) *J Neurosci* **23**, 3316–3324.

126 Corti, O., Hampe, C., Koutnikova, H., Darios, F., Jacquier, S., Prigent, A., Robinson, J.C., Pradier, L., Ruberg, M., Mirande, M., Hirsch, E., Rooney, T., Fournier, A. and Brice, A. (2003) *Hum Mol Genet* **12**, 1427–1437.

127 Jiang, H., Jiang, Q. and Feng, J. (2004) *J Biol Chem* **279**, 54380–54386.

128 Um, J.W., Min, D.S., Rhim, H., Kim, J., Paik, S.R. and Chung, K.C. (2006) *J Biol Chem* **281**, 3595–3603.

129 Yang, F., Jiang, Q., Zhao, J., Ren, Y., Sutton, M.D. and Feng, J. (2005) *J Biol Chem* **280**, 17154–17162 .

130 Murakami, T., Shoji, M., Imai, Y., Inoue, H., Kawarabayashi, T., Matsubara, E., Harigaya, Y., Sasaki, A., Takahashi, R. and Abe, K. (2004) *Ann Neurol* **55**, 439–442.

131 Sriram, S.R., Li, X., Ko, H.S., Chung, K.K., Wong, E., Lim, K.L., Dawson, V.L. and Dawson, T.M. (2005) *Hum Mol Genet* **14**, 2571–2586.

132 Yang, Y., Nishimura, I., Imai, Y., Takahashi, R. and Lu, B. (2003) *Neuron* **37**, 911–924.

133 Feany, M.B. and Pallanck, L.J. (2003) *Neuron* **38**, 13–16.

134 Dong, Z., Ferger, B., Paterna, J.C., Vogel, D., Furler, S., Osinde, M., Feldon, J. and Bueler, H. (2003) *Proc Natl Acad Sci USA* **100**, 12438–12443.

135 Springer, W., Hoppe, T., Schmidt, E. and Baumeister, R. (2005) *Hum Mol Genet* **14**, 3407–3423.

136 McNaught, K.S. and Olanow, C.W. (2003) *Ann Neurol* **53**(Suppl 3), S73–S84; discussion S84–S86.

137 Muqit, M.M., Davidson, S.M., Payne Smith, M.D., MacCormac, L.P., Kahns, S., Jensen, P.H., Wood, N.W. and Latchman, D.S. (2004) *Hum Mol Genet* **13**, 117–135.

138 Zhao, J., Ren, Y., Jiang, Q. and Feng, J. (2003) *J Cell Sci* **116**, 4011–4019.

139 Olanow, C.W., Perl, D.P., DeMartino, G.N. and McNaught, K.S. (2004) *Lancet Neurol* **3**, 496–503.

140 Higashi, Y., Asanuma, M., Miyazaki, I., Hattori, N., Mizuno, Y. and Ogawa, N. (2004) *J Neurochem* **89**, 1490–1497.

141 Darios, F., Corti, O., Lucking, C.B., Hampe, C., Muriel, M.P., Abbas, N., Gu, W.J., Hirsch, E.C., Rooney, T., Ruberg, M. and Brice, A. (2003) *Hum Mol Genet* **12**, 517–526.

142 Jiang, H., Ren, Y., Zhao, J. and Feng, J. (2004) *Hum Mol Genet* **13**, 1745–1754.

143 Oluwatosin-Chigbu, Y., Robbins, A., Scott, C.W., Arriza, J.L., Reid, J.D. and Zysk, J.R. (2003) *Biochem Biophys Res Commun* **309**, 679–684.

144 Kim, S.J., Sung, J.Y., Um, J.W., Hattori, N., Mizuno, Y., Tanaka, K., Paik, S.R., Kim, J. and Chung, K.C. (2003) *J Biol Chem* **278**, 41890–41899.

145 Lo Bianco, C., Schneider, B.L., Bauer, M., Sajadi, A., Brice, A., Iwatsubo, T. and Aebischer, P. (2004) *Proc Natl Acad Sci USA* **101**, 17510–17515.

146 Moore, D.J., Zhang, L., Troncoso, J., Lee, M.K., Hattori, N., Mizuno, Y., Dawson, T.M. and Dawson, V.L. (2005) *Hum Mol Genet* **14**, 71–84.

147 Smith, W.W., Pei, Z., Jiang, H., Moore, D.J., Liang, Y., West, A.B., Dawson, V.L., Dawson, T.M. and Ross, C.A. (2005) *Proc Natl Acad Sci USA* **102**, 18676–18681.

148 Gu, W.J., Corti, O., Araujo, F., Hampe, C., Jacquier, S., Lucking, C.B., Abbas, N., Duyckaerts, C., Rooney, T., Pradier, L., Ruberg, M. and Brice, A. (2003) *Neurobiol Dis* **14**, 357–364.

149 Matsuda, N., Kitami, T., Suzuki, T., Mizuno, Y., Hattori, N. and Tanaka, K. (2006) *J Biol Chem* **281**, 3204–3209.

150 Sakata, E., Yamaguchi, Y., Kurimoto, E., Kikuchi, J., Yokoyama, S., Yamada, S., Kawahara, H., Yokosawa, H., Hattori, N., Mizuno, Y., Tanaka, K. and Kato, K. (2003) *EMBO Rep* **4**, 301–306.

151 Dachsel, J.C., Lucking, C.B., Deeg, S., Schultz, E., Lalowski, M., Casademunt, E., Corti, O., Hampe, C., Patenge, N., Vaupel, K., Yamamoto, A., Dichgans, M., Brice, A., Wanker, E.E., Kahle, P.J. and Gasser, T. (2005) *FEBS Lett* **579**, 3913–3919.

152 Henn, I.H., Gostner, J.M., Lackner, P., Tatzelt, J. and Winklhofer, K.F. (2005) *J Neurochem* **92**, 114–122.

153 Cookson, M.R., Lockhart, P.J., McLendon, C., O'Farrell, C., Schlossmacher, M. and

Farrer, M.J. (2003) *Hum Mol Genet* **12**, 2957–2965.

154 Kahns, S., Lykkebo, S., Jakobsen, L.D., Nielsen, M.S. and Jensen, P.H. (2002) *J Biol Chem* **277**, 15303–15308.

155 Sato, S., Chiba, T., Sakata, E., Kato, K., Mizuno, Y., Hattori, N. and Tanaka, K. (2006) *EMBO J* **25**, 211–221.

156 Chung, K.K., Dawson, T.M. and Dawson, V.L. (2005) *Cell Mol Biol (Noisy-le-grand)* **51**, 247–254.

157 Yamamoto, A., Friedlein, A., Imai, Y., Takahashi, R., Kahle, P.J. and Haass, C. (2005) *J Biol Chem* **280**, 3390–3399.

158 Whitworth, A.J., Theodore, D.A., Greene, J.C., Benes, H., Wes, P.D. and Pallanck, L.J. (2005) *Proc Natl Acad Sci USA* **102**, 8024–8029.

159 Pesah, Y., Pham, T., Burgess, H., Middlebrooks, B., Verstreken, P., Zhou, Y., Harding, M., Bellen, H. and Mardon, G. (2004) *Development* **131**, 2183–2194.

160 Itier, J.M., Ibanez, P., Mena, M.A., Abbas, N., Cohen-Salmon, C., Bohme, G. A., Laville, M., Pratt, J., Corti, O., Pradier, L., Ret, G., Joubert, C., Periquet, M., Araujo, F., Negroni, J., Casarejos, M.J., Canals, S., Solano, R., Serrano, A., Gallego, E., Sanchez, M., Denefle, P., Benavides, J., Tremp, G., Rooney, T.A., Brice, A. and Garcia de Yebenes, J. (2003) *Hum Mol Genet* **12**, 2277–2291.

161 Jiang, H., Jiang, Q., Liu, W. and Feng, J. (2006) *J Biol Chem* **281**, 8591–8599.

162 Palacino, J.J., Sagi, D., Goldberg, M.S., Krauss, S., Motz, C., Klose, J. and Shen, J. (2004) *J Biol Chem* **279**, 18614–18622.

163 Von Coelln, R., Thomas, B., Savitt, J.M., Lim, K.L., Sasaki, M., Hess, E.J., Dawson, V.L. and Dawson, T.M. (2004) *Proc Natl Acad Sci USA* **101**, 10744–10749.

164 Perez, F.A. and Palmiter, R.D. (2005) *Proc Natl Acad Sci USA* **102**, 2174–2179.

Index

Protein Degradation, Vol. 4: The Ubiquitin-Proteasome System and Disease.
Edited by R. J. Mayer, A. Ciechanover, M. Rechsteiner

ISBN: 978-3-527-31436-2

v

w

y

z